Teubner Studienbücher Informatik

J. Berstel
Transductions and Context-Free Languages

# Leitfäden der angewandten Mathematik und Mechanik LAMM



Band 38

Die Lehrbücher dieser Reihe sind einerseits allen mathematischen Theorien und Methoden von grundsätzlicher Bedeutung für die Anwendung der Mathematik gewidmet; andererseits werden auch die Anwendungsgebiete selbst behandelt. Die Bände der Reihe sollen dem Ingenieur und Naturwissenschaftler die Kenntnis der mathematischen Methoden, dem Mathematiker die Kenntnisse der Anwendungsgebiete seiner Wissenschaft zugänglich machen. Die Werke sind für die angehenden Industrie- und Wirtschaftsmathematiker, Ingenieure und Naturwissenschaftler bestimmt, darüber hinaus aber sollen sie den im praktischen Beruf Tätigen zur Fortbildung im Zuge der fortschreitenden Wissenschaft dienen.

# Transductions and Context-Free Languages

By Dr. Jean Berstel
Professor at the Université P. et M. Curíe, Paris

With 32 figures, 158 exercises
and numerous examples

B. G. Teubner Stuttgart 1979

Prof. Jean Berstel

Born in 1941 at Nîmes (France). From 1961 to 1966 studies in mathematics at the University of Paris. 1967 Thèse de troisième cycle, 1972 Thèse d'Etat at the University of Paris. From 1970 to 1972 Chargé d'Enseignement at the University of Strasbourg. Since 1972, Maître de Conférence, then Professor at the Institut de Programmation, University of Paris VI. During the Summer Semester 1976 on leave at the University of Saarbrücken. Activity: Mathematics and Theoretical Computer Science.

CIP-Kurztitelaufnahme der Deutschen Bibliothek

**Berstel, Jean:**
Transductions and context-free languages / by
Jean Berstel. – Stuttgart : Teubner, 1979. –
(Leitfäden der angewandten Mathematik und
Mechanik ; Bd. 38) (Teubner Studienbücher :
Informatik)
ISBN 3-519-02340-7


Printed in Germany
Setting: The Universities Press (Belfast) Ltd., Belfast
Printer: J. Beltz, Hemsbach/Bergstr.
Cover design: W. Koch, Sindelfingen

# Preface

This book presents a theory of formal languages with main emphasis on rational transductions and their use for the classification of context-free languages. The level of presentation corresponds to that of beginning graduate or advanced undergraduate work. Prerequisites for this book are covered by a "standard" first-semester course in formal languages and automata theory: e.g. a knowledge of Chapters 1–3 of Ginsburg [1966], or Chapters 3–4 of Hopcroft and Ullman [1971], or Chapter 2 of Salomaa [1973], or Chapters 2 and 4 of Becker and Walter [1977] would suffice. The book is self-contained in the sense that complete proofs are given for all theorems stated, except for some basic results explicitly summarized at the beginning of the text. Chapter IV and Chapters V–VIII are independent from each other.
The subject matter is divided into two preliminary and six main chapters. The initial two chapters contain a general survey of the "classical" theory of regular and context-free languages with a detailed description of several special languages. Chapter III deals with the general theory of rational transductions, treated in an algebraic fashion along the lines of Eilenberg, and which will be used systematically in subsequent chapters. Chapter IV is concerned with the important special case of rational functions, and gives a full treatment of the latest developments, including subsequential transductions, unambiguous transducers and decision problems.
The study of families of languages (in the sense of Ginsburg) begins with Chapter V. There, the structures of "rational cone" and "full AFL" are introduced, and some general results are established. The chapter ends with the treatment of several examples of cones of linear languages. Chapter VI contains a theory of operators on families of languages, setting up an algebraic framework for transformation and comparison of families of languages. Other general results on cones and full AFLs are easily derived from a series of inequalities involving only operators. Chapter VII is concerned with the study of principal cones and full AFLs, that is families generated by one language only. Main interest is in subcones of the context-free languages. First, several languages are proved to generate the entire cone of context-free languages. Then S. Greibach's "Syntactic Lemma" is proved and used to exhibit nonprincipal cones. A detailed study of two important families follows, namely the family of one counter and the family of quasi-rational (nonexpansive, derivation-bounded $\cdots$) languages. Chapter VIII presents a general method (due to Boasson and Beauquier) for proving strict containment or incompara-

bility of cones of context-free languages, using iterative pairs and systems of iterative pairs. This method yields easy proofs of the existence of several wellknown and of new hierarchies of cones of context-free languages, showing thus the intimate connection between iterative pairs and the structure of context-free languages.

Main interest is in languages rather than in grammars or in acceptors. Indeed, a language (even a context-free one) exists independently from the grammars generating it, and a great number of context-free languages can be described by a combinatorial or an algebraic property, without any reference to a grammar. Moreover, grammatical characterizations of classes of languages usually require just the existence of one grammar of some special type. Thus to prove that a given language is n o t in the family, one must show that a l l grammars generating it violate some property. This is usually a very delicate proof. Finally there are results which are proved by the method of iterative pairs and which cannot be proved—up to now—by considering only grammars.

Rational cones are treated in greater detail than full AFLs. They are indeed in a natural relationship with rational transductions; further full AFLs are a secondary structure in the sense that any full AFL is the rational closure of some cone. Since we are mainly concerned with context-free languages, a description of AFLs and "trios" (in opposition to full AFLs and rational cones) seemed unnecessary, all the more as some fundamental results such as the Syntactic Lemma are still lacking for these families.

The notes from which this book derives were used in courses at the University of Paris and at the University of Saarbrücken. I want to thank Professor G. Hotz for the opportunity he gave me to stay with the Institut für angewandte Mathematik and Informatik, and for his encouragements to write this book. I am grateful to the following people for useful discussions or comments concerning various parts of the text: J. M. Autebert, J. Beauquier, Ch. Choffrut, G. Cousineau, K. Estenfeld, R. Linder, M. Nivat, D. Perrin, J. F. Perrot, J. Sakarovitch, M. Soria, M. Stadel, H. Walter. I am deeply indebted to M. P. Schützenberger for his constant interest in this book and for many fruitful discussions. Special thanks are due to L. Boasson whose comments have been of an invaluable help in the preparation of many sections of this book. I want to thank also J. Messerschmidt for his careful reading of the manuscript and for many pertinent comments, and Ch. Reutenauer for checking the galley proofs. I owe a special debt to my wife for her active contribution at each step of the preparation of the book, and to Bruno and Clara for their indulgence.

Paris, Spring 1978 J. Berstel

## Contents

**I Preliminaries**

I.1 Some Notations . . . . . . . . . . . . . . . . . . . 9
I.2 Monoids, Free Monoids . . . . . . . . . . . . . . . 9
I.3 Morphisms, Congruences. . . . . . . . . . . . . . . 12
I.4 Finite Automata, Regular Languages . . . . . . . . 15

**II Context-Free Languages**

II.1 Grammars, Languages, Equations . . . . . . . . . . 22
II.2 Closure Properties, Iteration . . . . . . . . . . . 31
II.3 Dyck Languages . . . . . . . . . . . . . . . . . . 35
II.4 Two Special Languages . . . . . . . . . . . . . . . 47

**III Rational Transductions**

III.1 Recognizable Sets . . . . . . . . . . . . . . . . 51
III.2 Rational Sets . . . . . . . . . . . . . . . . . . 55
III.3 Rational Relations . . . . . . . . . . . . . . . . 61
III.4 Rational Transductions. . . . . . . . . . . . . . 65
III.5 Examples . . . . . . . . . . . . . . . . . . . . . 72
III.6 Transducers. . . . . . . . . . . . . . . . . . . . 77
III.7 Matrix Representations. . . . . . . . . . . . . . 80
III.8 Decision Problems . . . . . . . . . . . . . . . . 87

**IV Rational Functions**

IV.1 Rational Functions . . . . . . . . . . . . . . . . 92
IV.2 Sequential Transductions . . . . . . . . . . . . . 96
IV.3 The Cross-Section Theorem . . . . . . . . . . . . 111
IV.4 Unambiguous Transducers . . . . . . . . . . . . . 114
IV.5 Bimachines. . . . . . . . . . . . . . . . . . . . 123
IV.6 A Decidable Property . . . . . . . . . . . . . . . 128

**V Families of Languages**

V.1 Definition. . . . . . . . . . . . . . . . . . . . . 134
V.2 Rational Equivalence, Rational Cones. . . . . . . . 135
V.3 Rationally Closed Families. . . . . . . . . . . . . 138
V.4 Full AFLs . . . . . . . . . . . . . . . . . . . . . 141

V.5 Substitution . . . . . . . . . . . . . . . . . . 145
V.6 Example: The Cone of Linear Languages . . . . . . . 150
V.7 Examples of Incomparable Languages. . . . . . . . 156

**VI Operators**

VI.1 Operators . . . . . . . . . . . . . . . . . . 162
VI.2 Examples of Operators . . . . . . . . . . . . 164
VI.3 Closure Operators . . . . . . . . . . . . . . 166
VI.4 Subcommutative Relations . . . . . . . . . . . 171
VI.5 Marked Substitution . . . . . . . . . . . . . 177

**VII Generators**

VII.1 Generators of the Context-Free Languages . . . . . 185
VII.2 The Syntactic Lemma. . . . . . . . . . . . . 194
VII.3 Substitution Closure . . . . . . . . . . . . . 197
VII.4 One Counter Languages. . . . . . . . . . . . 201
VII.5 Quasi-Rational Languages . . . . . . . . . . . 208

**VIII Iterative Pairs**

VIII.1 Types of Iterative Pairs. . . . . . . . . . . . 219
VIII.2 Grammatical Pairs . . . . . . . . . . . . . 223
VIII.3 Transfer of Iterative Pairs . . . . . . . . . . 227
VIII.4 Systems of Iterative Pairs . . . . . . . . . . 232
VIII.5 Grammatical Systems . . . . . . . . . . . . 237
VIII.6 Transfer of Systems . . . . . . . . . . . . . 246
VIII.7 Applications. . . . . . . . . . . . . . . . 251

**IX Open Problems, Further Developments** 265

**Bibliography** . . . . . . . . . . . . . . . . . . 268

**List of Symbols** . . . . . . . . . . . . . . . . . 274

**Index**. . . . . . . . . . . . . . . . . . . . . 275

# I Preliminaries

This chapter is a short review of some basic concepts used in the sequel. Its aim is to agree on notations and terminology. We first consider monoids, especially free monoids, and morphisms. Then a collection of definitions and results is given, dealing with finite automata and regular languages.

## I.1 Some Notations

$\mathbb{N} = \{0, 1, 2, \ldots .\}$ is the set of nonnegative integers. $\mathbb{Z} = \{\cdots -2, -1, 0, 1, \ldots\}$ is the set of integers. Let $E$ be a set. Then $\mathrm{Card}(E)$ is the number of its elements. The empty set is denoted by $\emptyset$. If $A$, $B$ are subsets of $E$, then we write $A \subset B$ iff $x \in A \Rightarrow x \in B$, and $A \subsetneq B$ iff $A \subset B$ and $A \neq B$. Further

$$A \backslash B = \{x \in E \mid x \in A \text{ and } x \notin B\}.$$

A singleton is a subset of $E$ consisting of just one element. If no confusion can arise, we shall not distinguish elements of $E$ from singletons. The set of all subsets of $E$, i.e. the powerset of $E$, is denoted by $\mathfrak{P}(E)$ or $2^E$. With the preceding convention, $E \subset \mathfrak{P}(E)$.

The domain $\mathrm{dom}(\alpha)$ of a partial function $\alpha: E \to F$ is the set of elements $x$ in $E$ for which $\alpha(x)$ is defined. $\alpha$ can be viewed as a (total) function from $E$ into $\mathfrak{P}(F)$, and with the convention $F \subset \mathfrak{P}(F)$, as a total function from $E$ into $F \cup \{\emptyset\}$. Then $\mathrm{dom}(\alpha) = \{x \in E \mid \alpha(x) \neq \emptyset\}$.

## I.2 Monoids, Free Monoids

A semigroup consists of a set $M$ and a binary operation on $M$, usually denoted by multiplication, and which is postulated to be associative: For any $m_1, m_2, m_3 \in M$, $m_1(m_2m_3) = (m_1m_2)m_3$. A neutral element or a unit is an element $1_M \in M$ (also noted 1 for short) such that $1_M m = m 1_M = m$ for all $m \in M$. A semi-group which has a neutral element is a monoid. The neutral element of a monoid is unique. Indeed, if $1'$ is another neutral element then $1' = 11' = 1$.

Given two subsets $A$, $B$ of a monoid $M$, the product $AB$ is defined by

$$AB = \{c \in M \mid \exists a \in A, \exists b \in B : c = ab\}. \qquad (2.1)$$

This definition converts $\mathfrak{P}(M)$ into a monoid with unit $\{1_M\}$. A subset $A$ of $M$ is a subsemigroup (submonoid) of $M$ if $A^2 \subset A$ ($1 \in A$ and $A^2 \subset A$). Given any subset $A$ of $M$, the sets

$$A^+ = \bigcup_{n \geq 1} A^n, \qquad A^* = \bigcup_{n \geq 0} A^n,$$

where $A^0 = \{1\}$ and $A^{n+1} = A^n A$, are a subsemigroup resp. a submonoid of $M$. In fact, $A^+$ (resp. $A^*$) is the least subsemigroup (resp. submonoid) for the order of set inclusion containing $A$. It is called the subsemigroup (submonoid) generated by $A$. If $M = A^*$ for some $A \subset M$, then $A$ is a system of generators of $M$. A monoid is finitely generated if it has a finite system of generators. The unary operations $A \mapsto A^+$ and $A \mapsto A^*$ on subsets of $M$ are called the (Kleene) plus and star operations. The formulas $A^+ = AA^* = A^*A$ and $A^* = 1 \cup A^+$ are readily verified.

For any set $X$, the free monoid $X^*$ generated by $X$ or with base $X$ is defined as follows. The elements of $X^*$ are $n$-tuples

$$u = (x_1, x_2, \ldots, x_n) \qquad (n \geq 0) \tag{2.2}$$

of elements of $X$. If $v = (y_1, \ldots, y_m)$ is another element of $X^*$, the product $uv$ is defined by concatenation, i.e.

$$uv = (x_1, x_2, \ldots, x_n, y_1, \ldots, y_m).$$

This produces a monoid with the only 0-tuple $1 = (\ )$ as neutral element. We shall agree to write $x$ instead of the 1-tuple $(x)$. Thus (2.2) may be written as

$$u = x_1 x_2 \cdots x_n.$$

Because of this, $u$ is called a word, $x \in X$ is called a letter and $X$ itself is called an alphabet. By the convention $x = (x)$, $X$ can be considered as a subset of $X^*$. This justifies the notation $X^*$, since indeed $X^*$ is the only submonoid of $X^*$ containing $X$. Further $X^+ = X^* \setminus 1$. In the sequel, and unless otherwise indicated, an alphabet will be supposed to be finite and nonempty.

We shall use the following terminology concerning a free monoid $X^*$ generated by an alphabet $X$. A (formal) language over $X$ is any subset of $X^*$. The length $|u|$ of a word $u \in X^*$ is the number of letters composing it. The neutral element of $X^*$ is called the empty word, and is noted 1 or $\varepsilon$. It is the only word of length 0. Clearly $|uv| = |u| + |v|$. If $Y \subset X$, then $|u|_Y$ is the number of occurrences of letters $y \in Y$ in $u$. Thus

$$|u| = \sum_{x \in X} |u|_x.$$

The reversal of a word $u = x_1 x_2 \cdots x_n$ ($n \geq 0, x_i \in X$) is denoted by $\tilde{u}$ or $u\tilde{}$ and is defined by $\tilde{u} = x_n x_{n-1} \cdots x_2 x_1$. Clearly $\tilde{\tilde{u}} = u$, $\tilde{1} = 1$, $(uv)\tilde{} = \tilde{v}\tilde{u}$ For $A \subset X^*$, $\tilde{A} = \{\tilde{u} \mid u \in A\}$. If $B \subset X^*$, then $(AB)\tilde{} = \tilde{B}\tilde{A}$, and $(A^*)\tilde{} = (\tilde{A})^*$.

Let $u \in X^*$. Then a word $v$ is a factor of $u$ if $u = avb$ for some $a, b \in X^*$. If $a = 1$, then $v$ is a left factor; if $b = 1$, then $v$ is a right factor of $u$. $v$ is a proper factor (left factor, right factor) of $u$ if further $v \neq u$. A word $v$ may occur at several places as a factor of $u$. A fixed occurrence of $v$ as a factor of $u$ is called a segment. This definition always refers to some previously defined factorization $u = avb$. If $u' = a'v'b'$ is another factorization, then the segment $v'$ is contained in the segment $v$ iff $a$ is a left factor of $a'$ and $b$ is a right factor of $b'$. Finally $v$ is a subword of $u$ iff $u = w_0 x_1 w_1 \cdots x_n w_n$, $(n \geq 0, x_1, \ldots, x_n \in X, w_0, \ldots, w_n \in X^*)$ and $v = x_1 \cdots x_n$.

Let $M$ be a submonoid of $X^*$. Then $A = (M \backslash 1) \backslash (M \backslash 1)^2$ is a system of generators of $M$, i.e. $A^* = M$. Further $A$ is minimal with this property, i.e. $B^* = M$ implies $B \supset A$. A submonoid $M$ of $X^*$ is free with base $C$ if any word $u \in M$ has one and only one factorization $u = c_1 c_2 \cdots c_n$, with $n \geq 0$ and $c_1, \ldots, c_n \in C$. The base of a free submonoid $M$ is unique and is equal to $(M \backslash 1) \backslash (M \backslash 1)^2$. Thus $X^*$ is free with base $X$. A base of a free submonoid is called a code. Examples of codes are supplied by prefix and suffix sets. A subset $A$ of $X^+$ is prefix iff $AX^+ \cap A = \emptyset$, i.e. if $A$ contains no proper left factor of some of its word, and $A$ is suffix iff $X^+ A \cap A = \emptyset$. $A$ is bifix if it is both prefix and suffix. Any prefix or suffix subset is a code.

Let $M$ be any monoid, and let $A, B \subset M$. The left and right quotients $B^{-1}A$ and $AB^{-1}$ are the sets

$$B^{-1}A = \{c \in M \mid \exists a \in A, \exists b \in B : a = bc\},$$

$$AB^{-1} = \{c \in M \mid \exists a \in A, \exists b \in B : a = cb\}.$$

If $M$ is a group and $u, v \in M$, then $v^{-1}u$ and $uv^{-1}$ are always singletons. If $M$ is a free monoid, then $uv^{-1}$ is non empty iff $v$ is a right factor of $u$; thus $uM^{-1}$ is the set of left factors of $u$.

Sometimes, we shall need the notion of semiring. A semiring consists of a set $S$ and of two binary operations, called addition and multiplication, noted $+$ and $\cdot$, and satisfying the following conditions:

(i) $S$ is a commutative monoid for the addition ($s + t = t + s$ for all $s, t \in S$) with neutral element 0;
(ii) $S$ is a monoid for multiplication;
(iii) the multiplication is distributive with respect to the addition:

$$s(t_1 + t_2) = st_1 + st_2; \qquad (t_1 + t_2)s = t_1 s + t_2 s \quad \text{for all} \quad s, t_1, t_2 \in S;$$

(iv) for all $s \in S$, $0 \cdot s = s \cdot 0 = 0$.

If $M$ is a monoid, then $\mathfrak{P}(M)$ is a semiring with set union for addition and the multiplication (2.1).

## Exercises

**2.1** Let $M_1$ and $M_2$ be monoids. Show that the cartesian product $M_1 \times M_2$ is a monoid when multiplication is defined by $(m_1, m_2)(m_1', m_2') = (m_1 m_1', m_2 m_2')$.

**2.2** Show that if $S$ is a semiring, then the set $S^{n \times n}$ of square matrices of size $n$ with elements in $S$ can be made a semiring, for addition and multiplication of matrixes induced by the operations in $S$.

**2.3** Let $M$ be a monoid, and let $A, B, C \subset M$. Prove the following formulas: $(AB)^{-1}C = B^{-1}(A^{-1}C)$; $(A^{-1}B)C^{-1} = A^{-1}(BC^{-1})$.

**2.4** Let $X$ be an alphabet, $A \subset X^+$, $A \neq \emptyset$. Show that $A$ is prefix iff $A^{-1}A = 1$.

**2.5** Let $X$ be an alphabet, and let $a, b \in X^+$. Show that the three following conditions are equivalent:
(i) $a = d^r$, $b = d^s$ for some word $d$ and $r, s \geqslant 1$;
(ii) $ab = ba$;
(iii) $a^m = b^n$ for some $m, n \geqslant 1$.

**2.6** Two words $a$ and $b$ are conjugate if $ac = cb$ for some word $c$. Show that this equation holds iff $a = uv$, $b = vu$, $c = (uv)^k u$ for some words $u$, $v$, and $k \geqslant 0$.

**2.7** A word $a$ is primitive iff it is not a nontrivial power of another word, i.e. if $a = d^n$ implies $n = 1$.
a) Show that any word $a \neq 1$ is a power of a unique primitive word.
b) Show that if $a$ and $b$ are conjugate, and $a$ is primitive, then $b$ is also primitive.
c) Show that if $ac = cb$ and $a \neq 1$, then there are unique primitive words $u$, $v$, and integers $p \geqslant 1$, $k \geqslant 0$, such that $a = (uv)^p$, $b = (vu)^p$, $c = (uv)^k u$.

## I.3 Morphisms, Congruences

If $M$, $M'$ are monoids, a (monoid) morphism $\alpha : M \to M'$ is a function satisfying

$$\alpha(m_1 m_2) = \alpha(m_1)\alpha(m_2) \qquad \text{for all} \quad m_1, m_2 \in M \tag{3.1}$$
$$\alpha(1_M) = 1_{M'}.$$

Then clearly $\alpha(M)$ is a submonoid of $M'$. If only (3.1) is postulated, then $\alpha$ is called a semigroup morphism and $\alpha(M)$ is a subsemigroup of $M'$. Unless otherwise indicated, morphism always means monoid morphism. A morphism $\alpha : X^* \to M'$, where $X$ is an alphabet, is completely defined by the values $\alpha(x)$ of the letters $x \in X$. We now review some formulas. Let $\alpha$: $M \to M'$ be a function, and let $A, B \subset M$, $A', B' \subset M'$. Then

$$\alpha(A \cup B) = \alpha(A) \cup \alpha(B); \qquad \alpha^{-1}(A' \cup B') = \alpha^{-1}(A') \cup \alpha^{-1}(B'),$$
$$\alpha^{-1}(A' \cap B') = \alpha^{-1}(A') \cap \alpha^{-1}(B'); \qquad \alpha(\alpha^{-1}(A') \cap B) = A' \cap \alpha(B).$$

Next if $\alpha$ is a semigroup morphism, then

$$\alpha(AB)=\alpha(A)\alpha(B), \qquad \alpha(A^+)=(\alpha(A))^+.$$

If $\alpha$ is a morphism, then $\alpha(A^*)=(\alpha(A))^*$.

Note that the formula $\alpha^{-1}(A'B')=\alpha^{-1}(A')\alpha^{-1}(B')$ is in general false. This observation leads to the definition of particular morphism, for which that formula holds.

Let $X$, $Y$ be alphabets, and let $\alpha: X^* \to Y^*$ be a morphism. Then $\alpha$ is called:
alphabetic if $\alpha(X) \subset Y \cup 1$;
strictly alphabetic if $\alpha(X) \subset Y$;
continuous or $\varepsilon$-free if $\alpha(X) \subset Y^+$;
a projection if $Y \subset X$, and if $\alpha(y)=y$ for $y \in Y$, $\alpha(x)=1$ for $x \in X \setminus Y$.

Thus projections are particular alphabetic morphisms.

If $\alpha: X^* \to Y^*$ is an alphabetic morphism, then

$$\alpha^{-1}(AB)=\alpha^{-1}(A)\alpha^{-1}(B), \qquad \alpha^{-1}(A^+)=(\alpha^{-1}(A))^+ \quad \text{for} \quad A, B \subset Y^*.$$

For the proof, it suffices to show that $\alpha^{-1}: Y^* \to \mathfrak{P}(X^*)$ is a semigroup morphism. Define

$$Z=\{x \in X \mid \alpha(x)=1\}=\alpha^{-1}(1) \cap X; \qquad Z_y=\{x \in X \mid \alpha(x)=y\}=\alpha^{-1}(y) \cap X$$

for $y \in Y$.

Then $\quad \alpha^{-1}(1)=Z^*, \qquad \alpha^{-1}(y)=Z^*Z_yZ^* \qquad (y \in Y)$.

If $g=y_1 \cdots y_n$ $(y_i \in Y)$, then

$$\alpha^{-1}(g)=Z^*Z_{y_1}Z^*Z_{y_2}Z^* \cdots Z^*Z_{y_n}Z^*.$$

Thus $\alpha^{-1}(g_1g_2)=\alpha^{-1}(g_1)\alpha^{-1}(g_2)$ for all $g_1, g_2 \in Y^*$. This completes the proof.
Note that the formula $\alpha^{-1}(A^*)=(\alpha^{-1}(A))^*$ is only true if further $\alpha$ is continuous, or if $1 \in A$ i.e. $A^*=A^+$.

We shall frequently use special morphisms called copies. Let $\alpha: X^* \to Y^*$ be an isomorphism. Then $\alpha(X)=Y$. For each subset $A$ of $X^*$, $\alpha(A)$ is called a copy of $A$ on $Y$.

Another class of particular morphisms are substitutions. A substitution $\sigma$ from $X^*$ into $Y^*$ is a (monoid) morphism from $X^*$ into $\mathfrak{P}(Y^*)$; thus $\sigma$ verifies: $\sigma(x) \subset Y^*$ for $x \in X$, and

$$\sigma(1)=1, \qquad \sigma(uv)=\sigma(u)\sigma(v) \qquad \text{for} \quad u, v \in X^*.$$

Thus if $\alpha: Y^* \to X^*$ is an alphabetic morphism, the function $\alpha^{-1}$ is a substitution iff $\alpha^{-1}(1)=1$. A substitution $\sigma$ is extended to $\mathfrak{P}(X^*)$ by the convention

$$\sigma(A)=\bigcup_{u \in A} \sigma(u) \qquad (A \subset X^*).$$

For sake of simplicity, we write $\sigma : X^* \to Y^*$ for a substitution from $X^*$ into $Y^*$. If $\tau : Y^* \to Z^*$ is another substitution, then the function $\tau \circ \sigma$ from $X^*$ into $\mathfrak{P}(Z^*)$ is a substitution.

Finally we note that any finitely generated monoid is a homomorphic image of a free monoid. Consider indeed a monoid $M$, and let $S = \{m_1, \ldots, m_k\}$ be a system of generators of $M$. Set $X = \{x_1, \ldots, x_k\}$, and define a morphism $\alpha : X^* \to M$ by $\alpha(x_i) = m_i$ for $i = 1, \ldots, k$. Then $\alpha(X^*) = (\alpha(X))^* = S^* = M$. Clearly this result remains true for any monoid if infinite alphabets are considered.

Let $E$ and $F$ be two sets. A relation over $E$ and $F$ is a subset $\theta$ of $E \times F$. For $(x, y) \in \theta$, we also write $x\theta y$, $x \sim y (\bmod\ \theta)$ or $x \equiv y (\bmod\ \theta)$, or simply $x \sim y$, $x \equiv y$ if no confusion can arise. If $E = F$, then $\theta$ is a relation over $E$. Relations are ordered by (set) inclusion.

Let $M$ be a monoid. A congruence over $M$ is an equivalence relation $\theta$ which is compatible with the monoid operation, i.e. which satisfies

$$m_1 \equiv m_1' (\bmod\ \theta),\ m_2 \equiv m_2' (\bmod\ \theta) \Rightarrow m_1 m_2 \equiv m_1' m_2' (\bmod\ \theta). \qquad (3.2)$$

For each $m \in M$, the class of $m$ mod $\theta$ is

$$[m]_\theta = \{m' \in M \mid m \equiv m' (\bmod\ \theta)\}.$$

Then (3.2) is equivalent to

$$[m_1]_\theta [m_2]_\theta \subset [m_1 m_2]_\theta.$$

If $\theta$ is a congruence, then the function which associates to each $m \in M$ its class $[m]_\theta$ is a morphism from $M$ onto the quotient monoid $M/\theta$. Conversely, if $\alpha : M \to M'$ is a morphism, then the relation $\theta$ defined by

$$m \sim m' (\bmod\ \theta) \quad \text{iff} \quad \alpha(m) = \alpha(m')$$

is a congruence. The number of equivalence classes of an equivalence relation $\theta$ is the index of $\theta$. The index is a positive integer or infinite.

Given a relation $\theta$ over a monoid $M$, the congruence $\hat{\theta}$ generated by $\theta$ is the least congruence containing $\theta$. The congruence $\hat{\theta}$ can be constructed as follows: Define a relation $\theta_1$ on $M$ by

$$m \sim m' (\bmod\ \theta_1) \quad \text{iff} \quad m = uav, \quad m' = ubv,$$
$$\text{and} \quad (a \sim b (\bmod\ \theta) \text{ or } b \sim a (\bmod\ \theta)).$$

Next define a relation $\theta_1^*$ by $m \equiv m' (\bmod\ \theta_1^*)$ iff there exist $k \geq 0$ and $m_0, \ldots, m_k \in M$ such that $m = m_0$, $m' = m_k$ and $m_i \sim m_{i+1} (\bmod\ \theta_1)$ for $i = 0, \ldots, k-1$. Then it is easily shown that $\theta_1^* = \hat{\theta}$.

**Example 3.1** Let $X$ be an alphabet, and define a relation over $X^*$ by

$$xy \sim yx \qquad \text{for} \quad x, y \in X, \quad x \neq y.$$

Let $\hat{\theta}$ be the congruence generated by this relation. Then $u \equiv v(\text{mod } \hat{\theta})$ iff $|u|_x = |v|_x$ for all $x \in X$. The quotient monoid $X^*/\hat{\theta}$ is denoted by $X^{\oplus}$ and is called the free commutative monoid generated by $X$.

**Exercises**

**3.1** Let $M$ be a group, $M'$ a monoid, and let $\alpha : M \to M'$ be a monoid morphism. Show that $\alpha(M)$ is a group and that $\alpha$ is a group morphism ($\alpha(m^{-1}) = \alpha(m)^{-1}$ for $m \in M$).

**3.2** Give examples of morphisms $\alpha : X^* \to Y^*$, such that $\alpha^{-1}(AB) \supsetneq \alpha^{-1}(A)\alpha^{-1}(B)$, and $\alpha^{-1}(A^*) \supsetneq (\alpha^{-1}(A))^*$.

**3.3** Let $X$, $Y$ be alphabets and let $\alpha : X^* \to Y^*$ be a morphism. Show that there are an alphabet $Z$, an injective morphism $\beta : X^* \to Z^*$ and a projection $\gamma : Z^* \to Y^*$ such that $\alpha = \gamma \circ \beta$.

**3.4** Let $X$ be an alphabet. Given $A \subset X^*$, the norm of $A$ is the number

$$\|A\| = 2^{-\omega(A)}, \qquad \text{where} \quad \omega(A) = \min\{|u| \mid u \in A\}, \quad \omega(\emptyset) = \infty.$$

If $B \subset X^*$, the distance $d(A, B)$ is the number

$$d(A, B) = \|A \backslash B \cup B \backslash A\|.$$

a) Show that $\| \ \|$ and $d$ are a norm and a distance in the usual topological sense, and that $d$ satisfies the ultrametric inequality: $d(A, B) \leqslant \max\{d(A, C), d(C, B)\}$.
b) Let $\alpha : X^* \to Y^*$ be a morphism. Show that the mapping from $\mathfrak{P}(X^*)$ into $\mathfrak{P}(Y^*)$ defined by $\alpha$ is continuous for this topology iff $\alpha(X) \subset Y^+$. (This is the reason why an $\varepsilon$-free morphism is called continuous.)

**3.5** Let $X$ be an alphabet, and let $L \subset X^*$. The syntactic congruence $\theta_L$ of $L$ is the coarsest (greatest) congruence over $X^*$ which saturates $L$, i.e. such that $u \in L$, $u \equiv v(\text{mod } \theta_L)$ implies $v \in L$. Show that

$$u \equiv v(\text{mod } \theta_L) \quad \text{iff for all} \quad f, g \in X^* : fug \in L \quad \Leftrightarrow \quad fvg \in L.$$

The quotient monoid $\text{Synt}(L) = X^*/\theta_L$ is called the syntactic monoid of $L$. Show that $\text{Synt}(L) = \text{Synt}(X^* \backslash L)$.

**3.6** Let $M$, $N$ be monoids, $\alpha : M \to N$ a morphism. Let $A \subset M$, $Q \subset N$, and set $B = \alpha(A)$, $P = \alpha^{-1}(Q)$. Show that $A^{-1}P = \alpha^{-1}(B^{-1} \cdot Q)$.

## I.4 Finite Automata, Regular Languages

In this section, we review some basic facts concerning finite automata, mainly in order to fix notations and to allow references in later chapters. When the proofs are omitted, they can be found in any of the books listed in the bibliography.

**Definition** A finite (deterministic) automaton $\mathbf{A}=\langle X, Q, q_-, Q_+, \delta\rangle$ consists of an alphabet $X$, a finite set $Q$ of states, an initial state $q_- \in Q$, a set of final states $Q_+ \subset Q$, and a next state function $\delta: Q \times X \to Q$.
If no confusion can arise, we denote $\delta$ by a dot, and we write

$$\mathbf{A}=\langle X, Q, q_-, Q_+\rangle$$

instead of the above notation. The next state function is extended to $Q \times X^*$ by setting

$$q \cdot 1 = q \tag{4.1}$$

$$q \cdot ux = (q \cdot u) \cdot x \qquad u \in X^*, x \in X. \tag{4.2}$$

Then the formula

$$q \cdot uv = (q \cdot u) \cdot v \qquad u, v \in X^* \tag{4.3}$$

is easily verified. A word $u \in X^*$ is recognized or accepted by $\mathbf{A}$ iff $q_- \cdot u \in Q_+$. The language recognized by $\mathbf{A}$ is

$$|\mathbf{A}| = \{u \in X^* \mid q_- \cdot u \in Q_+\}. \tag{4.4}$$

A language $L \subset X^*$ is recognizable or regular iff $L = |\mathbf{A}|$ for some finite automaton $\mathbf{A}$.

Finite automata can be represented by a graph in the following way. Each state $q$ is represented by a vertex, and an edge labelled $x$ is drawn from $q$ to $q'$ iff $q \cdot x = q'$. The initial state has an arrow entering in it. Final states are circled twice.

**Example 4.1** Let $\mathbf{A}$ be defined by $X=\{x, y\}$, $Q=\{1,2,3,4\}$, $q_-=1$, $Q_+=\{4\}$, and the next state function given by

| | $x$ | $y$ |
|---|---|---|
| 1 | 2 | 1 |
| 2 | 1 | 3 |
| 3 | 4 | 1 |
| 4 | 2 | 3 |

Then $\mathbf{A}$ is represented in Fig. I.1. A word is recognized by $\mathbf{A}$ iff it has $xyx$ as a right factor. Thus $|\mathbf{A}| = X^*xyx$.

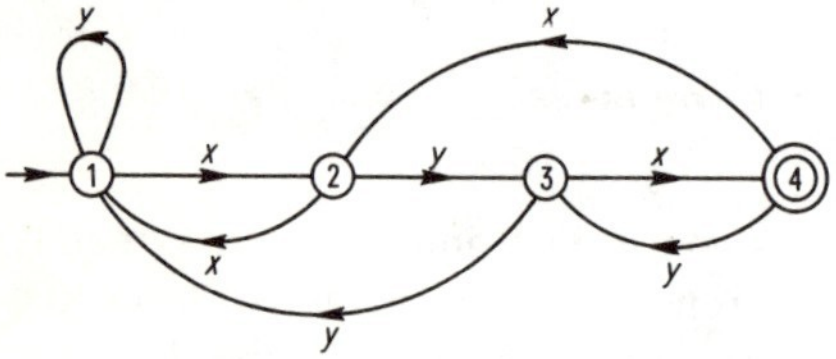


Fig. I.1

The following result is known as Kleene's Theorem.

**Theorem 4.1** (Kleene [1956]) *The family of regular languages over X is equal to the least family of languages over X containing the empty set and the singletons, and closed under union, product, and the star operation.*

We shall see another formulation of this theorem in Section III.2. Following closure properties can be proved for regular languages.

**Proposition 4.2** *Regular languages are closed under union, product, the star and the plus operation, intersection, complementation, reversal, morphism, inverse morphism, regular substitution.*

A substitution $\sigma : X^* \to Y^*$ is called regular iff $\sigma(x)$ is a regular language for all $x \in X$.

There are several variations for the definition of finite automata. Thus in a nondeterministic finite automaton, the next state function is a function from $Q \times X$ into the subsets of $Q$. Thus $q \cdot x \subset Q$ for $q \in Q$, $x \in X$. This notation is extended by defining first

$$Q' \cdot x = \bigcup_{q \in Q'} q \cdot x \quad \text{for} \quad Q' \subset Q. \tag{4.5}$$

Then the next state function can be defined on $Q \times X^*$ by (4.1) and (4.2), and (4.3) is easily seen to hold. The language recognized by $\mathbf{A}$ is then

$$|\mathbf{A}| = \{u \in X^* \mid q_- \cdot u \cap Q_+ \neq \emptyset\}.$$

Note that this definition agrees with (4.4) in the case where $\mathbf{A}$ is deterministic. Note next that (4.5) can also be considered as the definition of the next state function of a deterministic finite automaton $\boldsymbol{B} = \langle X, P, p_-, P_+\rangle$ where $P = \mathfrak{P}(Q)$. With $p_- = \{q_-\}$, and $P_+ = \{Q' \subset Q \mid Q' \cap Q_+ \neq \emptyset\}$, it is easily seen that $|\boldsymbol{B}| = |\mathbf{A}|$. Thus a language is regular iff it is recognized by some nondeterministic finite automaton. Nondeterministic automata are represented pictorially like deterministic automata, by drawing an edge labelled $x$ from $q$ to $q'$ whenever $q' \in q \cdot x$.

**Example 4.2** Fig. I.2 represents a nondeterministic finite automaton $\mathbf{A}$, with alphabet $X = \{x, y\}$. It is easily verified that $|\mathbf{A}| = X^*xyx$.

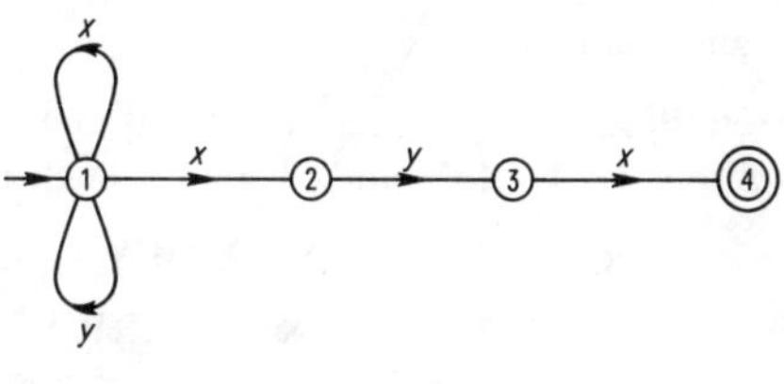


Fig. I.2

Let $\mathbf{A} = \langle X, Q, q_-, Q_+\rangle$ be a finite (deterministic) automaton. A state $q$ is called accessible if $q_- \cdot u = q$ for at least one $u \in X^*$. $\mathbf{A}$ is accessible if all its states are accessible. If $\mathbf{A}$ is not accessible, let $P \subset Q$ be the set of accessible states of $\mathbf{A}$. Then $q_- \in P$. Define $\boldsymbol{B} = \langle X, P, q_-, P_+\rangle$ by $P_+ = Q \cap Q_+$, and by taking as next state function the restriction to $P$ of the next state function of $\mathbf{A}$. Then $|\boldsymbol{B}| = |\mathbf{A}|$. $\boldsymbol{B}$ is called the accessible part of $\mathbf{A}$.

Given a finite automaton $\mathbf{A} = \langle X, Q, q_-, Q_+\rangle$, an equivalence relation called the Nerode equivalence, noted $\equiv$, is defined by

$$q \equiv q' \quad \text{iff for all} \quad u \in X^*, \qquad q \cdot u \in Q_+ \quad \Leftrightarrow \quad q' \cdot u \in Q_+.$$

This equivalence relation is easily seen to be right regular, i.e. to verify $q \equiv q'$, $w \in X^* \Rightarrow q \cdot w \equiv q' \cdot w$. Hence a next state function can be defined on the quotient set $Q/\equiv$ by $[q] \cdot x = [q \cdot x]$ ($[q]$ is the class of $q$ in the equivalence). Let $L = |\mathbf{A}|$, and let

$$\mathbf{A}/\equiv\; = \langle X, Q/\equiv, [q_-], \{[q] \mid q \in Q_+\}\rangle$$

be the quotient automaton with the next state function defined above. Then it can be shown that $|\mathbf{A}/\equiv| = |\mathbf{A}|$, and that the accessible part of $\mathbf{A}/\equiv$ is the unique automaton (up to a renaming of states) recognizing $L$ having a minimal number of states among all finite automata recognizing $L$. Therefore this automaton is called the minimal automaton of the language $L$.

Another useful concept is the notion of semiautomaton. A semiautomaton $\boldsymbol{S} = \langle X, Q, q_-\rangle$ is defined as a finite automaton, but without specifying the set of final states. There is a language recognized by $\boldsymbol{S}$ for any subset $Q' \subset Q$, defined by $|\boldsymbol{S}(Q')| = \{u \in X^* \mid q_- \cdot u \in Q'\}$. Semiautomata are used to recognize "simultaneously" several regular languages: Consider two (more generally any finite number) regular languages $A$, $B \subset X^*$, and let $\mathbf{A} = \langle X, Q, q_-, Q_+\rangle$, $\boldsymbol{B} = \langle X, P, p_-, P_+\rangle$ be finite automata with $|\mathbf{A}| = A$, $|\boldsymbol{B}| = B$. Define a semiautomaton $\boldsymbol{S} = \langle X, Q \times P, (q_-, p_-)\rangle$ by

$$(q, p) \cdot x = (q \cdot x, p \cdot x) \qquad x \in X, \quad (q, p) \in Q \times P.$$

Then $A = |\boldsymbol{S}(Q_+ \times P)|$ and $B = |\boldsymbol{S}(Q \times P_+)|$. Usually only the accessible part of $\boldsymbol{S}$ is conserved in this construction.

There exist several characterizations of regular languages. The first uses local regular languages.

**Definition** A language $K \subset X^*$ is a local regular language iff there are subsets $U$, $V$ of $X$ and $W$ of $X^2$ such that

$$K = (UX^* \cap X^* V) \backslash X^* W X^*$$

or $$K = 1 \cup (UX^* \cap X^* V) \backslash X^* W X^*.$$

Clearly, such a language is regular.

The terminology is justified by the following observation: In order to check that a word $w$ is in $K$, it suffices to verify that the first letter of $w$ is in $U$, the last letter of $w$ is in $V$, and that no couple of consecutive letters of $w$ is in $W$. These verifications are all of local nature. The set $W$ is called the set of forbidden transitions, and $X^2\backslash W$ is called the set of authorized transitions.

**Proposition 4.3** *A language $L \subset X^*$ is regular iff there are an alphabet $Z$, a local regular language $K \subset Z^*$, and a strictly alphabetic morphism $\alpha: Z^* \to X^*$ such that $\alpha(K) = L$.*

Proof. By Proposition 4.2 $\alpha(K)$ is regular for a regular language $K$. Conversely, let $\mathbf{A} = \langle X, Q, q_-, Q_+\rangle$ be a finite automaton such that $L = |\mathbf{A}|$. Define $Z$ by

$$Z = \{(q, x, q \cdot x) \mid q \in Q, x \in X\}$$

and define a morphism $\alpha: Z^* \to X^*$ by $\alpha((q, x, q \cdot x)) = x$. Then $\alpha$ is strictly alphabetic. Next let

$$U = \{(q, x, q \cdot x) \mid q = q_-\}, \qquad V = \{(q, x, q \cdot x) \mid q \cdot x \in Q_+\}$$
$$W = \{(q_1, x_1, q_1 \cdot x_1)(q_2, x_2, q_2 \cdot x_2) \mid q_1 \cdot x_1 \neq q_2\}$$

and set $K = (UZ^* \cap Z^*V)\backslash Z^*WZ^*$. Then for $n \geq 1$,

$$c = (q_1, x_1, q_1 \cdot x_1)(q_2, x_2, q_2 \cdot x_2) \cdots (q_n, x_n, q_n \cdot x_n) \in K \tag{4.6}$$

iff

$$q_1 = q_-, \qquad q_{i+1} = q_i \cdot x_i \qquad i = 1, \ldots, n-1, \qquad q_n \cdot x_n \in Q_+. \tag{4.7}$$

Consequently $\alpha(c) = x_1 x_2 \cdots x_n \in L$. Conversely, if $u = x_1 x_2 \cdots x_n \in L$, $(n \geq 1, x_i \in X)$, then there are states $q_1, \ldots, q_n$ such that (4.7) holds, and in view of (4.6), $u \in \alpha(K)$. Thus $L = \alpha(K)$ if $1 \notin L$. If $1 \in L$, the same equality holds if the empty word is added to $K$. ■

Another important characterization of regular languages is the following.

**Proposition 4.4** *A language $L \subset X^*$ is regular iff there exist a finite monoid $M$, a morphism $\alpha: X^* \to M$, and a subset $R \subset M$ such that $L = \alpha^{-1}(R)$.*

Proof. We first show that the condition is necessary. Consider a finite automaton $\mathbf{A} = \langle X, Q, q_-, Q_+\rangle$ such that $L = |\mathbf{A}|$. For each word $w$ define a mapping $\bar{w}: Q \to Q$ which associates to $q \in Q$ the state $q \cdot w$. For convenience, we write the function symbol on the right of the argument. Thus $(q)\bar{w} = q \cdot w$. Then

$$(q)\overline{ww'} = q \cdot ww' = (q \cdot w) \cdot w' = (q\bar{w})\bar{w}' \tag{4.8}$$
$$(q)\bar{1} = q \cdot 1 = q. \tag{4.9}$$

Let $\alpha$ be the function from $X^*$ into the (finite!) monoid $Q^Q$ of all functions from $Q$ into $Q$ defined by $\alpha(w) = \bar{w}$. Then $\alpha$ is a morphism in view of (4.8) and (4.9). Next, define $R \subset Q^Q$ by $R = \{m \in Q^Q \mid (q_-)m \in Q_+\}$. Then $w \in L$ iff $q_- \cdot w \in Q_+$, thus iff $\alpha(w) \in R$. Consequently $L = \alpha^{-1}(R)$.

Conversely, define a finite automaton $\mathbf{A}=\langle X, M, 1_M, R\rangle$ by setting

$$m \cdot x = m\alpha(x) \qquad m \in M, \quad x \in X.$$

Since $\alpha$ is a morphism, $m \cdot w = m\alpha(w)$ for all $w \in X^*$. Consequently $w \in |\mathbf{A}|$ iff $1_M\alpha(w)=\alpha(w)\in R$, thus iff $w \in \alpha^{-1}(R)$. ■

There exist several versions of the Iteration Lemma or Pumping Lemma for regular languages. The most general formulation is perhaps the analogue of an Iteration Lemma for context-free languages proved by Ogden (Lemma II.2.3). Let $X$ be an alphabet, and consider a word

$$f = x_1x_2 \cdots x_n \qquad (x_i \in X).$$

Then a position in $f$ is any integer $i \in \{1, \ldots, n\}$. Given a subset $I$ of $\{1, \ldots, n\}$, a position $i$ is called marked with respect to $I$ iff $i \in I$.

**Lemma 4.5** (Ogden's Iteration Lemma for Regular Languages) *Let $L \subset X^*$ be a regular language. Then there exists an integer $N \geqslant 1$ such that, for any word $f \in L$, and for any choice of at least $N$ marked positions in $f$, $f$ admits a factorization $f = aub$, $(a, u, b \in X^*)$ such that*
(i) *$u$ contains at least one and at most $N$ marked positions;*
(ii) *$au^*b \subset L$.*

Proof. Let $\mathbf{A}=\langle X, Q, q_-, Q_+\rangle$ be a finite automaton recognizing $L$, and set $N = \mathrm{Card}(Q)$. *Let* $f = x_1x_2 \cdots x_n$, $(x_j \in X)$ be a word in $L$, and consider a choice $I \subset \{1, \ldots, n\}$ of at least $N$ marked positions in $f$. Since $\mathrm{Card}(I) \geqslant N$, we have $n \geqslant N$. Let $1 \leqslant i_1 < i_2 < \cdots < i_N \leqslant n$ be the $N$ smallest elements of $I$, and define a factorization

$$f = g_0g_1 \cdots g_Ng_{N+1}$$

by $\quad g_0 = x_1 \cdots x_{i_1-1}, \qquad g_1 = x_{i_1}, \qquad g_k = x_{i_{k-1}+1} \cdots x_{i_k} \quad (k=2, \ldots, N),$

$$g_{N+1} = x_{i_N+1} \cdots x_n.$$

Then each $g_k (1 \leqslant k \leqslant N)$ contains exactly one marked position. Set

$$q_0 = q_- \cdot g_0, \qquad q_k = q_{k-1} \cdot g_k \quad (k=1, \ldots, N), \qquad q_+ = q_N \cdot g_{N+1}.$$

By assumption, $q_+ \in Q_+$. Next two among the $N+1$ states $q_0, \ldots, q_N$ are equal. Thus there exist $i, j$, $(0 \leqslant i < j \leqslant N)$, such that $q_i = q_j$. Define

$$a = g_0g_1 \cdots g_i, \qquad u = g_{i+1} \cdots g_j, \qquad b = g_{j+1} \cdots g_{N+1}.$$

Then $q_- \cdot a = q_- \cdot au = q_- \cdot au^m = q_j$ for all $m \geqslant 1$, whence $q_- \cdot au^mb = q_+$ for all $m \geqslant 0$, and $au^*b \subset L$. Next $u$ contains exactly $j-i$ marked positions. Since $0 < j-i \leqslant N$, this proves the lemma. ■

If the marked positions in $f$ are chosen to be consecutive, the same proof gives the following corollary.

**Corollary 4.6** *Let $L \subset X^*$ be a regular language. Then there exists an integer $N \geq 1$ such that for any word $f \in L$, and for any factorization $f = hgh'$ with $|g| \geq N$, $g$ admits a factorization $g = aub$ such that*
(i) $0 < |u| \leq N$;
(ii) $hau^*bh' \subset L$.

If $Y$ is a subset of $X$, and if the marked positions are chosen to be occurrences of letters in $Y$, we obtain

**Corollary 4.7** *Let $L \subset X^*$ be a regular language, and let $Y \subset X$. Then there is an integer $N \geq 1$ such that for any $f \in L$, and for any factorization $f = hgh'$ with $|g|_Y \geq N$, $g$ admits a factorization $g = aub$ such that*
(i) $0 < |u|_Y \leq N$;
(ii) $hau^*bh' \subset L'$.

## Exercises

**4.1** Let $K$ be a local regular language, and let $a$, $u$, $b$ be words. Show that if $aub$, $au^2b \in K$, then $au^+b \subset K$.

**4.2** Let $K \subset X^*$ be a regular language. Show that there are two integers $N, M$ such that if $au^kb \in K$ and $k \geq N$, then $au^k(u^M)^*b \subset K$.

**4.3** (continuation of Exercise 3.5). Let $L \subset X^*$, and assume that $L = \alpha^{-1}(P)$, where $\alpha$ is a morphism from $X^*$ onto a monoid $M$, and $P \subset M$. Show that there is a morphism $\beta : M \rightarrow \text{Synt}(L)$, and $R \subset \text{Synt}(L)$, such that $P = \beta^{-1}(R)$. Show that $L$ is regular iff $\text{Synt}(L)$ is finite. (Since $\text{Synt}(L) = \text{Synt}(X^* \setminus L)$, such a characterization cannot exist for context-free languages. See Perrot and Sakarovitch [1977].)

# II Context-Free Languages

The first section of this chapter contains the definitions of context-free or algebraic languages by means of context-free grammars and of systems of algebraic equations. In the second section, we recall without proof several constructions and closure properties of context-free languages. This section contains also the iteration lemmas for context-free languages. The third section gives a description of the various families of Dyck languages. They have two definitions, as classes of certain congruences, and as languages generated by some context-free grammars. The section ends with a proof of the Chomsky-Schützenberger Theorem. Two other languages, the Lukasiewicz language and the language of completely parenthesized arithmetic expressions, are studied in the last section.

## II.1 Grammars, Languages, Equations

In this section we define context-free grammars and context-free languages. We show how a system of equations can be associated to each context-free grammar in such a way that the languages generated by the grammar are precisely the minimal solution of the system of equations. For this reason, context-free languages and more generally context-free grammars are also called algebraic languages and algebraic grammars.

**Definition** A context-free or algebraic grammar $G=\langle V, X, P\rangle$ consists of an alphabet $V$ of variables or nonterminals, of an alphabet $X$, disjoint from $V$, of terminal letters, and of a finite set $P\subset V\times(V\cup X)^*$ of productions.

A production $(\xi, \alpha)\in P$ is usually written in the form

$$\xi\rightarrow\alpha.$$

If $\xi\rightarrow\alpha_1, \xi\rightarrow\alpha_2, \ldots, \xi\rightarrow\alpha_n$ are the productions of $G$ having the same left side $\xi$, they are grouped together by using one of the following notations

$$\xi\rightarrow\alpha_1\mid\alpha_2\mid\cdots\mid\alpha_n; \qquad \xi\rightarrow\alpha_1+\alpha_2+\cdots+\alpha_n; \qquad \xi\rightarrow\{\alpha_1, \alpha_2, \ldots, \alpha_n\}.$$

Clearly, the above definition is equivalent to another notation consisting of triple $T, X, P$, where $T$ is the total alphabet and $X$ is a subset of $T$. Then $V=T\backslash X$.

**Example 1.1** Let $V=\{\xi\}$, $X=\{a, b\}$, $P=\{\xi\rightarrow\xi\xi, \xi\rightarrow a\}$. Then the productions can be written as $\xi\rightarrow\xi\xi+a$.

**Example 1.2** Let $V=\{\xi, \xi_a, \xi_b\}$, $X=\{a, b\}$, and the set $P$ given by:

$$\xi \to 1 + a\xi_a b\xi + b\xi_b a\xi; \qquad \xi_a \to 1 + a\xi_a b\xi_a;$$
$$\xi_b \to 1 + b\xi_b a\xi_b.$$

Let $G=\langle V, X, P\rangle$ be a context-free grammar, and let $f, g$ be in $(V\cup X)^*$. Then we define

$$f \underset{G}{\to} g \tag{1.1}$$

iff there are factorizations $f=u\xi v$, $g=u\alpha v$ with $\xi\in V$, $u, \alpha, v\in(V\cup X)^*$ and $\xi\to\alpha\in P$. If no confusion can arise, we write $f\to g$ instead of (1.1). For any $p\geqslant 0$, define

$$f \overset{p}{\underset{G}{\to}} g$$

iff there exist $f_0, f_1, \ldots, f_p\in(V\cup X)^*$ such that

$$f=f_0, \qquad g=f_p, \qquad \text{and} \quad f_{i-1}\underset{G}{\to} f_i \qquad \text{for} \quad i=1,\ldots,p.$$

(In particular, $f \overset{0}{\underset{G}{\to}} f$ for any $f\in(V\cup X)^*$). The sequence

$$(f_0, f_1, \ldots, f_p)$$

is a derivation from $f$ into $g$, and $p$ is the length of the derivation. Finally, we define

$$f \overset{*}{\underset{G}{\to}} g \quad \text{iff} \quad f \overset{p}{\underset{G}{\to}} g \qquad \text{for some} \quad p\geqslant 0;$$

$$f \overset{+}{\underset{G}{\to}} g \quad \text{iff} \quad f \overset{p}{\underset{G}{\to}} g \qquad \text{for some} \quad p>0.$$

In the first case, we say that $f$ derives $g$ in $G$, in the second case that $f$ properly derives $g$ in $G$. If no confusion can arise, the index $G$ is dropped. For any variable $\xi\in V$, the language generated by $\xi$ in $G$ is

$$L_G(\xi)=\{w\in X^* \mid \xi \overset{*}{\to} w\}.$$

More generally, the language generated by $f\in(V\cup X)^*$ in $G$ is

$$L_G(f)=\{w\in X^* \mid f \overset{*}{\to} w\}.$$

Clearly, $L_G(f)=\{f\}$ for $f\in X^*$. The language of sentential forms generated by $f\in(V\cup X)^*$ in $G$ is

$$\hat{L}_G(f)=\{w\in(V\cup X)^* \mid f \overset{*}{\to} w\}.$$

Of course,

$$L_G(f)=\hat{L}_G(f)\cap X^*.$$

As noted by M. P. Schützenberger [1961a], there is a close relation between the derivations in a context-free grammar and derivations in an algebra. Recall that a derivation in an algebra $M$ is a linear function $\partial$ satisfying

$$\partial(xy)=\partial(x)\cdot y+x\cdot\partial(y) \qquad x, y\in M.$$

Given a grammar $G=\langle V, X, P\rangle$, define $\partial:(V\cup X)^*\to\mathfrak{P}((V\cup X)^*)$ by

$$g\in\partial(f) \quad \text{iff} \quad f\underset{G}{\to} g,$$

and for $A\subset(V\cup X)^*:\partial(A)=\bigcup_{f\in A}\partial(f)$. Then (see Lemma 1.1 below) we have

$$\partial(AB)=\partial(A)\cdot B\cup A\cdot\partial(B).$$

Further

$$\hat{L}_G(A)=\partial^*(A),$$

where $\partial^*(A)=\bigcup_{n\geqslant 0}\partial^n(A)$. Thus $L_G(f)=\partial^*(f)\cap X^*$.

A context-free grammar $G=\langle V, X, P\rangle$ generates a language $L\subset X^*$ iff $L=L_G(\xi)$ for some $\xi\in V$. Thus a grammar $G$ generates Card($V$) languages, not necessarily distinct.

**Definition** A language $L$ is context-free or algebraic if there is some grammar $G$ that generates $L$. The set of all context-free languages of $X^*$ is denoted by Alg($X^*$).

**Example 1.1** (continued). The language generated by $\xi$ is $L_G(\xi)=a^+$; further $\hat{L}_G(\xi)=\{a, \xi\}^+$.

**Example 1.2** (continued). The language generated by $\xi_a$ is the so-called restricted Dyck language $D_1'^*$ over $X$, with opening parenthesis $a$, and closing parenthesis $b$ (see also Section 3); $L_G(\xi_b)$ is obtained from $L_G(\xi_a)$ by exchanging $a$ and $b$. Finally, $L_G(\xi)$ is the Dyck set over $X$, consisting of all words $w$ such that $|w|_a=|w|_b$.

Let $G=\langle V, X, P\rangle$ be a context-free grammar. The following lemma is very useful.

**Lemma 1.1** *Let $f_1, f_2, g\in(V\cup X)^*$, and let $p\geqslant 0$ be an integer. Then*

$$f_1f_2\overset{p}{\to} g$$

*iff there are $g_1, g_2\in(V\cup X)^*$, $p_1, p_2\geqslant 0$ such that*

$$f_1\overset{p_1}{\to} g_1, \qquad f_2\overset{p_2}{\to} g_2, \qquad g=g_1g_2, \qquad p=p_1+p_2.$$

Proof. If $f_1 \xrightarrow{p_1} g_1$ and $f_2 \xrightarrow{p_2} g_2$, then

$$f_1f_2 \xrightarrow{p_1} g_1f_2 \xrightarrow{p_2} g_1g_2, \quad \text{thus} \quad f_1f_2 \xrightarrow{p} g.$$

Assume conversely $f_1f_2 \xrightarrow{p} g$. If $p=0$, there is nothing to prove. Arguing by induction on $p$, suppose $p>0$. Then

$$f_1f_2 \xrightarrow{p-1} h \rightarrow g$$

for some $h \in (V \cup X)^*$, and by induction there is a factorization

$$h = h_1h_2 \qquad \text{with} \quad f_1 \xrightarrow{q_1} h_1, \quad f_2 \xrightarrow{q_2} h_2, \quad q_1+q_2=p-1.$$

Since $h \rightarrow g$, there are $\xi \rightarrow \alpha \in P$, and words $u$, $v$ with

$$h = u\xi v, \qquad g = u\alpha v.$$

If $|u\xi| \leq |h_1|$, then $h_1 = u\xi\bar{u}$ for some word $\bar{u}$. Thus setting

$$g_1 = u\alpha\bar{u}, \qquad g_2 = h_2,$$

we obtain

$$g = g_1g_2, \qquad f_1 \xrightarrow{1+q_1} g_1, \qquad f_2 \xrightarrow{q_2} g_2.$$

Otherwise, $|\xi v| \leq |h_2|$, and a symmetric argument achieves the proof. ∎

**Corollary 1.2** *For any* $f_1, f_2 \in (V \cup X)^*$, $L_G(f_1f_2) = L_G(f_1)L_G(f_2)$.

Proof. By the preceding lemma, $w \in L_G(f_1f_2)$ iff there is a factorization $w = w_1w_2$ such that $w_1 \in L_G(f_1)$, $w_2 \in L_G(f_2)$. ∎

Since $L_G(1) = \{1\}$, the mapping $f \mapsto L_G(f)$ is a substitution from $(V \cup X)^*$ into itself. We denote it by $L_G$. The same is true for $\hat{L}_G$ (Exercise 1.2).

**Lemma 1.3** *Let* $G = \langle V, X, P \rangle$ *be a context-free grammar,* $\xi \in V$. *Then*

$$L_G(\xi) = \bigcup_{\xi \rightarrow \alpha \in P} L_G(\alpha) = L_G(\{\alpha \mid \xi \rightarrow \alpha \in P\}).$$

Proof. For $\xi \rightarrow \alpha \in P$, clearly $L_G(\alpha) \subset L_G(\xi)$. Conversely, let $w \in L_G(\xi)$. Since $w \in X^*$, we have $\xi \xrightarrow{+} w$. Thus there is a production $\xi \rightarrow \alpha \in P$ such that $\xi \rightarrow \alpha \xrightarrow{*} w$. Thus $w \in L_G(\alpha)$. ∎

Now we associate to each context-free grammar a system of equations. We shall see that the minimal solution of the system of equations is formed of the languages generated by the grammar. In certain special cases, the system has a unique solution. This gives a characterization of context-free languages by systems of equations.

**Definition** Let $V=\{\xi_1,\ldots,\xi_N\}$ and $X$ be two disjoint alphabets. A *system of algebraic equations* is a set

$$\xi_i=P_i \qquad i=1,\ldots,N \tag{1.2}$$

of equations, where $P_1,\ldots,P_N$ are *finite* subsets of $(V\cup X)^*$. The letters $\xi_i$ are called the *variables* of the system.

The terminology comes from the analogy with systems of algebraic equations over, say the field of real numbers. Such a system is given by a set of polynomial equations $Q_i(y_1, y_2,\ldots,y_N)=0$, $(i=1,\ldots,N)$. In the case where, in each $Q_i$, there is a monomial $y_i$, the system can be written in the form $y_i=Q_i'(y_1,y_2,\ldots,y_N)$, with each $Q_i'$ a polynomial. In the same manner, the sets $P_i$ of (1.2) can be considered as "polynomials" by writing

$$P_i=\sum_{\alpha\in P_i}\alpha$$

with coefficients in the boolean semiring. The theory of systems of algebraic equations over arbitrary semirings allows in particular to take into account the ambiguity of a grammar. This is beyond the scope of the book. See Salomaa and Soittola [1978] and Eilenberg [1978].

The correspondence between systems of algebraic equations and context-free grammars is established as follows. Given a context-free grammar $G=\langle V, X, P\rangle$, number the nonterminals such that $V=\{\xi_1,\ldots,\xi_N\}$ with $N=\text{Card}(V)$, and define

$$P_i=\{\alpha\mid \xi_i\to\alpha\in P\} \qquad i=1,\ldots,N.$$

Then (1.2) is the system of equations *associated* to $G$.
Conversely, the context-free grammar *associated* to (1.2) has as set of productions

$$P=\{\xi_i\to\alpha\mid \alpha\in P_i,\ 1\le i\le N\}.$$

We now define a solution of (1.2) as a vector $A=(A_1,\ldots,A_N)$ of languages such that the substitution in $P_i$ of the language $A_j$ to each occurrence of $\xi_j$ yields precisely the language $A_i$.
Formally, given (1.2) let $A=(A_1,\ldots,A_N)$ with $A_i\subset(V\cup X)^*$ for $i=1,\ldots,N$. Define a substitution $\underset{\sim}{A}$ from $(V\cup X)^*$ into itself by

$$\underset{\sim}{A}(x)=\{x\} \qquad x\in X;$$
$$\underset{\sim}{A}(\xi_i)=A_i \qquad i=1,\ldots,N.$$

**Definition** The vector $A=(A_1,\ldots,A_N)$ is a solution of the system of equations (1.2) iff

$$\underline{A}(P_i)=A_i \qquad i=1,\ldots,N. \tag{1.3}$$

**Example 1.1** (continued). The equations $\xi=\xi\xi+a$ has the solution $a^+$ since $a^+=a^+a^+\cup a$ and also the solutions $X^*$ and $(V\cup X)^*$, since $X^*=X^*X^*\cup a$ and similarly for the second set.

**Example 1.2** (continued). As will be shown below, the vector $(L_G(\xi), L_G(\xi_a), L_G(\xi_b))$ is the unique solution of the system

$$\xi=1+a\xi_a b\xi+b\xi_b a\xi; \qquad \xi_a=1+a\xi_a b\xi_a; \qquad \xi_b=1+b\xi_b a\xi_b.$$

A system of equations (1.2) may have several, and even an infinity of solutions. We order the solutions by setting, for $A=(A_1,\ldots,A_N)$, $B=(B_1,\ldots,B_N)$, $A\subset B$ iff $A_i\subset B_i$ for $i=1,\ldots,N$.

**Theorem 1.4** *Let G be a context-free grammar, and let* (1.2) *be the system of algebraic equations associated to G. The vector $L_G=(L_G(\xi_1),\ldots,L_G(\xi_N))$ is the minimal solution of* (1.2).

This result contains a converse statement: given a system of algebraic equations, the components of the minimal solution are context-free languages. For this reason, context-free languages are called algebraic languages. Note that only the components of the minimal solution are claimed to be context-free. There are solutions of systems which are not context-free (Exercise 1.4).

Proof. By definition, we have $\underline{L}_G(f)=L_G(f)$ for all $f\in(V\cup X)^*$. We shall verify that the substitution $L_G$ satisfies (1.3). Indeed, in view of Lemma 1.3,

$$L_G(P_i)=\bigcup_{\alpha\in P_i} L_G(\alpha)=L_G(\xi_i) \qquad i=1,\ldots,N.$$

This shows that $L_G=(L_G(\xi_1),\ldots,L_G(\xi_N))$ is a solution of (1.2). Next, let $A=(A_1,\ldots,A_N)$ be another solution of (1.2). We show that

$$L_G(f)\subset\underline{A}(f) \qquad f\in(V\cup X)^* \tag{1.4}$$

by induction on the length of the derivation of the words of $L_G(f)$. Let $w\in L_G(f)$. If $f\xrightarrow{0}w$, then $f=w\in X^*$ and $w\in\underline{A}(f)$. Assume now $f\xrightarrow{p}w$ and $p>0$. There exist a word $g$ such that

$$f\to g\xrightarrow{p-1}w,$$

and factorizations $f=u\xi_i v$, $g=u\alpha v$ such that $\xi_i\to\alpha\in P$. Since $w\in\underline{A}(g)$ by induction, it follows that

$$w \in \underline{A}(g) = \underline{A}(u)\underline{A}(\alpha)\underline{A}(v) \subset \underline{A}(u)\underline{A}(P_i)\underline{A}(v).$$

Since $\underline{A}(P_i) = A_i = \underline{A}(\xi_i)$, we have

$$\underline{A}(u)\underline{A}(P_i)\underline{A}(v) = \underline{A}(u\xi_i v) = \underline{A}(f).$$

Thus $w \in \underline{A}(f)$. From (1.4), we obtain

$$L_G(\xi_i) \subset \underline{A}(\xi_i) = A_i \qquad i = 1, \ldots, N.$$ ∎

We now show that in some cases, a system of algebraic equations has a unique solution.

**Definition** An algebraic grammar $G = \langle V, X, P\rangle$ is strict iff for each production $\xi \to \alpha \in P$, either $\alpha = 1$ or $\alpha$ contains at least one terminal letter, thus iff

$$\alpha \in 1 \cup (V \cup X)^* X (V \cup X)^*.$$

A system of equations is strict if the associated grammar is strict.

By Greibach's Normal Form Theorem, a strict grammar can be supplied for any context-free language (see the books listed in the bibliography).

**Theorem 1.5** *Let $G = \langle V, X, P\rangle$ be a context-free grammar. If $G$ is strict, then $L_G = (L_G(\xi_1), \ldots, L_G(\xi_N))$ is the unique solution of the system of equations associated to $G$.*

Proof. Let (1.2) be the system of equations associated to $G$, and let $A = (A_1, \ldots, A_N)$, $B = (B_1, \ldots, B_N)$ be two solutions of this system. We prove:

$$\text{for} \quad i = 1, \ldots, N, \quad w \in A_i, \quad |w| \leq n \quad \text{implies} \quad w \in B_i \tag{1.5}$$

by induction on $n$. This shows that $A \subset B$, and $A = B$ by symmetry.

If $1 \in A_i = \underline{A}(P_i)$ then $1 \in \underline{A}(\alpha)$ for some $\alpha \in P_i$, and since $G$ is strict, this implies $\alpha = 1$. Thus $1 \in P_i$ and $1 \in \underline{B}(P_i) = B_i$. Assume $w \in A_i$ and $|w| = n > 0$. As before, $w \in \underline{A}(\alpha)$ for some $\alpha \in P_i$. If $\alpha \in X^*$, then $\alpha = w \in P_i$ and $w \in B_i$. Thus suppose the contrary. Then

$$\alpha = u_0 \xi_{i_1} u_1 \cdots u_{r-1} \xi_{i_r} u_r,$$

with $r \geq 1$, $u_0, \ldots, u_r \in X^*$, $\xi_{i_1}, \ldots, \xi_{i_r} \in V$. Therefore

$$w = u_0 v_1 u_1 \cdots u_{r-1} v_r u_r,$$

with $v_k \in \underline{A}(\xi_{i_k}) = A_{i_k}$ for $k = 1, \ldots, r$. Now $u_0 u_1 \cdots u_r \neq 1$ since $G$ is strict. Since $r \geq 1$, $|v_k| < n$ for all $k = 1, \ldots, r$, and by the induction hypothesis, $v_k \in B_{i_k} = \underline{B}(\xi_{i_k})$ for $k = 1, \ldots, n$. Thus

$$w \in u_0 \underline{B}(\xi_{i_1}) u_1 \cdots u_{r-1} \underline{B}(\xi_{i_r}) u_r = \underline{B}(\alpha) \subset B_i.$$ ∎

Note that the finiteness of the sets $P_i$ was used in none of the proofs of Theorems 1.4 or 1.5. Thus these remain true if the sets $P_i$ are infinite, provided grammars with infinite sets of productions are allowed or alternatively, if the connections with grammars are dropped in the statements. Thus especially Theorem 1.5 can be used to proof unicity of the solution of equations (Exercise 1.5).

We conclude this section with a result that permits transformations of system of equations without changing the set of solutions. This is used later to show that systems of equations which are not strict have a unique solution by transforming them into strict systems.

**Definition** Two systems of equations

$$\xi_i = P_i \quad (1 \leq i \leq N) \qquad \text{and} \qquad \xi_i = Q_i \quad (1 \leq i \leq N)$$

with the same set of variables are equivalent if they have the same set of solutions.

**Proposition 1.6** (Substitution Lemma) *Let*

$$\xi_i = P_i \qquad i = 1, \ldots, N \tag{1.6}$$

*be a system of equation. Assume that* $\alpha = u\xi_j v \in P_k$ *for some* $j, k \in \{1, \ldots, N\}$ *and some words* $u, v$. *Define* $Q_i = P_i$ *for* $i \neq k$, *and* $Q_k = (P_k \backslash \alpha) \cup uP_j v$. *Then* (1.6) *is equivalent to*

$$\xi_i = Q_i \qquad i = 1, \ldots, N. \tag{1.7}$$

**Example 1.1** (continued). Starting with $\xi = P$, where $P = a + \xi\xi$, we single out $\alpha = \xi\xi$, and form $Q = P \backslash \alpha \cup \xi P = a + \xi\xi\xi + \xi a$. The substitution lemma claims that the equation $\xi = a + \xi a + \xi\xi\xi$ is the equivalent to the initial one.

**Example 1.3** Let $X = \{a, b\}$, and consider the system

$$\xi = 1 + \eta\xi; \qquad \eta = a\xi b.$$

Taking $\alpha = \eta\xi$, and replacing $\eta$ by $a\xi b$, yields the strict system

$$\xi = 1 + a\xi b\xi; \qquad \eta = a\xi b.$$

By the Substitution Lemma, the first system has a unique solution.

For the proof of Proposition 1.6, we need a technical lemma.

**Lemma 1.7** *Let* $Y$ *be an alphabet, and let* $A, B, C, L$ *and* $M$ *be subsets of* $Y^*$. *If*

$$L = A \cup BMC \quad \text{and} \quad M = A \cup BLC,$$

*then* $L = M$.

Proof. If $B = \emptyset$ or $C = \emptyset$, then $L = M = A$. Next, if $1 \in B$ and $1 \in C$, then by the first equation $M \subset BMC \subset L$, and similarly $L \subset M$, hence $L = M$. Thus we may assume that $1 \notin BC$. Then $1 \in L$ iff $1 \in A$, hence iff $1 \in M$. Arguing by induction on the length of words, consider $w \in Y^*$, $|w| = p > 0$, and assume $w \notin A$. Then $w \in L$ iff $w \in BMC$, hence iff $w = bw'c$ with $b \in B$, $c \in C$, $w' \in M$ and $|w'| < p$. Thus $w' \in L$ and $w \in BLC \subset M$. Similarly, $w \in M$ implies $w \in L$. This proves the lemma. ■

Proof of Proposition 1.6. Let $A = (A_1, \ldots, A_N)$ by a solution of (1.6). By definition, $\underline{A}(P_i) = A_i$ for $i = 1, \ldots, N$. Thus

$$\underline{A}(Q_i) = A_i \qquad i = 1, \ldots, N, i \neq k,$$

$$\underline{A}(Q_k) = \underline{A}(P_k \backslash \alpha) \cup \underline{A}(u)\underline{A}(P_j)\underline{A}(v) = \underline{A}(P_k \backslash \alpha) \cup \underline{A}(\alpha) = A_k,$$

showing that $A$ is a solution of (1.7).

Conversely, let $B = (B_1, \ldots, B_N)$ be a solution of (1.7). Then

$$\underline{B}(P_i) = B_i \qquad i = 1, \ldots, N, i \neq k,$$

$$\begin{aligned} \underline{B}(P_k) &= \underline{B}(P_k \backslash \alpha) \cup \underline{B}(\alpha) = \underline{B}(P_k \backslash \alpha) \cup \underline{B}(u)\underline{B}(\xi_j)\underline{B}(v) \\ &= \underline{B}(P_k \backslash \alpha) \cup \underline{B}(u)\underline{B}(Q_j)\underline{B}(v). \end{aligned} \tag{1.8}$$

If $j \neq k$, then $Q_j = P_j$. Thus

$$\underline{B}(P_k) = \underline{B}(P_k \backslash \alpha \cup uP_jv) = \underline{B}(Q_k) = B_k,$$

and $B$ is a solution of (1.6). If $j = k$, then by (1.8)

$$\underline{B}(P_k) = \underline{B}(P_k \backslash \alpha) \cup \underline{B}(u)\underline{B}(Q_k)\underline{B}(v);$$

$$\underline{B}(Q_k) = \underline{B}(P_k \backslash \alpha) \cup \underline{B}(u)\underline{B}(P_k)\underline{B}(v)$$

by definition. In view of Lemma 1.7, we have

$$\underline{B}(P_k) = \underline{B}(Q_k) = B_k;$$

thus $B$ is a solution of (1.6). ■

## Exercises

**1.1** A context-free grammar $G = \langle V, X, P \rangle$ is called proper if each production $\xi \to \alpha \in P$ verifies $\alpha \notin 1 \cup V$. A vector $A = (A_1, \ldots, A_N)$ of languages is called proper if $1 \notin A_i$ for $i = 1, \ldots, N$. Prove that the system of equations associated to a proper grammar has a unique proper solution.

**1.2** Show that Corollary 1.2 remains true if $L_G$ is replaced by $\hat{L}_G$, and that Lemma 1.3 becomes false.

**1.3** Show that the languages of sentential forms of a context-free grammar are context-free.

**1.4** Show that there are non context-free languages among the solutions of the equation of Example 1.1

**1.5** Use Theorem 1.5 to show that the equation $\xi = A \cup \xi B$ has the unique solution $AB^*$ provided $1 \notin B$.

## II.2 Closure Properties, Iteration

We recall here some closure properties of the family of context-free languages, and also the iteration lemmas for these languages. We give no proof; we just recall some constructions that will be used later. Proofs of the results stated here can be found in standard books on formal languages (see Bibliography).

**Theorem 2.1** *Context-free languages are closed under union, product, star operation, reversal, morphism, inverse morphism, intersection with regular sets, context-free substitution.*

A context-free substitution is a substitution $\theta : X^* \to Y^*$ such that $\theta(x)$ is a context-free language for each $x \in X$.

We now recall the usual constructions employed to prove closure under morphism, inverse morphism, intersection with regular sets and substitution. Let $L \subset X^*$ be an algebraic language, let $G = \langle V, X, P \rangle$ be an algebraic language and let $\sigma \in V$ be such that $L = L_G(\sigma)$.

**a) Morphism** Let $\psi : X^* \to Y^*$ be a morphism. Extend $\psi$ to a morphism from $(V \cup X)^*$ into $(V \cup Y)^*$ by setting $\psi(\xi) = \xi$ for $\xi \in V$. Define a grammar

$$\psi G = \langle V, Y, \psi P \rangle$$

by $\psi P = \{\xi \to \psi(\alpha) \mid \xi \to \alpha \in P\}.$

Then it is readily shown that

$$\psi L_G(\xi) = L_{\psi G}(\xi) \qquad \xi \in V.$$

Thus $\psi L = L_{\psi G}(\sigma)$ is a context-free language.

**b) Inverse alphabetic morphism** Let $\varphi : Y^* \to X^*$ be an alphabetic morphism. As above, extend $\varphi$ to $(V \cup Y)^*$ by setting $\varphi(\xi) = \xi$ for $\xi \in V$. Define $Z = \{y \in Y \mid \varphi(y) = 1\}$ and $T = Y \backslash Z$. Finally, let $\omega$ be a new letter ($\omega \notin V \cup X \cup Y$). Define a grammar

$$\varphi^{-1}G = \langle \omega \cup V, Y, P' \rangle$$

where $P' = P'' \cup \left\{\omega \to 1 + \sum_{z \in Z} \omega z\right\}$

and where $P''$ is defined as follows: For $\xi \in V$, $k \geqslant 0$, $y_1, y_2, \ldots, y_k \in V \cup T$

$$\xi \to \omega y_1 \omega y_2 \omega \cdots \omega y_k \omega \in P'' \quad \Leftrightarrow \quad \xi \to \varphi(y_1 y_2 \cdots y_k) \in P.$$

Since the restriction of $\varphi$ to $(V\cup T)^*$ is strictly alphabetic, $P''$ is finite. Next, it is easy to prove that

$$\varphi^{-1}L_G(\xi)=L_{\varphi^{-1}(G)}(\xi) \qquad \xi\in V,$$

and $L_{\varphi^{-1}(G)}(\omega)=Z^*$. Thus $\varphi^{-1}L$ is a context-free language. Any inverse morphism can be factorized into an alphabetic morphism, followed by the intersection with a regular language, followed by a morphism. Thus closure under arbitrary inverse morphism can be deduced from the above and from the following construction.

**c) Intersection with a regular language** Let $K\subset X^*$ be a regular language, and $\mathbf{A}=\langle X, Q, q_-, Q_+\rangle$ be a finite automaton such that $K=|\mathbf{A}|$. Let $\hat\sigma$ be a new symbol, and define a grammar

$$G_K=\langle\hat\sigma\cup(Q\times V\times Q), X, P_K\rangle$$

where $P_K=P'\cup P''$,

with $P''=\{\hat\sigma\to(q_-,\sigma,q_+)\mid q_+\in Q_+\}$,

and $P'$ composed of the following productions:
For $k\geqslant 0$, $\xi,\eta_1,\ldots,\eta_k\in V$, $u_0,\ldots,u_k\in X^*$, $q,q',q_1,\ldots,q_k,q'_1,\ldots,q'_k\in Q$,

$$(q,\xi,q')\to u_0(q_1,\eta_1,q'_1)u_1(q_2,\eta_2,q'_2)\cdots(q_k,\eta_k,q'_k)u_k\in P'$$

if and only if

$$\xi\to u_0\eta_1u_1\eta_2\cdots\eta_ku_k\in P$$

and $\quad q\cdot u_0=q_1, \qquad q'_i\cdot u_i=q_{i+1} \quad (i=1,\ldots,k-1), \qquad q'_k\cdot u_k=q'.$

It is not difficult to show that

$$L_{G_K}(q,\xi,q')=L_G(\xi)\cap K_{q,q'} \qquad (q,q'\in Q,\ \xi\in V)$$

where $K_{q,q'}=\{f\in X^*\mid q\cdot f=q'\}$.

Thus $\quad L\cap K=L_{G_K}(\hat\sigma)$.

**d) Context-free substitution** Let $\theta$: $X^*\to Y^*$ be a context-free substitution. For each $x\in X$, let

$$G_x=\langle V_x, Y, P_x\rangle$$

be a context-free grammar such that $\theta(x)=L_{G_x}(\sigma_x)$ for some $\sigma_x\in V_x$. Clearly the alphabets $V_x$ may be assumed pairwise disjoint and disjoint from $V$. Define a copy morphism $\gamma:(V\cup X)^*\to(V\cup\{\sigma_x:x\in X\})^*$ by $\gamma(\xi)=\xi$ for $\xi\in V$, $\gamma(x)=\sigma_x$ for $x\in X$, and let $\gamma G=\langle V,\{\sigma_x:x\in X\},\gamma P\rangle$ be defined as in a). Let

$$H=\langle W, Y, Q\rangle$$

be the grammar with $W=V\cup\bigcup_{x\in X}V_x$, $Q=\gamma P\cup\bigcup_{x\in X}P_x$. Then it can be shown that

$$L_H(\xi)=\theta(L_G(\xi)) \qquad \xi\in V.$$

Consequently $\theta(L)=L_H(\sigma)$.

The iteration lemmas for context-free languages are not as accurate as the corresponding lemmas for regular sets. It can be shown (Exercise 2.1) that a strict analog of the iteration lemmas for regular languages does not exist. The most frequently used iteration lemma is due to Bar-Hillel, Perles and Shamir.

**Lemma 2.2** (Iteration Lemma for Algebraic Languages) *Let $L\subset X^*$ be an algebraic language. There exists an integer $N\geqslant 1$ such that any word $f\in L$ with $|f|\geqslant N$ admits a factorization $f=aubvc$ ($a, u, b, v, c\in X^*$) satisfying*

(i) $au^nbv^nc\in L$ *for all* $n\geqslant 0$;

(ii) $0<|uv|\leqslant N$.

There is some difficulty in the use of this lemma arising from the fact that the position of the segments $u$ and $v$ in $f$ cannot be predicted. The following refinement of the above lemma states that at least the position of one of the two segments can be located with some precision. The notion of marked position is the same as in Section I.5.

**Lemma 2.3** (Ogden's Iteration Lemma for Algebraic Languages) *Let $L\subset X^*$ be an algebraic language. There exists an integer $N\geqslant 1$ such that for any $f\in L$, and for any choice of at least $N$ marked positions in $f$, $f$ admits a factorization $f=aubvc$ ($a, u, b, v, c\in X^*$) satisfying*

(i) $au^nbv^nc\in L$ *for all* $n\geqslant 0$;

(ii) *(each of $a$ and $u$ and $b$) or (each of $b$ and $v$ and $c$) contains at least one marked position;*

(iii) *$uv$ contains at most $N$ marked positions.*

If all positions of $f$ are marked, we obtain Lemma 2.2. Assume that $N$ consecutive positions are marked, hence that a segment $g$ of $f$ has been distinguished. Then (ii) implies that either $u$ or $v$ is a segment of $g$. Thus we have:

**Corollary 2.4** *Let $L\subset X^*$ be an algebraic language. Then there exists an integer $N$ such that for any word $f\in L$, and for any factorization $f=hgh'$ ($h, g, h'\in X^*$)*

*with* $|g| \geq N$, *f admits a factorization* $f = aubvc$ $(a, u, b, v, c \in X^*)$ *satisfying*

(i) $au^n bv^n c \in L$ *for all* $n \geq 0$;

(ii) *either u is a segment of g and* $|u| > 0$ *or v is a segment of g and* $|v| > 0$.

We do not prove these lemmas (see Ogden [1968], Aho and Ullman [1972], Autebert and Cousineau [1976]). The proof is on derivation trees and the lemmas are in fact results on derivations in algebraic grammars. We give this version of the lemma for later use.

**Lemma 2.5** (Ogden's Iteration Lemma for Algebraic Grammars) *Let* $G = \langle V, X, P \rangle$ *be an algebraic grammar. Then there exists an integer N such that, for any derivation* $\xi \xrightarrow{*} f$ *with* $\xi \in V$, $f \in (V \cup X)^*$, *and for any choice of N marked positions in f, there is a factorization*
$f = aubvc$ $(a, u, b, v, c \in X^*)$ *and* $\eta \in V$ *such that*

(i) $\xi \xrightarrow{*} a\eta c$, $\eta \xrightarrow{*} u\eta v$, $\eta \xrightarrow{*} b$;

(ii) (*a and u and b*) *or* (*b and v and c*) *contain at least one marked position;*

(iii) *uv contains at most N marked positions.*

Note that the integer $N$ is independent of the nonterminal $\xi$. Note also that the lemma is true for sentential forms as well as for words in $X^*$.

A context-free grammar $G = \langle V, X, P \rangle$ can be "reduced" in several ways. Let $\sigma \in V$. Then $G$ is

reduced in $\sigma$, if for each $\xi \in V$, $L_G(\xi) \neq \emptyset$, and $\sigma \xrightarrow{*} u\xi v$ for some words $u, v \in X^*$,

strictly reduced in $\sigma$, if $G$ is reduced in $\sigma$, and if further $L_G(\xi)$ is infinite for each $\xi \in V$.

**Lemma 2.6** *Let* $L \subset X^*$ *be a context-free language. If L is nonempty, then* $L = L_G(\sigma)$ *for some context-free grammar* $G = \langle V, X, P \rangle$ *which is reduced in* $\sigma$. *If L is infinite, then G can be assumed to be strictly reduced in* $\sigma$.

We only give the construction. Let $L = L_G(\sigma)$ for some grammar $G = \langle V, X, P \rangle$, and let $V'$ be the set of $\xi \in V$ such that $L_G(\xi) \neq \emptyset$ and $\sigma \xrightarrow{*} u\xi v$ for some $u, v \in X^*$. If $L \neq \emptyset$, then $\sigma \in V'$. Let $G' = \langle V', X, P' \rangle$ where $P' = \{\xi \to \alpha \in P \mid \xi \in V', \alpha \in (V' \cup X)^*\}$. Then $G'$ is reduced in $\sigma$, and $L_G(\xi) = L_{G'}(\xi)$ for each $\xi \in V'$.

Next let $V'' = \{\xi \in V' \mid L_{G'}(\xi)$ is infinite$\}$, and define a grammar $G'' = \langle V'', X, P'' \rangle$ as follows. Let $\theta : (V' \cup X)^* \to (V'' \cup X)^*$ be the substitution given by

$$\theta(x) = x \ (x \in X), \quad \theta(\xi) = \xi \ (\xi \in V''), \quad \theta(\xi) = L_{G'}(\xi) \ (\xi \in V' \setminus V''),$$

and set

$$P''=\{\xi\to\beta \mid \xi\to\alpha\in P',\quad \beta\in\theta(\alpha)\}.$$

If $L$ is infinite, then $\sigma\in V''$, $G''$ is strictly reduced in $\sigma$, and $L_{G''}(\xi)=L_{G'}(\xi)$ for each $\xi\in V''$.

**Exercise**

**2.1** Show that Corollary 2.4 cannot be strengthened to assert the existence, for each factorization $f=hgh'g'h''$ of $f\in L$, of a factorization $f=aubvc$ satisfying (i) and such that both $u$ is a segment of $g$ and $v$ is a segment of $g'$.

## II.3 Dyck Languages

The Dyck sets are among the most frequently cited context-free languages. In view of the Chomsky-Schützenberger Theorem proved below, they are also the most "typical" context-free languages. In Chapter VII, we shall see another formulation of this fact: The Dyck languages are, up to four exceptions, generators of the rational cone of context-free languages.

A Dyck language consists of "well-formed" words over a finite number of pairs of parentheses. There are two (and in fact even four) families of Dyck languages defined by different constraints on the use of parenthesis. The restricted Dyck languages $D_n'^*$, $(n\geq 1)$ are formed of the words over $n$ pairs of parentheses which are "correct" in the usual sense. Thus

$$(\,[\,(\,)\,(\,)\,]\,\{\,\}\,[\,]\,)\,(\,)$$

is a word of $D_3'^*$. For the Dyck languages $D_n^*$, the interpretation of the parentheses is different. Two parentheses of the same type are rather considered as formal inverse for each other. A word is considered as "correct" iff successive deletion of factors of associated parentheses (say of the form $x\bar{x}$ and $\bar{x}x$) yields the empty word. Thus

$$\bar{x}x\bar{x}\bar{x}\bar{y}y\bar{x}x$$

is a word of $D_2^*$. This interpretation is used for the construction of free groups. Finally, $D_n$ and $D_n'$ are the sets of Dyck-primes and restricted Dyck-primes, that is the words of $D_n^*$ (resp. $D_n'^*$) which are not product of two nonempty words of $D_n^*$ (resp. $D_n'^*$).

The appropriate framework to formalize the definitions of $D_n'^*$ and $D_n^*$ are congruences. We first give this definition, and prove then that the four families consist of context-free languages. The section ends with a proof of the Chomsky-Schützenberger Theorem.

Let $n\geq 1$ be an integer, and let $X_n=\{x_1,\ldots,x_n\}$, $\bar{X}_n=\{\bar{x}_1,\ldots,\bar{x}_n\}$ be two alphabets of $n$ letters. Each couple $x_k,\bar{x}_k$ can be considered as a pair of parentheses of the same type. Define $Z_n = X_n \cup \bar{X}_n$. We introduce the following useful notation. For $z \in Z_n$, let

$$\bar{z} = \begin{cases} \bar{x}_k & \text{if} \quad z = x_k; \\ x_k & \text{if} \quad z = \bar{x}_k. \end{cases}$$

Thus $\bar{\bar{z}} = z$.

**Definition** The restricted Dyck congruence $\delta'_n$ is the congruence of $Z_n^*$ generated by

$$x_k\bar{x}_k \sim 1 \qquad k = 1, \ldots, n. \tag{3.1}$$

The Dyck congruence $\delta_n$ is the congruence generated by (3.1) and by

$$\bar{x}_k x_k \sim 1 \qquad k = 1, \ldots, n \tag{3.2}$$

Thus two words $w$ and $w'$ are congruent modulo $\delta'_n$ or modulo $\delta_n$, and we write

$$w \equiv w' (\text{mod } \delta'_n) \quad \text{or} \quad w \equiv w' (\text{mod } \delta_n)$$

iff $w'$ can be obtained from $w$ by a finite number of insertions or deletions of factors of the form $x_k\bar{x}_k$ (resp. $x_k\bar{x}_k$ or $\bar{x}_k x_k$).

**Definition** The restricted Dyck language $D_n'^*$ is the class of 1 in the congruence $\delta'_n : D_n'^* = [1]_{\delta'_n}$. The Dyck language $D_n^*$ is the class of 1 in the congruence $\delta_n : D_n^* = [1]_{\delta_n}$.
Clearly by definition both $D_n'^*$ and $D_n^*$ are submonoids of $Z_n^*$.

**Definition** The set $D'_n$ of restricted Dyck primes is

$$D'_n = (D_n'^* \backslash 1) \backslash (D_n'^* \backslash 1)^2.$$

The set of Dyck primes is $D_n = (D_n^* \backslash 1) \backslash (D_n^* \backslash 1)^2$.

The notation is consistent since $D'_n$ and $D_n$ indeed generate the submonoids $D_n'^*$ and $D_n^*$ (see Section I.2). In fact, we shall see that $D'_n$ and $D_n$ are bifix codes; thus $D_n'^*$ and $D_n^*$ are free submonoids of $Z_n^*$. In order to give a unified treatment, we follow an idea of M. P. Schützenberger and introduce a more general family of congruences and languages. It will appear that the restricted and the general Dyck languages are just extremal cases in the new formulation.

**Definition** Let $I$ be a subset of $\{1, \ldots, n\}$. The congruence $\delta_I$ is the congruence generated by

$$x_k\bar{x}_k \sim 1 \quad (k = 1, \ldots, n) \qquad \text{and} \qquad \bar{x}_i x_i \sim 1 \quad (i \in I).$$

The language $D_I^*$ is the class of the empty word in the congruence $\delta_I$.
Clearly $D_I^*$ is a submonoid of $Z_n^*$, justifying thus the notation. Anyone of the $2^n$ subsets of $\{1, \ldots, n\}$ defines a "Dyck-like" language. If $I = \emptyset$, then $\delta_I = \delta'_n$ and $D_I^* = D_n'^*$; if $I = \{1, \ldots, n\}$, then $\delta_I = \delta_n$ and $D_I^* = D_n^*$.

Our aim is to prove that $D_I^*$ and $D_I = (D_I^* \backslash 1) \backslash (D_I^* \backslash 1)^2$ are context-free languages for any $I \subset \{1, \ldots, n\}$. For this we first introduce a new relation. Let $u, v \in Z_n^*$, and set $u \underset{I}{\vdash} v$ iff there are $f, g \in Z_n^*$ such that $u = f\alpha g$, $v = fg$, and either $\alpha = x_k \bar{x}_k$ for some $k \in \{1, \ldots, n\}$ or $\alpha = \bar{x}_i x_i$ for some $i \in I$. The reflexive and transitive closure $\overset{*}{\underset{I}{\vdash}}$ of $\underset{I}{\vdash}$ is called the Dyck reduction. Clearly, if $u \overset{*}{\underset{I}{\vdash}} v$, then $|u| \geqslant |v|$, and $|u| = |v|$ implies $u = v$. The congruence $\delta_I$ and the reduction $\overset{*}{\underset{I}{\vdash}}$ are linked by

$$u \equiv v \pmod{\delta_I}$$

iff there are $k \geqslant 0$, $u_0, \ldots, u_k \in Z_n^*$ such that

$$u_0 = u, \qquad u_k = v,$$

and $\quad (u_p \underset{I}{\vdash} u_{p+1} \quad \text{or} \quad u_{p+1} \underset{I}{\vdash} u_p) \quad p = 0, \ldots, k-1.$

Thus $u \overset{*}{\underset{I}{\vdash}} v$ implies $u \equiv v (\text{mod } \delta_I)$, but the converse is false.
A word $u$ is reduced mod $\delta_I$ iff it contains no factor of the form $x_k \bar{x}_k$ or $\bar{x}_i x_i$, $(i \in I)$. Thus $u$ is reduced iff $\{v \mid u \underset{I}{\vdash} v\} = \emptyset$. For any word $w \in Z_n^*$ there is at least one reduced word $u$ congruent to $w (\text{mod } \delta_I)$. Usually, there are several ways to compute a reduced word. We shall prove that all computations lead to the same reduced word which is unique. We follow Autebert and Cousineau [1976] rather than the standard exposition as treated in Magnus, Karrass and Solitar [1966]. Indeed, the presentation below is closer to the extensions to more general congruences over free monoids as considered by Cochet and Nivat [1971], Benois and Nivat [1970].

**Example 3.1** For $n = 2$, $I = \{1, 2\}$, consider the word $w = \bar{x}x\bar{x}x\bar{y}y\bar{x}x$. It can be reduced to the empty word in at least the two following fashions:

$$\begin{array}{ll}
w = \underline{\bar{x}x}\bar{x}x\bar{y}y\bar{x}x & \qquad w = \bar{x}\underline{x\bar{x}}x\bar{y}y\bar{x}x \\[1ex]
w_1 = \underline{\bar{x}x}\bar{y}y\bar{x}x & \qquad w'_1 = \bar{x}x\underline{\bar{y}y}\bar{x}x \\[1ex]
w_2 = \underline{\bar{y}y}\bar{x}x & \qquad w'_2 = \bar{x}\underline{x\bar{x}}x \\[1ex]
w_3 = \underline{\bar{x}x} & \qquad w'_3 = \underline{\bar{x}x} \\[1ex]
w_4 = 1 & \qquad w'_4 = 1.
\end{array}$$

From now on, we write $\vdash$ and $\overset{*}{\vdash}$ instead of $\underset{I}{\vdash}$ and $\overset{*}{\underset{I}{\vdash}}$.

**Lemma 3.1** (Confluence Lemma) *If $w \overset{*}{\vdash} u_1$ and $w \overset{*}{\vdash} u_2$, then there exists a word $v$ such that $u_1 \overset{*}{\vdash} v$, $u_2 \overset{*}{\vdash} v$.*

Thus the lemma asserts the existence of a word $v$ such that the following diagram holds (Fig. II.1).

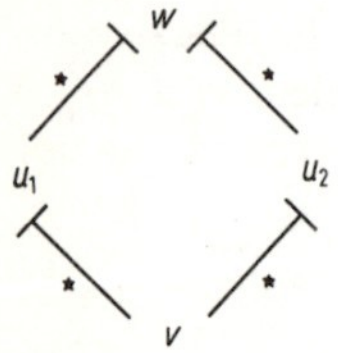


Fig. II.1

We first prove the lemma in a special case.

**Lemma 3.2** *If $w \vdash u_1$ and $w \vdash u_2$, then there exists a word $v$ such that $u_1 \overset{*}{\vdash} v$ and $u_2 \overset{*}{\vdash} v$.*

Proof. There are words $f_1, g_1, f_2, g_2 \in Z_n^*$ and $a_1, a_2 \in Z_n^2$ such that

$$w = f_1 a_1 g_1 = f_2 a_2 g_2, \qquad u_1 = f_1 g_1, \qquad u_2 = f_2 g_2.$$

If $|f_1| = |f_2|$, then $u_1 = u_2$ and there is nothing to prove. Assume for instance $|f_1| < |f_2|$. We distinguish two cases.

a) $|f_1| + 2 \leqslant |f_2|$. Then $f_2 = f_1 a_1 h$ for some word $h$, thus $w = f_1 a_1 h a_2 g_2$, and $v = f_1 h g_2$ satisfies $u_1 \vdash v$, $u_2 \vdash v$.

b) $|f_1| + 1 = |f_2|$. Then $f_2 = f_1 z$ for some letter $z$, hence $a_1 = z\bar{z}$, and $a_2 = \bar{z} z$. (This implies that $z = x_i$ or $z = \bar{x}_i$ with $i \in I$.) Thus $w = f_1 z \bar{z} z g_2$, and $u_1 = f_1 z g_2 = u_2$. Hence $v = u_1$ satisfies the conditions. ■

Proof of Lemma 3.1 By induction on $|w|$. If $|w| = 0$, then $u_1 = w = u_2 = 1$, and $v = w$ satisfies the lemma. Assume $|w| = p > 0$. If $|u_1| = |w|$, then $u_1 = w$ and the lemma is verified for $v = u_2$. Thus we may suppose $|u_1| < p$ and similarly $|u_2| < p$. There exist two words $v_1, v_2$ with $|v_1| = |v_2| = p - 2$ such that

$$w \vdash v_1 \overset{*}{\vdash} u_1, \qquad w \vdash v_2 \overset{*}{\vdash} u_2.$$

Thus in view of Lemma 3.8, there is a word $t$ such that

$$v_1 \overset{*}{\vdash} t \quad \text{and} \quad v_2 \overset{*}{\vdash} t.$$

Since $v_1 \overset{*}{\vdash} u_1$, $v_1 \overset{*}{\vdash} t$ and $|v_1| < p$, by induction there is a word $w_1$ such that

$$u_1 \overset{*}{\vdash} w_1, \qquad t \overset{*}{\vdash} w_1;$$

similarly, there is a word $w_2$ such that

$$u_2 \overset{*}{\vdash} w_2, \qquad t \overset{*}{\vdash} w_2.$$

Now $t \overset{*}{\vdash} w_1$, $t \overset{*}{\vdash} w_2$ and $|t| < p$. Using the induction hypothesis once more, there exists a word $v$ such that

$$w_1 \overset{*}{\vdash} v, \qquad w_2 \overset{*}{\vdash} v.$$

This shows that $u_1 \overset{*}{\vdash} w_1 \overset{*}{\vdash} v$, $u_2 \overset{*}{\vdash} w_2 \overset{*}{\vdash} v$. ∎

The construction of the proof is reflected in Fig. II.2.

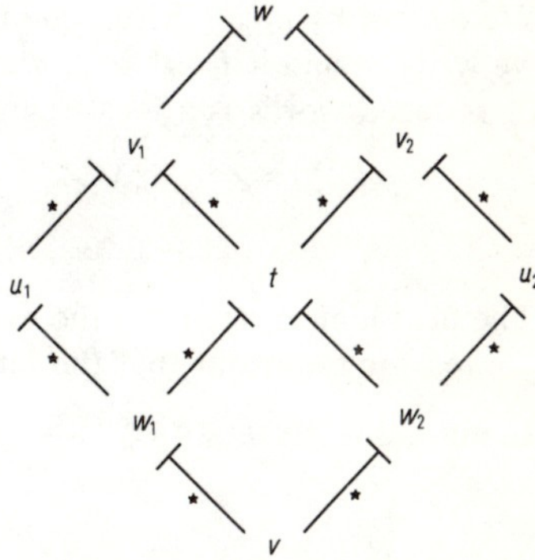


Fig. II.2

**Corollary 3.3** *Let $u, v \in Z_n^*$; then $u \equiv v (\mathrm{mod}\ \delta_I)$ iff there exists a word $w$ such that $u \overset{*}{\vdash} w$ and $v \overset{*}{\vdash} w$.*

Proof. Assume $u \equiv v (\mathrm{mod}\ \delta_I)$. Then there are $k \geqslant 0$, $u_0, \ldots, u_k \in Z_n^*$ such that $u_0 = u$, $u_k = v$ and

$$u_p \vdash u_{p+1} \quad \text{or} \quad u_{p+1} \vdash u_p \qquad \text{for} \quad p = 0, \ldots, k-1.$$

If $k = 0$, then $u = v$ and there is nothing to prove. Arguing by induction on $k$, there is a word $w_1$ such that

$$u_1 \overset{*}{\vdash} w_1 \quad \text{and} \quad v \overset{*}{\vdash} w_1.$$

If $u_0 \vdash u_1$, then $u_0 \overset{*}{\vdash} w_1$ and the corollary is true with $w = w_1$. If $u_1 \vdash u_0$, there exists, by the Confluence Lemma, a word $w_2$ such that $u_0 \overset{*}{\vdash} w_2$ and $w_1 \overset{*}{\vdash} w_2$, whence $v \overset{*}{\vdash} w_2$. The converse is obvious. ∎

Remark. The Confluence Lemma can be considered as a property of some binary relations: Let $\overset{*}{\dashv}$ be the relation opposed to $\overset{*}{\vdash}$. Then the congruence $\delta_I$ is the least congruence containing $\overset{*}{\dashv}$ and $\overset{*}{\vdash}$. Corollary 3.3 states that $\delta_I$ is the product (of relations) of $\overset{*}{\vdash}$ by $\overset{*}{\dashv}$; the Confluence Lemma asserts the existence of a weak commutativity property: the product of $\overset{*}{\dashv}$ by $\overset{*}{\vdash}$ is contained in the product of $\overset{*}{\vdash}$ by $\overset{*}{\dashv}$.

We list two other corollaries.

**Corollary 3.4** *If $v$ is reduced and $u \equiv v (\mathrm{mod}\ \delta_I)$, then $u \overset{*}{\vdash} v$.*

Proof. By Corollary 3.3, there is a word $w$ such that $u \overset{*}{\vdash} w$ and $v \overset{*}{\vdash} w$. Since $v$ is reduced, $v = w$. ∎

**Corollary 3.5** *Any class of the congruence $\delta_I$ contains exactly one reduced word.*

Proof. It is clear that any class contains at least one reduced word. Assume $u, v$ are reduced and $u \equiv v (\mathrm{mod}\ \delta_I)$. Then by Corollary 3.4, $u \overset{*}{\vdash} v$ and $v \overset{*}{\vdash} u$, thus $u = v$. ∎

We denote by $\rho_I(w)$ the unique reduced word congruent to $w$ mod $\delta_I$. If $I = \emptyset$, we write $\rho'$ and if $I = \{1, \ldots, n\}$, we write $\rho$ instead of $\rho_I$. The language $\rho_I(Z_n^*)$ of reduced words is a local regular set, since

$$\rho_I(Z_n^*) = Z_n^* \backslash Z_n^* V_I Z_n^*$$

with $\quad V_I = \{x_k \bar{x}_k : k = 1, \ldots, n\} \cup \{\bar{x}_i x_i : i \in I\}$.

The next lemma describes the words which reduce to a given word. It is the key lemma for the proof that the languages $D_I^*$ are context-free.

**Lemma 3.6** *Let $f, w \in Z_n^*$, $w = z_1 z_2 \cdots z_m (z_p \in Z_n^*)$. Then*

$$f \overset{*}{\vdash} w$$

*iff there exist $d_0, d_1, \ldots, d_m \in Z_n^*$ such that*

$$f = d_0 z_1 d_1 z_2 \cdots d_{m-1} z_m d_m,$$

*and* $\quad d_0 \overset{*}{\vdash} 1,\ d_1 \overset{*}{\vdash} 1, \ldots, d_m \overset{*}{\vdash} 1.$

Proof. The conditions are clearly sufficient. The proof of the converse is by induction on $|f| - |w|$. If $|f| = |w|$, then $f = w$; if $|f| > |w|$, then there exists $f'$ such that $|f'| = |f| - 2$ and $f \vdash f' \overset{*}{\vdash} w$. By induction,

$$f' = d_0 z_1 d_1 z_2 \cdots d_{m-1} z_m d_m$$

for some words $d_r$ with $d_r \overset{*}{\vdash} 1$, $(r = 0, \ldots, m)$. Next, there is a factorization

$$f = gah, \quad \text{with} \quad f' = gh,$$

and $a = x_k \bar{x}_k$ for some $k \in \{1, \ldots, n\}$ or $a = \bar{x}_i x_i$ for some $i \in I$. Hence there is an integer $j$, $(0 \leq j \leq m)$ and a factorization $d_j = d'd''$, $(d', d'' \in Z_n^*)$ such that

$$g = d_0 z_1 \ldots z_j d', \qquad h = d'' z_{j+1} \cdots z_m d_m.$$

Set $e_j = d'ad''$. Then $e_j \vdash d_j \overset{*}{\vdash} 1$ and

$$f = d_0 z_1 \cdots z_j e_j z_{j+1} \cdots z_m d_m.$$ ∎

**Theorem 3.7** *The language $D_I^*$ is context-free. More precisely, $D_I^*$ is the language generated by the grammar $G_I$ with productions:*

$$\xi \to 1 + \sum_{k=1}^{n} x_k \xi \bar{x}_k \xi + \sum_{i \in I} \bar{x}_i \xi x_i \xi. \tag{3.3}$$

First, we introduce the notation

$$Y_I = X_n \cup \{\bar{x}_i : i \in I\}.$$

Thus $Y_I = X_n$ if $I = \emptyset$, and $Y_I = Z_n$ if $I = \{1, \ldots, n\}$.

Proof. The grammar $G_I$ is strict. Thus it suffices to show that $D_I^*$ is a solution of the equation associated to (3.3). Assume $w = zw'\bar{z}w''$ with $w' \equiv w'' \equiv 1 \pmod{\delta_I}$ and $z \in Y_I$. Then $w \equiv z\bar{z} \equiv 1 \pmod{\delta_I}$ and $w \in D_I^*$. This shows the inclusion

$$D_I^* \supset 1 \cup \bigcup_{1 \leq k \leq n} x_k D_I^* \bar{x}_k D_I^* \cup \bigcup_{i \in I} \bar{x}_i D_I^* x_i D_I^*.$$

Conversely, let $w \in D_I^*$, $w \neq 1$. Since 1 is reduced, $w \overset{*}{\vdash} 1$ by Corollary 3.4. Since $w \neq 1$, there is a letter $z \in Y_I$ such that $w \overset{*}{\vdash} z\bar{z}$. By Lemma 3.6, $w$ factorizes in

$$w = d_0 z d_1 \bar{z} d_2$$

with $d_0, d_1, d_2 \in D_I^*$. If $d_0 = 1$, then $w \in zD_I^* \bar{z} D_I^*$. If $d_0 \neq 1$, then $|d_0| < |w|$, and arguing by induction, $d_0 \in yD_I^* \bar{y} D_I^*$ for some $y \in Y_I$. Thus $w \in yD_I^* \bar{y} D_I^* z D_I^* \bar{z} D_I^* \subset yD_I^* \bar{y} D_I^*$. This achieves the proof. ■

We now investigate the language

$$D_I = (D_I^* \backslash 1) \backslash (D_I^* \backslash 1)^2.$$

For $z \in Z_n$, define

$$D_{I,z} = D_I \cap zZ_n^*.$$

**Proposition 3.8** (i) *The language $D_I$ is bifix;*
(ii) *$D_{I,z} \neq \emptyset$ iff $z \in Y_I$;*
(iii) *if $D_{I,z} \neq \emptyset$, then any $w \in D_{I,z}$ admits a unique factorization*

$$w = zu_1 u_2 \cdots u_m \bar{z} \qquad \text{with} \qquad m \geq 0, \quad u_1, \ldots, u_m \in D_I \backslash D_{I,\bar{z}}.$$

Proof. (i) Let $w \in D_I$, and assume $w = uv$ with $u \in D_I$, $v \in Z_n^*$. Then $1 \equiv uv \equiv v \pmod{\delta_I}$, thus $v \in D_I^*$. Thus $v = 1$ by the definition of $D_I$. This shows that $D_I$ is prefix. A symmetric argument shows that $D_I$ is suffix.
(ii) Let $w \in D_{I,z}$. Then $w = zu$ for some word $u$. Since $w \overset{*}{\vdash} 1$, there is a letter $y \in Y_I$ such that $w \overset{*}{\vdash} y\bar{y}$. In view of Lemma 3.6, $w$ factorizes in $w = d_0 y d_1 \bar{y} d_2$,

with $d_0$, $d_1$, $d_2 \in D_I^*$. Since $w \in D_I$, $d_0 = d_2 = 1$ and $y = z$. This proves the assertion, since clearly $D_{I,z} \neq \emptyset$ if $z \in Y_I$.

(iii) We just have seen that a word $w \in D_{I,z}$ factorizes in $w = zd_1\bar{z}$ with $d_1 \in D_I^*$. Thus either $d_1 = 1$, or $d_1 = u_1u_2 \cdots u_m$ with $u_1, \ldots, u_m \in D_I$. Assume that $u_p \in D_{I,\bar{z}}$ for some $p$. Then $u_p = \bar{z}dz$ for some $d \in D_I^*$, and

$$w = (zu_1 \cdots u_{p-1}\bar{z})d(zu_{p+1} \cdots u_m\bar{z}) \in (D_I^*\backslash 1)^2$$

contrary to the definition of $D_I$. The unicity is immediate since $D_I$ is a code. ■

By Proposition 3.8,

$$D_I = \bigcup_{z \in Y_I} D_{I,z} \tag{3.4}$$

$$D_{I,z} = z\Delta_{I,z}\bar{z} \qquad (z \in Y_I) \tag{3.5}$$

where $\Delta_{I,z}$ is the submonoid of $Z_n^*$ generated by $D_I \backslash D_{I,\bar{z}}$:

$$\Delta_{I,z} = \Big(\bigcup_{y \in Y_I \backslash \bar{z}} D_{I,y}\Big)^*. \tag{3.6}$$

Finally, since $D_I$ generates $D_I^*$, we have:

$$D_I^* = 1 \cup D_I D_I^* \tag{3.7}$$

From equations (3.4)–(3.7), we deduce the following grammar $H_I = \langle V_I, Z_n, P_I \rangle$. Set

$$V_I = \{\xi, \eta\} \cup \{\xi_z, \eta_z : z \in Y_I\},$$

and let $P_I$ contain the productions:

$$\xi \to 1 + \eta\xi; \qquad \eta \to \sum_{z \in Y_I} \eta_z;$$

$$\eta_z \to z\xi_z\bar{z}; \qquad \xi_z \to 1 + \Big(\sum_{y \in Y_I \backslash \bar{z}} \eta_y\Big)\xi_z \qquad (z \in Y_I).$$

By the Substitution Lemma, the system of equations associated to $H_I$ is equivalent to:

$$\xi = 1 + \sum_{z \in Y_I} z\xi_z\bar{z}\xi; \qquad \xi_z = 1 + \sum_{y \in Y_I \backslash \bar{z}} y\xi_y\bar{y}\xi_z \qquad (z \in Y_I); \tag{3.8}$$

$$\eta = \sum_{z \in Y_I} \eta_z; \qquad \eta_z = z\xi_z\bar{z} \qquad (z \in Y_I).$$

The equations (3.8) are strict, thus the system associated to $H_I$ has a unique solution, and equations (3.4)–(3.7) show that $H_I$ generates the various languages related to the Dyck sets:

$$D_I^* = L_{H_I}(\xi); \qquad D_I = L_{H_I}(\eta); \qquad D_{I,z} = L_{H_I}(\eta_z), \qquad \Delta_{I,z} = L_{H_I}(\xi_z)$$

**Corollary 3.9** *The languages $D_I$ are context-free.* ∎

If $I = \emptyset$, the grammar $H_I$ reduces, after short-cutting the $\eta_z$, to the grammar with the productions:

$$\xi \to 1 + \eta\xi; \qquad \eta \to \sum_{k=1}^{n} x_k \xi \bar{x}_k.$$

Thus we have, for $D_n'^*$, $D_n'$, $D_n^*$, $D_n$ the following formulas:

$$D_n'^* = 1 \cup D_n' D_n'^*; \qquad D_n' = \bigcup_{1 \le k \le n} x_k D_n'^* \bar{x}_k;$$

$$D_n^* = 1 \cup D_n D_n^*; \qquad D_n = \bigcup_{z \in Z_n} D_{n,z};$$

$$D_{n,z} = z \Delta_{n,z} \bar{z}; \quad (z \in Z_n); \qquad \Delta_{n,z} = \Big( \bigcup_{y \in Z_n \setminus \bar{z}} D_{n,y} \Big)^* \quad (z \in Z_n).$$

It can be shown (Exercise 3.1) that $D_n'^*$ is also generated by the grammar with productions

$$\xi \to \xi\xi + \sum_{k=1}^{n} x_k \xi \bar{x}_k + 1,$$

and that $D_n^*$ is generated by the grammar with productions

$$\xi \to \xi\xi + \sum_{k=1}^{n} x_k \xi \bar{x}_k + \sum_{k=1}^{n} \bar{x}_k \xi x_k + 1.$$

Let $X_n^{(*)} = Z_n^* / \delta_n$ be the quotient monoid; we denote by $\delta_n$ the canonical morphism from $Z_n^*$ onto $X_n^{(*)}$ defined by $\delta_n(w) = [w]_{\delta_n}$. For $w = z_1 z_2 \cdots z_m \in Z_n^*$, $(z_i \in Z_n)$, define $\bar{w} = \bar{z}_m \bar{z}_{m-1} \cdots \bar{z}_1$. Since

$$w\bar{w} \equiv \bar{w}w \equiv 1 (\text{mod } \delta_n),$$

$X_n^{(*)}$ is a group, and $\delta_n(\bar{w}) = (\delta_n(w))^{-1}$. In particular, $\delta_n(\bar{z}) = (\delta_n(z))^{-1}$ for $z \in Z_n$. It can even be shown that $X_n^{(*)}$ is a free group (see Magnus, Karrass and Solitar [1966]). $X_n^{(*)}$ is called the free group generated by $X_n$. Since each class $[w]_{\delta_n}$ contains exactly one reduced word $\rho(w)$, there is a bijection from $X_n^{(*)}$ onto $\rho(Z_n^*)$ which associates to any $u \in X_n^{(*)}$ the unique reduced word $w$ such that $u = \delta_n(w)$. If no confusion can arise, the index $n$ will be omitted in the above notations.

Note that any word in $X_n^*$ is already reduced. Thus $X_n^* \subset \rho(Z_n^*)$. It is sometimes convenient to identify $X_n^*$ with its image in $X_n^{(*)}$. This identification allows use of inverses and may simplify considerably certain formulations. However, it is important not to confuse the product $f^{-1}g$ in $X_n^{(*)}$, where $f, g \in X_n^*$, with the left quotient operation defined in Section I.2: Viewed as an operation in $X_n^{(*)}$, $f^{-1}g$

is always a well defined element of $X_n^{(*)}$, and $f^{-1}g = h \in X_n^*$ iff $fh = g$. Viewed as an operation in $X_n^*$, $f^{-1}g$ is either the empty set or a word in $X_n^*$, according to $f$ is not, or is a left factor of $g$. The embedding of $X_n^*$ into $X_n^{(*)}$ will be used only in Sections IV.2 and IV.6. In all other circumstances, $f^{-1}g$ should be interpreted as the left quotient defined in Section I.2.

The Dyck languages are known by the Chomsky-Schützenberger Theorem. We prove the following

**Theorem 3.10** (Chomsky-Schützenberger Theorem) *Let $L \subset Y^*$ be an algebraic language. Then there are an integer $n \geq 1$, an alphabetic morphism $\varphi: Z_n^* \to Y^*$, and a local regular language $K$ such that*

$$L = \varphi(D_n^* \cap K) = \varphi(D_n'^* \cap K) = \varphi(D_n \cap K) = \varphi(D_n' \cap K).$$

Proof. Assume the theorem proved in the case where $1 \notin L$. Then $1 \notin K$. Thus setting $K' = K \cup 1$, $K'$ is still a local language and the theorem holds for $L \cup 1$. Thus we may assume that $1 \notin L$.

The idea of the proof is simple: each production in a grammar generating $L$ is bracketed by a distinct pair of parentheses, and new letters are added to make the new grammar generate a subset of $D_n^*$, and in fact of $D_n'$. Thus it has only to be shown that none of the generated words is in $D_n^* \backslash D_n'$.

We assume that $L$ is generated by a grammar $G = \langle V, Y, P \rangle$ in quadratic form, i.e. that each production $\xi \to \alpha \in P$ satisfies $\alpha \in Y \cup V^2$. Such a grammar can always be obtained (see e.g. the books listed in the bibliography). We set

$$V = \{\xi_1, \ldots, \xi_N\}, \qquad Y = \{y_1, \ldots, y_q\},$$

and define

$$X_n = Y \cup \{a_{i,j,k}, b_{i,j,k} \mid i, j, k = 1, \ldots, N\} \cup \{d_{i,s} : i = 1, \ldots, N, s = 1, \ldots, q\},$$

where the $a_{i,j,k}, b_{i,j,k}, d_{i,s}$ are new letters. Thus $n = 2N^3 + Nq + q$. Set $\bar{X}_n = \{\bar{x} \mid x \in X_n\}$ and $Z_n = X_n \cup \bar{X}_n$.

Let $H = \langle V, Z_n, Q \rangle$ be the grammar with following productions:

For $i, j, k \in \{1, \ldots, N\}$,

$$\xi_i \to a_{ijk} b_{ijk} \xi_j \bar{b}_{ijk} \xi_k \bar{a}_{ijk} \in Q \tag{3.9}$$

if and only if

$$\xi_i \to \xi_j \xi_k \in P.$$

Further, for $i \in \{1, \ldots, N\}$, $s = 1, \ldots, q$,

$$\xi_i \to d_{is} y_s \bar{y}_s \bar{d}_{is} \in Q \tag{3.10}$$

if and only if

$$\xi_i \to y_s \in P.$$

For $i = 1, \ldots, N$, set $M_i = L_H(\xi_i)$, and let $\varphi: Z_n^* \to Y^*$ be the projection. Then clearly

$$\varphi(M_i) = L_G(\xi_i).$$

We shall prove that

$$M_i = D_n^* \cap K_i = D_n'^* \cap K_i = D_n \cap K_i = D_n' \cap K_i \quad i = 1, \ldots, N \tag{3.11}$$

where $K_i = (A_i Z_n^* \cap Z_n^* \bar{A}_i) \backslash Z_n^* B Z_n^*$

is the local regular set defined by:

$$A_i = \{a_{ijk} \mid j, k = 1, \ldots, N\} \cup \{d_{i,s} \mid s = 1, \ldots, q\}, \qquad \bar{A}_i = \{\bar{x} \mid x \in A_i\};$$

$$Z_n^2 \backslash B = W_1 \cup W_2 \cup W_3,$$

with

$$W_1 = \{a_{ijk} b_{ijk} \mid i, j, k = 1, \ldots, N\}; \tag{3.12}$$

$$W_2 = \{d_{is} y_s, \bar{y}_s \bar{d}_{is} \mid i = 1, \ldots, N, s = 1, \ldots, q\} \cup \{y_s \bar{y}_s \mid s = 1, \ldots, q\}; \tag{3.13}$$

$$W_3 = \bigcup_{i,j,k} b_{ijk} A_j \cup \bar{b}_{ijk} A_k \cup \bar{A}_j \bar{b}_{ijk} \cup \bar{A}_k \bar{a}_{ijk}. \tag{3.14}$$

a) $M_i \subset D_n^* \cap K_i$, $(i = 1, \ldots, N)$.

Let indeed $w \in M_i$. Then either, by (3.10),

$$w = d_{is} y_s \bar{y}_s \bar{d}_{is}$$

for some $s \in \{1, \ldots, q\}$, and clearly $w \in D_n^* \cap K_i$, or by (3.9)

$$w = a_{ijk} b_{ijk} u \bar{b}_{ijk} v \bar{a}_{ijk}$$

for some $j, k \in \{1, \ldots, N\}$ and $u \in M_j$, $v \in M_k$. Arguing by induction, $u \in D_n^* \cap K_j$, $v \in D_n^* \cap K_k$, thus $w \in D_n^*$ and, in view of (3.12) and (3.14), $u \in K_i$.

b) $D_n^* \cap K_i \subset D_n' \cap K_i$, $(i = 1, \ldots, N)$.

First, we verify

$$D_{n,\bar{x}} \cap Z_n^* \backslash Z_n^* B Z_n^* = \emptyset \quad \text{for} \quad \bar{x} \in \bar{X}_n. \tag{3.15}$$

Assume the contrary, and let $w \in D_{n,\bar{x}} \cap Z_n^* \backslash Z_n^* B Z_n^*$ be of minimal length. Then $|w| > 2$ since $\bar{x}x \in B$. In view of Proposition 3.8(iii), $w = \bar{x} u_1 \cdots u_m x$, with $u_p \in D_n \cap X_n Z_n^* \bar{X}_n$, $(p = 1, \ldots, m)$ by the minimality of $w$. Since the first letter of $u_1$ is not barred, $\bar{x} = \bar{b}_{ijk}$ from some indices $i, j, k$ by (3.14). Thus, by (3.12), the last letter of $u_m$ is $a_{ijk}$ and $u_m \notin D_n \cap X_n Z_n^* \bar{X}_n$. This proves (3.15).

Now let $w \in D_n^* \cap K_i$, $w = w_1 w_2 \cdots w_r$ with $w_p \in D_n \cap X_n Z_n^* \bar{X}_n$, for $p = 1, \ldots, r$ by (3.15). Then $w_1 \in A_i Z_n^* \bar{A}_i$ for some $i$, thus if $r > 1$, the first letter of $w_2$ would be barred by (3.14). Thus $r = 1$ and $w \in D_n$. Next if $w$ begins with a letter $d_{is}$, then $w = d_{is} y_s \bar{y}_s \bar{d}_{is}$ by (3.13) and $w \in D_n'$. Finally, if $w$ begins with the letter $a_{ijk}$, then $w=a_{ijk}u\bar{a}_{ijk}$ for some $u\in D_n^*$, and in view of (3.12), $u=b_{ijk}v_1\bar{b}_{ijk}v_2$ for some $v_1, v_2\in D_n^*$. In view of (3.14), $v_1\in K_j$, $v_2\in K_k$, and arguing by induction, $v_1\in D'_n\cap K_j$, $v_2\in D'_n\cap K_k$. Thus $w\in D'_n$.

c) $D'_n\cap K_i\subset M_i$, $(i=1,\ldots,N)$.

Let $w\in D'_n\cap K_i$. If $w$ begins with a letter $d_{is}$, then by (3.13), $w=d_{is}y_s\bar{y}_s\bar{d}_{is}$ and $w\in M_i$ by (3.10). Otherwise, $w=a_{ijk}u\bar{a}_{ijk}$ for some indices $j$, $k$, and $u\in D_n'^*$. By (3.12), $u=b_{ijk}v_1\bar{b}_{ijk}v_2$ for some $v_1, v_2\in D_n'^*$. Moreover, $v_1\in K_j$ and $v_2\in K_k$. Thus $v_1\in D_n'^*\cap K_j\subset D'_n\cap K_j$, and similarly $v_2\in D'_n\cap K_k$ by part b) of the proof. Therefore by induction $v_1\in M_j$, $v_2\in M_k$ and $w\in M_i$ by (3.9).

Thus we proved

$$M_i\subset D_n^*\cap K_i\subset D'_n\cap K_i\subset M_i \qquad i=1,\ldots,N,$$

and (3.11) follows. ■

## Exercises

**3.1** Show that for any $I\subset\{1,\ldots,n\}$, $D_I^*$ is the language generated by the grammar with productions

$$\xi\to\xi\xi+\sum_{k=1}^{n} x_k\xi\bar{x}_k+\sum_{i\in I}\bar{x}_i\xi x_i+1.$$

**3.2** Same question as in 3.1, for the grammar

$$\xi\to\xi\xi+\sum_{k=1}^{n} x_k\xi\xi\bar{x}_k+\sum_{i\in I}\bar{x}_i\xi\xi x_i+1.$$

**3.3** (Magnus, Karrass and Solitar [1966]) Define a function $\theta_I: Z_n^*\to Z_n^*$ inductively as follows: $\theta_I(1)=1$, $\theta_I(z)=z$ for $z\in Z_n$, and if $\theta_I(w)=z_1z_2\cdots z_m$, $(z_i\in Z_n)$, then

$$\theta_I(wz)=\begin{cases} z_1z_2\cdots z_{m-1} & \text{if } z_m\in Y_I \text{ and } z_m=\bar{z};\\ z_1z_2\cdots z_mz & \text{otherwise.}\end{cases}$$

Show that $\theta_1=\rho_I$.

**3.4.** Show that $ww'\in D_n^*\Rightarrow w'w\in D_n^*$.

**3.5** Show that for each $w\in Z_n^*$, the class $[w]_{\delta_I}$ of $w$ is a context-free language.

**3.6** For $w\in Z_n^*$, define

$$\|w\|=|w|_{X_n}-|w|_{\bar{X}_n}=\sum_{k=1}^{n}|w|_{x_k}-|w|_{\bar{x}_k}.$$

Show the following assertions:

a) $w\in D_n^*\Rightarrow\|w\|=0$.
b) $w\in D_n'^*\Rightarrow\|w'\|\geq 0$ for each left factor $w'$ of $w$.
c) $w\in D'_n\Rightarrow\|w'\|>0$ for each proper non empty left factor $w'$ of $w$.
d) $w\in D_1^* \Leftrightarrow \|w\|=0$.

**3.7** (Requires knowledge in ambiguity.) Show that the grammars $H_I$ are unambiguous.

**3.8** Assume that the grammar $G=\langle V, Y, P\rangle$ for $L$ in the proof of the Chomsky-Schützenberger Theorem is in Greibach Normal Form, i.e. $\xi\to\alpha$ implies $\alpha\in Y\cup YV\cup YVV$.

a) Show that $G$ can be transformed in such a way that for any two productions $\xi\to y\beta$, $\xi\to y'\beta'(y, y'\in Y)$, if $y\neq y'$ then $\beta\neq\beta'$.
b) Replace the productions of the form

$$\xi_i\to y\xi_j\xi_k \quad \text{by} \quad \xi_i\to a_{ijk}\xi_j\bar{a}_{ijk}\xi_k$$

$$\xi_i\to y\xi_j \quad \text{by} \quad \xi_i\to b_{ij}\bar{b}_{ij}\xi_j$$

$$\xi_i\to y \quad \text{by} \quad \xi_i\to d_i\bar{d}_i,$$

and prove that $L=\varphi(D_n'^*\cap K)$ where $K$ is a local regular set and where $\varphi$ erases barred letters, and replaces unbarred letters according to the above rules.
c) Show that each word in $D_n'^*\cap K$ ends by exactly one barred letter, and that no word in $D_n'^*\cap K$ contains a factor of more than two barred letters.
d) Show that any context-free language $L$ can be represented in the form $L=\varphi(D_m'^*\cap R)$ with $R$ local and $\varphi$ $\varepsilon$-limited on $R$ (i.e. $k\cdot|\varphi(w)|\geq|w|$ for all $w$ in $R$ and for some $k>0$).

## II.4 Two Special Languages

We present some properties of the Lukasiewicz language, and of the language of completely parenthesized arithmetic expressions.

**a) The Lukasiewicz language** Ł over $X=\{a, b\}$ is the language generated by the grammar with productions

$$\xi\to a\xi\xi+b.$$

Thus Ł is the unique language satisfying

$$Ł=aŁŁ\cup b. \tag{4.1}$$

The first words of Ł are

*b, abb, aabbb, ababb, aaabbbb, aababbb, ....*

The language of Lukasiewicz is the simplest of a family of languages constructed in order to write arithmetic expressions without parentheses (prefix or "polish" notation). The letter $a$ represents a binary operation, say +, and $b$ represents the operand. Thus the word $abb$ represents the expression $b+b$, and $aababbb$ represents the expression $((b+(b+b))+b)$.
For $w\in X^*$, define

$$\|w\|=|w|_a-|w|_b.$$

Clearly $\|ww'\|=\|w\|+\|w'\|$.

**Proposition 4.1** *Let* $w \in X^*$. *Then* $w \in$ Ł *iff* $w$ *satisfies the two following conditions:*

(i) $\|w\| = -1$;

(ii) $\|w'\| \geqslant 0$ *for any proper left factor* $w'$ *of* $w$.

Clearly, Proposition 4.1 implies that Ł is prefix.

Proof. Let $w \in$ Ł. If $w = b$, then (i) and (ii) are satisfied. Assume $|w| > 1$. Then by (4.1), $w = auv$ with $u, v \in$ Ł. Thus $\|w\| = 1 + \|u\| + \|v\| = -1$. Next, let $w'$ be a proper left factor of $w$. If $w' = a$, or if $w'$ is a left factor of $au$, then clearly $\|w'\| \geqslant 0$. If $w' = auv'$ and $v'$ is a proper left factor of $v$, then $\|w'\| = \|v'\| \geqslant 0$. Conversely, let $w$ be a word satisfying (i) and (ii). If $|w| = 1$, then $w = b \in$ Ł. Arguing by induction on $|w|$, assume $|w| > 1$. First note that by (ii), $w$ begins with the letter $a$. Thus $w = aw'$ for some $w'$. Next, since $\|w\| = -1$, there exists a shortest non empty left factor $u$ of $w'$ such that $\|au\| = 0$. Set $w = auv$. Then $\|u\| = -1$, and for any proper left factor $u'$ of $u$, $\|u'\| \geqslant 0$ by the minimality assumption on $u$. Thus $u \in$ Ł. Next $\|v\| = \|w\| = -1$, and $\|v'\| = \|auv'\| \geqslant 0$ for any proper left factor $v'$ of $v$ since $w$ satisfies (ii). Thus $v \in$ Ł and $w \in$ Ł by (4.1). ■

Proposition 4.1 can be used to draw a pictorial representation of a word $w$ in Ł. This is given by the graph of the function $w' \mapsto \|w'\|$, where $w'$ ranges over the left factors of $w$. Thus, for $w = aabaabbabbabaaabbbb$, we obtain Fig. II.3.

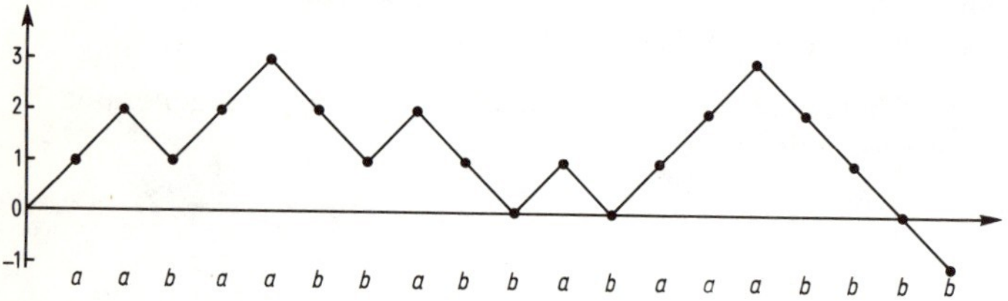


Fig. II.3

Next, consider the restricted Dyck language $D_1'^*$ over $X$, i.e. with $x_1 = a$, $\bar{x}_1 = b$; Then $D_1'^*$ is defined by

$$D_1'^* = 1 \cup aD_1'^* bD_1'^*.$$

Multiply this equation by $b$ on the right. This gives

$$D_1'^* b = b \cup aD_1'^* bD_1'^* b.$$

Thus $D_1'^* b$ is solution of (4.1), and therefore $D_1'^* b =$ Ł.

**Corollary 4.2** *Let* $w \in X^*$. *Then* $w \in D_1'^*$ *iff* $w$ *satisfies:*

(i) $\|w\| = 0$;

(ii) $\|w'\| \geqslant 0$ *for any left factor* $w'$ *of* $w$. ■

Recall that $u$ and $v$ are conjugate iff $u = fg$ and $v = gf$ for some words $f$, $g$.

**Proposition 4.3** *Let $u \in X^*$ with $\|u\| = -1$. Then there exists one and only one word $v$ conjugate to $u$ such that $v \in Ł$.*

Proof. We show first the unicity. Assume $u = fg$, $v = gf \in Ł$, $g \neq 1$. Then by Proposition 4.1 $\|f\| \geqslant 0$, thus $\|u\| = -1 = \|f\| + \|g\| \geqslant \|g\|$, and $g$ cannot be a proper left factor of $v$. Thus $f = 1$ and $u = v$.

Next, let $p = \min\{\|u'\| : u' \text{ proper left factor of } u\}$. If $p \geqslant 0$, then $u \in Ł$. Assume $p < 0$, and let $f$ be the shortest left factor of $u$ such that $\|f\| = p$. Write $u = fg$. Then

$$\|f'\| \geqslant p + 1 \qquad \text{for any proper left factor } f' \text{ of } f \tag{4.2}$$

by the minimality of $f$. Next

$$\|g'\| \geqslant 0 \quad \text{for any proper left factor } g' \text{ of } g \tag{4.3}$$

since $p \leqslant \|fg'\| = p + \|g'\|$ by definition of $p$ and

$$\|g\| = -1 - p \geqslant 0. \tag{4.4}$$

Let $v = gf$. Then $\|v\| = \|u\| = -1$. Let $v'$ be a proper left factor of $v$. If $v'$ is a left factor of $g$, then $\|v'\| \geqslant 0$ by (4.3) and (4.4). Otherwise, $v' = gf'$ where $f'$ is a proper left factor of $f$, and $\|v'\| = -1 - p + \|f'\| \geqslant 0$ by (4.2). In view of Proposition 4.1, $v \in Ł$. ∎

**b) The language of completely parenthesized arithmetic expressions** $E$ over $X = \{a, b, c, d\}$ is the language generated by the grammar with productions

$$\xi \to a\xi b\xi c + d.$$

Thus $E$ is the unique language satisfying

$$E = aEbEc \cup d. \tag{4.5}$$

The first words of $E$ are

$$d, adbdc, aadbdcbdc, adbadbdcc, aadbdcbadbdcc, \ldots.$$

The terminology is from M. Nivat [1967]. Write indeed "(" for "$a$", ")" for "$c$", "+" for "$b$" and "$i$" for "$d$". The words listed above become

$$i, (i+i), ((i+i)+i), (i+(i+i)), ((i+i)+(i+i)), \ldots.$$

Consider the morphism that erases $c$ and $d$. Then by (4.5) the image of $E$ is the language $D_1'^*$ over $\{a, b\}$. If $b$ and $c$ are erased, then the image of $E$ is the language $Ł$ over $\{a, d\}$. Thus $E$ is closely related to these languages. In fact, we shall prove later (Chapter VII) that the language $E$ is a generator of the cone of context-free languages.

**Lemma 4.4** *Let $w \in E$. Then*

(i) $|w|_a = |w|_b = |w|_c = |w|_d - 1$;

(ii) *If $w'$ is a proper, non empty left (resp. right) factor of $w$, then $|w'|_a > |w'|_c$ (resp. $|w'|_a < |w'|_c$).*

The easy proof is left to the reader. Note that (ii) implies that the language $E$ is bifix.

Let $\eta$ be the congruence over $X^*$ generated by the relation

$$adbdc \sim d.$$

**Theorem 4.5** *The language $E$ is equal to the class of $d$ in the congruence $\eta: E=[d]_\eta$.*

Proof. Clearly $d \in [d]_\eta$. Let $w \in E$, $w \neq d$. Then $w = aubvc$ for some $u, v \in E$ by (4.5). Arguing by induction, $u \equiv v \equiv d(\text{mod } \eta)$, thus $w \equiv adbdc \equiv d(\text{mod } \eta)$. This shows the inclusion $E \subset [d]_\eta$.

To show the converse conclusion, it suffices to prove that for any two words $w = udv$ and $w' = uadbdcv$,

$$w \in E \quad \Leftrightarrow \quad w' \in E. \tag{4.6}$$

We verify (4.6) by induction on $|w| = |w'| - 4$. If $u = 1$, then $w \in E$ iff $w = d$, and $w' \in E$ iff $w' = adbdc$, since $E$ is prefix. Thus we may assume $u \neq 1$. Suppose $w \in E$. Then $w = aw_1bw_2c$ for some $w_1, w_2 \in E$. Then either $|ud| \leqslant |aw_1|$ or $|dv| \leqslant |w_2c|$. In the first case, $w_1 = u_1dv_1$ for $u_1$ and $v_1$ defined by $au_1 = u$, $v = v_1bw_2c$. By induction, $w'_1 = u_1adbdcv_1$ belongs to $E$. Thus $aw'_1bw_2c = w' \in E$ by (4.5). The second case handles in the same way. Conversely, suppose $w' \in E$. Then

$$w' = uadbdcv = aw'_1bw'_2c$$

for $w'_1, w'_2 \in E$. If $|aw'_1| = |uad|$, then $w'_1 = d$ since $E$ is suffix, hence $u = 1$ contrary to the assumption. Thus either $|uad| < |aw'_1|$ or $|dcv| < |w'_2c|$. It suffices to consider the first case. Clearly, it implies that $|uadbdc| \leqslant |aw'_1|$, thus $w'_1 = u'_1adbdcv'_1$ with $au'_1 = u$ and $v = v'_1bw'_2c$. By induction, $w_1 = u'_1dv'_1 \in E$, hence

$$aw_1bw'_2c = w \in E.$$

■

Theorem 4.5 admits the following

**Corollary 4.6** *Let $u, u' \in E$. Then $fug \in E$ iff $fu'g \in E$.*

Proof. $fug \equiv fu'g(\text{mod } \eta)$. Thus $fug \equiv d(\text{mod } \eta)$ iff $fu'g \equiv d(\text{mod } \eta)$. ■

## Exercises

**4.1** Show that $Ł = [b]_\lambda$, where $\lambda$ is the congruence over $\{a, b\}^*$ generated by the relation $abb \sim b$.

**4.2** Let $p_n = \text{Card}(X^{2n+1} \cap Ł)$. Show that $p_n = \frac{1}{n+1}\binom{2n}{n}$. Show that $p_n = q_n$, where $q_n = \text{Card}(X^{4n+1} \cap E)$.

# III Rational Transductions

Rational transductions are defined by rational relations, i.e. rational subsets of the product of two free monoids. The chapter therefore begins with two sections concerned with recognizable and rational subsets of an arbitrary monoid. The next two sections contain the definition and basic properties of rational relations and rational transductions. Examples of rational transductions are given in Section 5. Then the machines realizing rational transductions are introduced. Matrix representations of rational transductions are investigated in Section 7. In the last section we show that most of the usual decision problems are unsolvable for rational transductions.

## III.1 Recognizable Sets

Kleene's Theorem gives a characterization of the regular languages of a finitely generated free monoid, but the theorem cannot be extended to arbitrary monoids. Therefore one can try to investigate the class of monoids where Kleene's Theorem remains true. An example of such a monoid was given by Amar and Putzolu [1965]. A wider family of semigroups, where Kleene's Theorem is partially true is formed by the equidivisible semigroups of McKnight and Storey [1969]. S. Eilenberg had the idea, formulated for instance in [1967], to distinguish in each monoid two families of subsets, called the recognizable and the rational subsets. These two families are of distinct nature and Kleene's Theorem precisely asserts that they coincide in finitely generated free monoids. Properties of regular languages like closure properties can be proved some for the recognizable subsets, others for the rational subsets of a monoid. This gives also insight in the structure of regular languages by showing from which of their two aspects originate their properties.

This section deals with recognizable, the second section with rational subsets of a monoid. We are mainly interested in properties which are of later use for rational transductions, but we also touch slightly on properties of rational subsets of groups.

We want recognizable sets to be, in free monoids, exactly the languages recognized by finite automata. Instead of a generalization of finite automata, we prefer to use as definition the characterization via a morphism into a finite monoid. This simplifies the exposition.

**Definition** Let $M$ be a monoid. A subset $A$ of $M$ is recognizable if there exist a finite monoid $N$, a morphism $\alpha$ from $M$ into $N$ and a subset $P$ of $N$ such that $A=\alpha^{-1}(P)$.

If this holds, then $\alpha(A)=P\cap\alpha(M)$ and consequently $A=\alpha^{-1}(\alpha(A))$. Next, $\alpha$ considered as a morphism onto $\alpha(M)$ is surjective, and $A=\alpha^{-1}(Q)$ with $Q=P\cap\alpha(M)$. Thus we may assume that $\alpha$ is surjective in the above definition. An equivalent condition for $A\subset M$ to be recognizable is the existence of a congruence relation $\theta$ on $M$ of finite index such that $A$ is saturated for $\theta$, i.e. $A$ is a union of equivalence classes of $\theta$.

The set of all recognizable subsets of $M$ is denoted by $\mathrm{Rec}(M)$.

**Example 1.1** Let $M$ be any monoid, and let $N=\{1\}$ be the monoid consisting of a single element. Let $\alpha$ be the unique morphism from $M$ onto $N$. Then $\emptyset=\alpha^{-1}(\emptyset)$ and $M=\alpha^{-1}(N)$. Thus $M, \emptyset\in\mathrm{Rec}(M)$ for any monoid $M$.

**Example 1.2** If $M$ is a finite monoid, then any subset of $M$ is recognizable.

**Example 1.3** If $M=X^*$ and $X$ is an alphabet, then $A\in\mathrm{Rec}(X^*)$ iff $A$ is recognized by a finite automaton (Proposition I.4.4).

**Example 1.4** Consider the additive group $\mathbb{Z}$ of integers. Let $\alpha$ be a morphism from $\mathbb{Z}$ onto a finite monoid $N$. Then $\alpha$ is a group morphism and $N=\alpha(\mathbb{Z})$ is a finite group, thus $N=\mathbb{Z}/n\mathbb{Z}$ for some integer $n\geqslant 1$ (Exercise I.3.1). Consequently $N$ can be identified with the set $\{0,1,\ldots,n-1\}$, and for $p\in N$, $\alpha^{-1}(p)=p+n\mathbb{Z}$. Thus for $P\subset N$, $\alpha^{-1}(P)=\bigcup_{p\in P} p+n\mathbb{Z}$. Conversely, any subset of $\mathbb{Z}$ of this form is recognizable. It follows that $A\in\mathrm{Rec}(\mathbb{Z})$ iff $A$ is a finite union of arithmetic progressions. In particular, any nonempty recognizable subset of $\mathbb{Z}$ is infinite.

**Proposition 1.1** *Let $M$ be a monoid. Then* $\mathrm{Rec}(M)$ *is closed under union, intersection and complementation.*

Since $\emptyset$, $M\in\mathrm{Rec}(M)$, it follows that $\mathrm{Rec}(M)$ is a boolean algebra.

Proof. Let $A\in\mathrm{Rec}(M)$, let $N$ be a finite monoid, let $\alpha: M\to N$ be a surjective morphism and let $P$ be a subset of $N$ such that $A=\alpha^{-1}(P)$. Then $M\backslash A=\alpha^{-1}(N\backslash P)$. Thus $M\backslash A\in\mathrm{Rec}(M)$. This proves the closure under complementation.

Next, let $B\in\mathrm{Rec}(M)$, $B=\beta^{-1}(Q)$, where $\beta$ is a morphism from $M$ onto some finite monoid $N'$ and $Q\subset N'$. Let $N''=N\times N'$ be the product monoid and define $\gamma: M\to N''$ by $\gamma(m)=(\alpha(m),\beta(m))$, $m\in M$. Then $\gamma$ is a morphism. Further $\gamma(m)\in P\times Q$ iff $\alpha(m)\in P$ and $\beta(m)\in Q$, thus iff $m\in\alpha^{-1}(P)\cap\beta^{-1}(Q)$. Consequently $A\cap B=\gamma^{-1}(P\times Q)$, and since $N''$ is finite, $A\cap B\in\mathrm{Rec}(M)$. Thus $\mathrm{Rec}(M)$ is closed under intersection. Closure under union follows from de Morgan's rule. ∎

**Corollary 1.2** *If $A$, $B \in \mathrm{Rec}(M)$, then $A \backslash B \in \mathrm{Rec}(M)$.* ■

An important property of recognizable sets is the closure under inverse morphisms.

**Proposition 1.3** *Let $M$ and $M'$ be monoids, and let $\gamma: M \to M'$ be a morphism. If $A' \in \mathrm{Rec}(M')$, then $\gamma^{-1}(A') \in \mathrm{Rec}(M)$.*

Proof. Let $\alpha: M' \to N$ be a surjective morphism onto a finite monoid $N$, and let $P \subset N$ such that $A' = \alpha^{-1}(P)$. Then $\gamma^{-1}(A') = \beta^{-1}(P)$, with $\beta = \alpha \circ \gamma$. Thus $\gamma^{-1}(A') \in \mathrm{Rec}(M)$. ■

If the monoid $M'$ of Proposition 1.3 is finitely generated, then $M$ can be chosen to be the free monoid generated by an alphabet. It follows that $\gamma^{-1}(A')$ is a regular language.

**Corollary 1.4** *Let $\gamma$ be an isomorphism from $M$ onto $M'$. Then $A \in \mathrm{Rec}(M)$ iff $\gamma(A) \in \mathrm{Rec}(M')$.* ■

The following example shows that the homomorphic image of a recognizable set is not recognizable in general.

**Example 1.5** Let $X = \{x, y\}$, and let $\gamma: X^* \to \mathbb{Z}$ be the morphism defined by $\gamma(f) = |f|_x - |f|_y$ $(f \in X^*)$. Then $\{1\} \in \mathrm{Rec}(X^*)$, and $\gamma(\{1\}) = \{0\}$. In view of Example 1.4, $\{0\} \notin \mathrm{Rec}(\mathbb{Z})$. This can also be seen by applying Proposition 1.3. Assume indeed $\{0\} \in \mathrm{Rec}(\mathbb{Z})$. Then $\gamma^{-1}(0)$ is a recognizable subset of $X^*$, i.e. a regular language. Since $\gamma^{-1}(0) = D_1^{*}$, the Dyck language over $X$ (Exercise II.3.6), this yields a contradiction.

In general, the family $\mathrm{Rec}(M)$ is closed neither under product nor under star operation. This is shown by the following example which is credited to S. Winograd by Eilenberg [1974].

**Example 1.6** Consider the additive group $\mathbb{Z}$, and add to $\mathbb{Z}$ two new elements $\varepsilon$, $a$. The set $M = \mathbb{Z} \cup \{\varepsilon, a\}$ is a commutative monoid with addition extended as follows:

$$\varepsilon + m = m \quad (m \in M), \qquad a + a = 0, \quad a + x = x \quad (x \in \mathbb{Z}).$$

Thus $\varepsilon$ is the neutral element of $M$. We first show that $\{\varepsilon\}$, $\{a\} \in \mathrm{Rec}(M)$. Consider indeed the commutative monoid $N = \{\bar{\varepsilon}, \bar{a}, \bar{0}\}$ with neutral element $\bar{\varepsilon}$, and with addition defined by $\bar{0} + \bar{0} = \bar{0} + \bar{a} = \bar{a} + \bar{a} = \bar{0}$. Then $\alpha: M \to N$ given by $\alpha(\varepsilon) = \bar{\varepsilon}$, $\alpha(a) = \bar{a}$, $\alpha(x) = \bar{0}$, $(x \in \mathbb{Z})$ is a morphism, and $\{\varepsilon\} = \alpha^{-1}(\bar{\varepsilon})$, $\{a\} = \alpha^{-1}(\bar{a})$. Next if $A \in \mathrm{Rec}(M)$, then $A \cap \mathbb{Z} \in \mathrm{Rec}(\mathbb{Z})$. Let indeed $\beta$ be a morphism from $M$ onto a finite monoid $N'$, let $\beta_1$ be the restriction of $\beta$ on $\mathbb{Z}$, and set $N_1 = \beta_1(\mathbb{Z})$.

If $A = \beta^{-1}(P)$ for $P \subset N'$, then

$$\beta_1^{-1}(P \cap N_1) = \beta^{-1}(P \cap N_1) \cap \mathbb{Z} = \beta^{-1}(P) \cap \mathbb{Z} = A \cap \mathbb{Z}.$$

Consequently $A \cap \mathbb{Z} \in \mathrm{Rec}(\mathbb{Z})$. Define now $A = \{a\}$. Then $A \in \mathrm{Rec}(M)$, and $A + A = \{0\}$, $A^+ = \{0, a\}$, $A^* = \{0, \varepsilon, a\}$. None of these subsets is in $\mathrm{Rec}(M)$, since otherwise their intersection with $\mathbb{Z}$, that is $\{0\}$ would be a recognizable subset of $\mathbb{Z}$ in contradiction with Example 1.4.

The following theorem gives a description of the recognizable subsets of the product of two monoids. Eilenberg [1974] attributes it to Mezei.

**Theorem 1.5** (Mezei) *Let $M_1$, $M_2$ be monoids and $M = M_1 \times M_2$. Then $B \in \mathrm{Rec}(M)$ iff $B$ is a finite union of sets of the form $A_1 \times A_2$, with $A_1 \in \mathrm{Rec}(M_1)$ and $A_2 \in \mathrm{Rec}(M_2)$.*

Proof. The condition is sufficient. Let indeed $\pi_i : M \to M_i$, $(i = 1, 2)$ be the canonical projections. If $A_1 \subset M_1$, $A_2 \subset M_2$, then

$$A_1 \times A_2 = (A_1 \times M_2) \cap (M_1 \times A_2) = \pi_1^{-1}(A_1) \cap \pi_2^{-1}(A_2).$$

Thus if $A_1 \in \mathrm{Rec}(M_1)$, $A_2 \in \mathrm{Rec}(M_2)$, then $A_1 \times A_2 \in \mathrm{Rec}(M)$ in view of Propositions 1.3 and 1.1. Since $\mathrm{Rec}(M)$ is closed under union, $B \in \mathrm{Rec}(M)$.

Conversely, assume $B \in \mathrm{Rec}(M)$. Then there exist a finite monoid $N$, a morphism $\beta : M \to N$, and a subset $P$ of $N$ such that $B = \beta^{-1}(P)$. Consider the morphisms $\alpha_i : M_i \to N$ defined by

$$\alpha_1(m_1) = \beta(m_1, 1), \qquad \alpha_2(m_2) = \beta(1, m_2)$$

and let $\gamma : M \to N \times N$ be the morphism defined by

$$\gamma(m_1, m_2) = (\alpha_1(m_1), \alpha_2(m_2)).$$

In $N \times N$ consider the set

$$Q = \{(n_1, n_2) \mid n_1 n_2 \in P\}.$$

Then $\gamma(m_1, m_2) \in Q$ iff $\alpha_1(m_1)\alpha_2(m_2) \in P$. Since

$$\alpha_1(m_1)\alpha_2(m_2) = \beta(m_1, m_2),$$

and since $(m_1, m_2) \in B$ iff $\beta(m_1, m_2) \in P$, it follows that $B = \gamma^{-1}(Q)$. Next $\gamma^{-1}(n_1, n_2) = \alpha_1^{-1}(n_1) \times \alpha_2^{-1}(n_2)$, whence

$$B = \bigcup_{(n_1, n_2) \in Q} \gamma^{-1}(n_1, n_2) = \bigcup_{(n_1, n_2) \in Q} \alpha_1^{-1}(n_1) \times \alpha_2^{-1}(n_2).$$

Since the sets $\alpha_i^{-1}(n_i)$ are recognizable subsets of $M_i$, $(i = 1, 2)$, the required decomposition of $B$ is obtained. ∎

**Exercises**

**1.1** Let $M'$ be a monoid, $M$ a submonoid of $M'$. Show that if $A' \in \mathrm{Rec}(M')$ then $A' \cap M \in \mathrm{Rec}(M)$. Give an example showing that $\mathrm{Rec}(M)$ is in general not contained in $\mathrm{Rec}(M')$, even if $M \in \mathrm{Rec}(M')$. (Hint (Perrin). Consider $M = (xy^+)^* \in \mathrm{Rec}(\{x, y\}^*)$.)

**1.2** Let $M$ be a monoid. Define a finite automaton $\mathbf{A}$ over $M$ by a finite set of states $Q$, an initial state $q_-$, a set of final states $Q_+$, and a next state function $Q \times M \to Q$ satisfying the following conditions:

$$q \cdot 1 = q \qquad (q \in Q)$$

$$q \cdot mm' = (q \cdot m) \cdot m' \qquad (q \in Q, m, m' \in M).$$

The subset of $M$ recognized by $\mathbf{A}$ is by definition $|\mathbf{A}| = \{m \in M \mid q_- \cdot m \in Q_+\}$. Show that $A \in \mathrm{Rec}(M)$ iff $A$ is recognized by a finite automaton over $M$. (For further discussion on these lines, see Walljasper [1970] and Vogel [1972].)

**1.3** Let $G$ be a group. Show that $A \in \mathrm{Rec}(G)$ iff there exists an invariant subgroup $H$ of $G$ of finite index (i.e. $G/H$ is finite) such that $A$ is a union of cosets of $H$. Show that a subgroup of $G$ is recognizable iff it is of finite index.

**1.4** Let $M$ be a monoid, $A \in \mathrm{Rec}(M)$. Show that for any $B \subset M$, $B^{-1}A = \{m \mid Bm \cap A \neq \emptyset\}$ is a recognizable subset of $M$. (Hint. Use Exercise I.3.6.)

## III.2 Rational Sets

In this section, we study the rational subsets of a monoid and their relation to recognizable subsets.

**Definition** Let $M$ be a monoid. The family $\mathrm{Rat}(M)$ of rational subsets of $M$ is the least family $\mathcal{R}$ of subsets of $M$ satisfying the following conditions:

(i) $\emptyset \in \mathcal{R}; \{m\} \in \mathcal{R}$ for all $m \in M$; (2.1)

(ii) if $A, B \in \mathcal{R}$, then $A \cup B, AB \in \mathcal{R}$; (2.2)

(iii) if $A \in \mathcal{R}$, then $A^+ = \bigcup_{n \geq 1} A^n \in \mathcal{R}$. (2.3)

In presence of (i) and (ii), the condition (iii) is equivalent to:

(iii′) $A \in \mathcal{R} \Rightarrow A^* \in \mathcal{R}$.

Assume indeed $A^+ \in \mathcal{R}$. Since $\{1\} \in \mathcal{R}$ by (i), it follows by (ii) that $A^* = \{1\} \cup A^+ \in \mathcal{R}$. Conversely, if $A$, $A^* \in \mathcal{R}$, then by (ii) $A^+ = AA^* \in \mathcal{R}$.

Any subset $A$ of $M$ obtained from the singletons by a finite number of unions, products and plus or star operations is in $\mathrm{Rat}(M)$. Moreover, the family of subsets of $M$ obtained in that way, together with the empty set satisfies conditions (i)–(iii), and therefore is the family $\mathrm{Rat}(M)$. Thus a rational subset of $M$ is either empty or can be expressed, starting with singletons, by a finite number of unions, products, and plus or stars. Such an expression is called a rational expression. It is the simplest way to show that a given set is rational.

**Example 2.1** Any subset of a finite monoid is rational.

**Example 2.2** Let $X$ be an alphabet, and $X^{\oplus}$ be the free commutative monoid generated by $X$. We claim that $A$ is a rational subset of $X^{\oplus}$ iff $A$ is a finite union of sets of the form

$$ab_1^*b_2^*\cdots b_n^* \qquad (n \geq 0, a, b_1, \ldots, b_n \in X^{\oplus}). \tag{2.4}$$

Unions of sets (2.4) are also called semilinear. Clearly, any set of the form (2.4) is rational, thus any semilinear set is rational. Next if $A$, $B \subset X^{\oplus}$, then $(A \cup B)^* = A^*B^*$ and $(ab_1^*b_2^* \cdots b_n^*)^* = a^*b_1^* \cdots b_n^*$. This shows that semilinear sets are closed under star operation. The empty set and the singletons are semilinear, further semilinear sets are obviously closed under union and product. Thus any rational set is semilinear. This proves that the semilinear sets are exactly the rational subsets of $X^{\oplus}$.

**Example 2.3** If $X$ is an alphabet, then the rational subsets are, according to Kleene's Theorem, exactly the languages recognized by finite automata.

Thus Kleene's Theorem can be formulated as follows:

**Theorem 2.1** (Kleene) *Let $X$ be a (finite) alphabet. Then* $\mathrm{Rat}(X^*) = \mathrm{Rec}(X^*)$.

In view of this theorem, we also call regular languages indistinctly rational or recognizable languages.

We now prove that rational sets are closed under morphism.

**Proposition 2.2** *Let $M$, $M'$ be monoids, and let $\alpha: M \to M'$ be a morphism. If $A \in \mathrm{Rat}(M)$, then $\alpha(A) \in \mathrm{Rat}(M')$. Further if $\alpha$ is surjective, then for any $A' \in \mathrm{Rat}(M')$ there is a set $A \in \mathrm{Rat}(M)$ such that $\alpha(A) = A'$.*

Proof. Let $\mathcal{R}$ be the family of subsets $A$ of $M$ such that $\alpha(A) \in \mathrm{Rat}(M')$. Then $\emptyset \in \mathcal{R}$ and $\{m\} \in \mathcal{R}$ for $m \in M$. Next

$$\alpha(A \cup B) = \alpha(A) \cup \alpha(B), \qquad \alpha(AB) = \alpha(A)\alpha(B), \qquad \alpha(A^+) = (\alpha(A))^+ \tag{2.5}$$

for any subsets $A$, $B$ of $M$. Thus $A$, $B \in \mathcal{R}$ implies that $A \cup B$, $AB$, $A^+ \in \mathcal{R}$. Thus $\mathcal{R}$ satisfies conditions (2.1), (2.2), (2.3). Consequently $\mathcal{R} \supset \mathrm{Rat}(M)$ and the first statement is proved. Consider now the family $\mathcal{S}$ of subsets $A'$ of $M'$ such that $A' = \alpha(A)$ for some $A \in \mathrm{Rat}(M)$. Since $\alpha$ is surjective, $\{m'\} \in \mathcal{S}$ for all $m' \in M'$. Obviously $\emptyset \in \mathcal{S}$. In view of (2.5), $\mathcal{S}$ is closed under union, product and the plus operation. Thus $\mathcal{S} \supset \mathrm{Rat}(M')$. ∎

**Corollary 2.3** *Let $\alpha$ be an isomorphism from $M$ onto $M'$. Then $A \in \mathrm{Rat}(M)$ iff $\alpha(A) \in \mathrm{Rat}(M')$.* ∎

Note that the second part of Proposition 2.2 only claims the existence of a rational set $A$ such that $\alpha(A) = A' \in \mathrm{Rat}(M')$. Obviously this does not imply that any subset $A$ of $M$ with $\alpha(A) \in \mathrm{Rat}(M')$ is rational. In particular, the inverse image $\alpha^{-1}(A')$ is generally not rational for rational subsets $A'$ of $M'$.

**Example 2.4** Consider as in Example 1.5 the alphabet $X=\{x, y\}$ and the morphism $\gamma: X^*\to \mathbb{Z}$ defined by $\gamma(f)=|f|_x-|f|_y$ $(f\in X^*)$. Then $\{0\}\in \mathrm{Rat}(\mathbb{Z})$, and $\gamma^{-1}(0)=\{f\in X^* \mid |f|_x=|f|_y\}=D_1^{*}\notin \mathrm{Rat}(X^*)$.

Although Kleene's Theorem is not true in arbitrary monoids, there is a weakened version for finitely generated monoids.

**Proposition 2.4** (McKnight (1964)] *Let M be a finitely generated monoid. Then* $\mathrm{Rec}(M)\subset \mathrm{Rat}(M)$.

Proof. Since $M$ is finitely generated, there exist an alphabet $X$ and a surjective morphism $\alpha: X^*\to M$. Let $A\in \mathrm{Rec}(M)$. Then $\alpha^{-1}(A)\in \mathrm{Rec}(X^*)$ by Proposition 1.3. By Kleene's Theorem, $\alpha^{-1}(A)\in \mathrm{Rat}(X^*)$. In view of Proposition 2.2, $\alpha(\alpha^{-1}(A))=A\in \mathrm{Rat}(M)$. ■

Proposition 2.4 is not true in monoids which are not finitely generated. Consider indeed such a monoid $M$. Then $M\in \mathrm{Rec}(M)$, but $M\notin \mathrm{Rat}(M)$ in view of the following lemma.

**Lemma 2.5** *Let M be a monoid. For any* $A\in \mathrm{Rat}(M)$, *there exists a finitely generated submonoid* $M_1$ *of M such that* $A\subset M_1$.

Proof. Let $\mathcal{R}$ be the family of subsets $A$ of $M$ contained in some finitely generated submonoid of $M$. Obviously, $\emptyset\in\mathcal{R}$ and $\{m\}\in\mathcal{R}$ for $m\in M$. Next let $A, B\in\mathcal{R}$, and let $R, S$ be finite subsets of $M$ such that $A\subset R^*$, $B\subset S^*$. Then $A\cup B$, $AB\subset(R\cup S)^*$ and $A^*\subset R^*$. Consequently $A\cup B$, $AB$, $A^*\in\mathcal{R}$ and $\mathcal{R}\supset \mathrm{Rat}(M)$. ■

**Proposition 2.6** *Let M be a monoid. If* $A\in \mathrm{Rat}(M)$ *and* $B\in \mathrm{Rec}(M)$, *then* $A\cap B\in \mathrm{Rat}(M)$.

Proof. Let $A$ be a rational subset of $M$. Then there exists a finitely generated submonoid $M_1$ of $M$ such that $A\subset M_1$, and consequently $A\in \mathrm{Rat}(M_1)$. Next there is an alphabet $X$ and a morphism $\alpha: X^*\to M$ that maps $X^*$ onto $M_1$. Thus by the second part of Proposition 2.2, there is a rational language $A'\subset X^*$ such that $\alpha(A')=A$. Let $B$ be a recognizable subset of $M$. Then $B'=\alpha^{-1}(B)$ is a recognizable subset of $X^*$ by Proposition 1.3. In view of Kleene's Theorem, $C'=A'\cap B'$ is a regular, thus a rational language, and $\alpha(C')\in \mathrm{Rat}(M)$ by Proposition 2.2. Since

$$\alpha(C')=\alpha(A'\cap\alpha^{-1}(B))=\alpha(A')\cap B=A\cap B,$$

it follows that $A\cap B\in \mathrm{Rat}(M)$. ■

The following example shows that the intersection of two rational sets is not necessarily rational.

**Example 2.5** Let $M=\{x\}^*\times\{y,z\}^*$, and consider the sets

$$A=(x,y)^*(1,z)^*=\{(x^n,y^nz^k):n,\ k\geq 0\},$$
$$B=(1,y)^*(x,z)^*=\{(x^n,y^kz^n):n,\ k\geq 0\}.$$

Clearly, $A,\ B\in \mathrm{Rat}(M)$. Suppose that

$$C=A\cap B=\{(x^n,y^nz^n):n\geq 0\}$$

is rational, and define a morphism $\pi:M\rightarrow\{y,z\}^*$ by $\pi(x,1)=1$, $\pi(1,y)=y$, $\pi(1,z)=z$. Then $\pi(C)=\{y^nz^n:n\geq 0\}$ would be a rational subset of $\{y,z\}^*$ by Proposition 2.2. Thus $C$ is not rational.

Sometimes the notion of starheight of a rational set is useful. Let $M$ be a monoid, and define inductively sets $\mathrm{Rat}_0(M)\subset \mathrm{Rat}_1(M)\subset\cdots$ by:

$A\in \mathrm{Rat}_0(M)$ iff $A$ is a finite subset of $M$;
$A\in \mathrm{Rat}_{h+1}(M)$ iff $A$ is a finite union of sets of the form $B_1B_2\cdots B_n$,

where either $B_i$ is a singleton or $B_i=C_i^*$ for some $C_i\in \mathrm{Rat}_h(M)$. It is readily shown (Exercise 2.1) that

$$\mathrm{Rat}(M)=\bigcup_{h\geq 0}\mathrm{Rat}_h(M).$$

The sets in $\mathrm{Rat}_h\backslash \mathrm{Rat}_{h-1}$ are said to have starheight $h$.

We use starheight in the proof of the following result which gives an interpretation of rational sets in groups.

**Theorem 2.7** (Anissimov and Seifert [1975]) *Let $G$ be a group, and let $H$ be a subgroup of $G$. Then $H$ is finitely generated iff $H$ is a rational subset of $G$.*

Proof. For any subset $A$ of $G$, let $\langle A\rangle$ denote the subgroup generated by $A$, and let $A^{-1}=\{x^{-1}\mid x\in A\}$. Then $\langle A\rangle=(A\cup A^{-1})^*$. This shows that a finitely generated subgroup of $G$ is rational.

In order to prove the converse we first consider the following situation. Let $A$ be a subset of $G$ such that

$$A=x_1T_1^*x_2T_2^*\cdots x_nT_n^*x_{n+1} \tag{2.6}$$

with $x_1,\ldots,x_{n+1}\in G$, $T_1,\ldots,T_n\subset G$, and define

$$y_i=x_1x_2\cdots x_i \qquad i=1,\ldots,n+1 \tag{2.7}$$
$$S_i=y_iT_iy_i^{-1} \qquad i=1,\ldots,n$$
$$A'=y_{n+1}\cup S_1\cup\cdots\cup S_n. \tag{2.8}$$

Then we claim:

$$\langle A\rangle=\langle A'\rangle. \tag{2.9}$$

Indeed, observe that by (2.6), $y_{n+1}, y_{n+1}^{-1} \in \langle A\rangle$. Further

$$S_i = (x_1 \cdots x_i T_i x_{i+1} \cdots x_{n+1}) y_{n+1}^{-1}.$$

Thus $S_i \subset \langle A\rangle$, whence $A' \subset \langle A\rangle$ and $\langle A'\rangle \subset \langle A\rangle$. Next

$$S_i^* = (y_i T_i y_i^{-1})^* = y_i T_i^* y_i^{-1}.$$

Since $x_1 = y_1$ and $x_i = y_{i-1}^{-1} y_i$, $(2 \leqslant i \leqslant n+1)$,

$$A = y_1 T_1^* y_1^{-1} y_2 T_2^* y_2^{-1} \cdots y_n T_n^* y_n^{-1} y_{n+1} = S_1^* S_2^* \cdots S_n^* y_{n+1}.$$

Thus $A \subset \langle A'\rangle$, whence $\langle A\rangle \subset \langle A'\rangle$. This proves (2.9).

Consider now a subgroup $H$ of $G$ such that $H \in \mathrm{Rat}(G)$. Since $H = \langle H\rangle$, $H$ has a rational set of generators. We have to show that $H$ has a system of generators of starheight 0. Let $R$ be a rational set of generators of minimal starheight $h$, and assume $h > 0$. Then

$$R = A_1 \cup A_2 \cup \cdots \cup A_r,$$

where each $A_k$, $(1 \leqslant k \leqslant r)$ has the form (2.6), and at least one $A_k$ has starheight $h$. Set

$$R' = A_1' \cup A_2' \cup \cdots \cup A_r',$$

where each $A_k'$ is deduced from $A_k$ by (2.7) and (2.8). Then clearly $R'$ has starheight $h-1$. By (2.9), each $A_k$ is contained in $\langle R'\rangle$, and conversely each $A_k'$ is contained in $R$. Thus $\langle R\rangle = \langle R'\rangle = H$, and $R'$ is a system of generators of $H$ of starheight $h-1$, in contradiction with the minimality of $h$. Thus $h = 0$ and the theorem is proved. ■

In the case of free groups, a more precise description of rational sets can be given. Consider an alphabet $X = \{x_1, \ldots, x_n\}$, let $\bar{X} = \{\bar{x} \mid x \in X\}$ and set $Z = X \cup \bar{X}$. Let $X^{(*)}$ be the free group generated by $X$ (see Section II.3), and let $\delta : Z^* \to X^{(*)}$ be the canonical morphism. As already mentioned, there exists an injection $\iota : X^{(*)} \to Z^*$ which associates to each element $u \in X^{(*)}$ the unique reduced word $\iota(u) = f \in \rho(Z^*)$ such that $\delta(f) = u$. The following result describes a property of the mapping $\rho$.

**Proposition 2.8** (Benois [1969]) *Let $K \subset Z^*$ be a regular language. Then the language $\rho(K)$ is also regular.*

This theorem yields the following characterization of rational subsets of $X^{(*)}$.

**Theorem 2.9** (Benois [1969]) *Let $K \subset X^{(*)}$. Then $K \in \mathrm{Rat}(X^{(*)})$ iff $\iota(K)$ is a regular language.*

Proof. Let $K \in \mathrm{Rat}(X^{(*)})$. In view of Proposition 2.2, there exists a regular language $K' \subset Z^*$ such that $\delta(K') = K$. Thus $K'' = \rho(K')$ is regular by Proposition 2.8. Now $\rho = \iota \circ \delta$, whence $K'' = \iota(K)$. Thus $\iota(K)$ is regular. Conversely, assume $\iota(K) \in \mathrm{Rat}(Z^*)$. Then the homomorphic image $\delta(\iota(K)) = K$ is in $\mathrm{Rat}(X^{(*)})$ by Proposition 2.2. ■

The following corollary is interesting:

**Corollary 2.10** (Fliess [1971]) $\mathrm{Rat}(X^{(*)})$ *is closed under intersection and*

Proof. It suffices to show closure under complementation. Let $K \in \mathrm{Rat}(X^{(*)})$. Then $\iota(K) \in \mathrm{Rat}(Z^*)$ by Theorem 2.9. Next $\rho(Z^*)$ is regular, thus $\rho(Z^*) \backslash \iota(K)$ is regular. Since $\iota(K) \subset \rho(Z^*)$, it follows that $\delta(\rho(Z^*) \backslash \iota(K)) = X^{(*)} \backslash K \in \mathrm{Rat}(X^{(*)})$ by Proposition 2.2. ■

It remains to prove Proposition 2.8. For this, we first establish a lemma derived from Fliess [1971]. Consider an alphabet $Y$, and let $A \subset Y^*$ be an arbitrary language. Define a function $\lambda_A$ from $Y^*$ into the subsets of $Y^*$ as follows. For $w, w' \in Y^*$, $w' \in \lambda_A(w)$ iff there exists a factorization

$$w = a_0 x_1 a_1 x_2 \cdots a_{r-1} x_r a_r,$$

with $r \geq 0$, $a_0, a_1, \ldots, a_r \in A$, $x_1, \ldots, x_r \in Y$ such that

$$w' = x_1 x_2 \cdots x_r.$$

Thus $\lambda_A(w)$ consists of all subwords of $w$ obtained by deleting, in $w$, factors in $A$ which are separated by letters.

**Lemma 2.11** *For any $A \subset Y^*$, and for any regular language $K \subset Y^*$, $\lambda_A(K)$ is a regular language.*

Proof. Let $\mathbf{A} = \langle Y, Q, q_-, Q_+ \rangle$ be a finite automaton recognizing $K$. Set $K_{p,q} = \{f \in Y^* \mid p \cdot f = q\}$ for $p, q \in Q$. Let $s \notin Q$, and let $\boldsymbol{B} = \langle Y, Q \cup s, s, Q' \rangle$ be the nondeterministic finite automaton with next state function defined by

$$q \in p \cdot x \quad \text{iff} \quad xA \cap K_{p,q} \neq \emptyset \qquad x \in Y, \quad p, q \in Q;$$

$$q \in s \cdot x \quad \text{iff} \quad AxA \cap K_{q_-q} \neq \emptyset \qquad x \in Y, \quad q \in Q.$$

Next let

$$Q' = \begin{cases} Q_+ & \text{if} \quad A \cap K = \emptyset; \\ s \cup Q_+ & \text{if} \quad A \cap K \neq \emptyset. \end{cases}$$

Then clearly

$$\lambda_A(K) = |\boldsymbol{B}| = \{f \in Y^* \mid s \cdot f \cap Q' \neq \emptyset\}.$$ ■

Proof of Proposition 2.8. Choose in Lemma 2.11 $A = D_n^* = \rho^{-1}(1)$, and $Y = Z$. Then for $f \in Z^*$

$$\rho(f) = \lambda_{D_n^*}(f) \cap \rho(Z^*).$$

Consequently, for $K \in \mathrm{Rat}(Z^*)$, $\rho(K) = \lambda_{D_n^*}(K) \cap \rho(Z^*)$. Since $\rho(Z^*)$ is regular, $\rho(K)$ is a regular language. ■

## Exercises

**2.1** Show that $\operatorname{Rat}(M) = \bigcup_{h \geq 0} \operatorname{Rat}_h(M)$. Compute $\operatorname{Rat}_h(X^{\oplus})$ for $h \geq 0$.

**2.2** Let $G$ be a group. Show that $K \in \operatorname{Rat}(G)$ implies $K^{-1} \in \operatorname{Rat}(G)$.

**2.3** Prove the following group theoretic result: let $G$ be a finitely generated group, and let $H$ be a subgroup of $G$. If $H$ is of finite index, then $H$ is finitely generated (Hint. Use Exercise 1.3.)

**2.4** (Anissimov and Seifert [1975]). Prove the following theorem of Howson: The intersection of two finitely generated subgroups of a free group is again a finitely generated subgroup.

**2.5** Show that for any rational subset $K$ of $X^{(*)}$, $\delta^{-1}(K)$ is a context-free language. (Hint (Sakarovitch [1977]). Write $K = (K^{-1})^{-1} \cdot 1$ and use Exercise I.3.6.)

**2.6** Show that in Proposition 2.8 and in the following statements, $\rho$ can be replaced by $\rho'$ and in fact by $\rho_I$ as defined in Section II.3.

## III.3 Rational Relations

A relation can be considered as a subset of the cartesian product of two sets, or as mapping from the first set into the set of subsets of the second. For the exposition of rational transductions we use in this section the first, "static" aspect, and in the next section the second, more "dynamic" point of view. Rational transductions (more precisely relations) are defined as rational subsets of the product of two monoids. Several characterizations are given. The examples are grouped in Section 5.

**Definition** Let $X$ and $Y$ be alphabets. A rational (resp. recognizable) relation over $X$ and $Y$ is a rational (resp. recognizable) subset of the monoid $X^* \times Y^*$.

The family $\operatorname{Rec}(X^* \times Y^*)$ of recognizable relations is described by Mezei's Theorem 1.5. More precisely we have

**Proposition 3.1** (i) $\operatorname{Rec}(X^* \times Y^*) \subsetneq \operatorname{Rat}(X^* \times Y^*)$;
(ii) *if* $A, B \in \operatorname{Rec}(X^* \times Y^*)$, *then* $AB \in \operatorname{Rec}(X^* \times Y^*)$.

Thus recognizable relations are closed under product. It follows from the proof below that they are not closed under star operation.

Proof. (i) Since $X^* \times Y^*$ is a finitely generated monoid, the inclusion $\operatorname{Rec}(X^* \times Y^*) \subset \operatorname{Rat}(X^* \times Y^*)$ follows from Proposition 2.4. To show that the inclusion is proper, let $x \in X$, $y \in Y$ and consider $A = (x, y)^* = \{(x^n, y^n) : n \geq 0\}$. Clearly $A$ is a rational relation. Assume $A$ is recognizable, let $Z = \{\bar{x}, \bar{y}\}$, and consider the morphism $\gamma : Z^* \to X^* \times Y^*$ defined by $\gamma(\bar{x}) = (x, 1)$, $\gamma(\bar{y}) = (1, y)$.

Then $\gamma^{-1}A = \{f \in Z^* \mid |f|_{\bar{x}} = |f|_{\bar{y}}\}$ is recognizable, thus a regular language. This yields the contradiction.
(ii) In view of Mezei's Theorem,

$$A = \bigcup_{i=1}^{n} R_i \times S_i, \qquad B = \bigcup_{j=1}^{m} R'_j \times S'_j,$$

with $R_i$, $R'_j \in \mathrm{Rat}(X^*)$, $S_i$, $S'_j \in \mathrm{Rat}(Y^*)$. Consequently

$$AB = \bigcup_{i=1}^{n} \bigcup_{j=1}^{m} R_i R'_j \times S_i S'_j,$$

and $R_i R'_j \in \mathrm{Rat}(X^*)$, $S_i S'_j \in \mathrm{Rat}(Y^*)$. By Mezei's Theorem, $AB \in \mathrm{Rec}(X^* \times Y^*)$. ■

We extend the notion of copy defined in Section I.3 as follows: $X^* \times Y^*$ is a copy of $X'^* \times Y'^*$ if $X^*$ is a copy of $X'^*$ and $Y^*$ is a copy of $Y'^*$. Then $X^* \times Y^*$ and $X'^* \times Y'^*$ are isomorphic, and recognizable and rational relations are preserved through the isomorphism by Corollaries 1.4 and 2.3.

The following characterizations of rational relations are fundamental.

They allow to express rational relations by means of regular languages and morphisms of free monoid, and thus rely the algebraic definition to more combinatorial notions. Further, we shall see later that in view of the theorem, a family of languages is closed under rational transduction iff it is closed under morphism, inverse morphism and intersection with regular sets.

**Theorem 3.2** (Nivat [1968]) *Let X and Y be alphabets. The following conditions are equivalent*:
(i) $A \in \mathrm{Rat}(X^* \times Y^*)$;
(ii) *There exist an alphabet Z, two morphisms* $\varphi: Z^* \to X^*$, $\psi: Z^* \to Y^*$ *and a regular language* $K \subset Z^*$ *such that*

$$A = \{(\varphi h, \psi h) : h \in K\};$$

(iii) *There exist an alphabet Z, two alphabetic morphisms* $\alpha: Z^* \to X^*$, $\beta: Z^* \to Y^*$ *and a regular language* $K \subset Z^*$ *such that*

$$A = \{(\alpha h, \beta h) : h \in K\};$$

(iv) *There exist an alphabet Z, two alphabetic morphisms* $\alpha: Z^* \to X^*$, $\beta: Z^* \to Y^*$ *and a local regular language* $K \subset Z^*$ *such that*

$$A = \{(\alpha h, \beta h) : h \in K\};$$

*if* $X \cap Y = \emptyset$, *then* (i) *is equivalent to*
(v) *There exists a regular language* $K \subset (X \cup Y)^*$ *such that*

$$A = \{(\pi_X h, \pi_Y h) : h \in K\},$$

*where* $\pi_X$ *and* $\pi_Y$ *are the projections of* $(X \cup Y)^*$ *onto* $X^*$ *and* $Y^*$ *respectively.*

A couple $(\varphi, \psi)$ of morphisms $\varphi: Z^* \to X^*$ and $\psi: Z^* \to Y^*$ is called a bimorphism.

Proof. The implications (iv)$\Rightarrow$(iii)$\Rightarrow$(ii) are obvious. We prove (ii)$\Rightarrow$(i). Define $\gamma: Z^* \to X^* \times Y^*$ by $\gamma h = (\varphi h, \psi h)(h \in Z^*)$. Then $\gamma$ is a morphism and $\gamma(K) = A$. Since $K \in \text{Rat}(Z^*)$, $A \in \text{Rat}(X^* \times Y^*)$ by Proposition 2.2. Next, we prove (iii)$\Rightarrow$(iv). There exist an alphabet $Z'$, an alphabetic morphism $\gamma: Z'^* \to Z^*$ and a local regular language $K' \subset Z'^*$ such that $\gamma(K') = K$ (see for instance Section I.4). Thus $A = \{(\alpha(\gamma h'), \beta(\gamma h')) : h' \in K'\}$ and the morphisms $\alpha \circ \gamma$, $\beta \circ \gamma$ are alphabetic.

Assume now $X \cap Y = \emptyset$, and define $\pi: (X \cup Y)^* \to X^* \times Y^*$ by $\pi h = (\pi_X h, \pi_Y h)$. Obviously $\pi$ is a surjective morphism. Thus if $A \in \text{Rat}(X^* \times Y^*)$ there exists, by the second part of Proposition 2.2, a regular language $K \subset (X \cup Y)^*$ such that $\pi(K) = A$. This proves (i)$\Rightarrow$(v).

Conversely, if (v) holds, then $\pi(K) = A \in \text{Rat}(X^* \times Y^*)$ by the first part of Proposition 2.2.

Finally we prove (i)$\Rightarrow$(iii). Assume $A \in \text{Rat}(X^* \times Y^*)$. If $X \cap Y = \emptyset$, then (iii) follows from (v). Otherwise, let $X'^* \times Y'^*$ be a copy of $X^* \times Y^*$with $X' \cap Y' = \emptyset$, let $\omega_X: X^* \to X'^*$, $\omega_Y: Y^* \to Y'^*$ be the copy isomorphisms and set $A' = \{(\omega_X f, \omega_Y g) : (f, g) \in A\}$. Then $A'$ is a rational relation, and in view of (v) $A' = \{(\pi_{X'} h, \pi_{Y'} h) : h \in K\}$ for some regular language $K \subset (X' \cup Y')^*$. Consequently, $A = \{((\omega_X^{-1} \circ \pi_{X'})h, (\omega_Y^{-1} \circ \pi_{Y'})h) : h \in K\}$. ■

Theorem 3.2 can be used to derive an iteration lemma for rational relations.

**Lemma 3.3** (Iteration Lemma for Rational Relations) *Let $A \subset X^* \times Y^*$ be a rational relation. There exists an integer $N \geq 1$ such that any $(f, f') \in A$ with $|f| + |f'| \geq N$ admits a factorization*

$$(f, f') = (a, a')(u, u')(b, b') \qquad a, u, b \in X^*, \quad a', u', b' \in Y^*$$

*such that*

(i) $\quad 0 < |u| + |u'| \leq N$;

(ii) $\quad (a, a')(u, u')^*(b, b') \subset A$.

Proof. After a copy, we may assume $X \cap Y = \emptyset$, and by Theorem 3.2(v), $A = \{(\pi_X h, \pi_Y h) : h \in K\}$ for some regular language $K$. Since $|h| = |\pi_X h| + |\pi_Y h|$, the lemma follows directly from the iteration lemmas for regular languages (see Section I.4) applied to $K$. ■

Remark. Several versions of the iteration lemma for regular languages can be transposed to rational relations. Thus we may assume that in addition to (i) and (ii), the following holds:

$$|a| + |a'| + |u| + |u'| \leq N.$$

The definition of rational relations holds also for arbitrary monoids:

**Definition** Let $M$ and $M'$ be monoids. A rational relation over $M$ and $M'$ is a rational subsets of $M \times M'$.

We shall see in the next section how this definition can be used to define interesting rational transductions. Here we just note the following:

**Proposition 3.4** *Let $M$, $M'$ be monoids. Then $A$ is a rational relation over $M$ and $M'$ iff there exist an alphabet $Z$, two morphisms $\alpha: Z^* \to M$, $\beta: Z^* \to M'$ and a regular language $K \subset Z^*$ such that $A = \{(\alpha h, \beta h): h \in K\}$.*

Proof. Let $A \in \mathrm{Rat}(M \times M')$. Then $A \subset N$, in view of Lemma 2.5, where $N$ is a finitely generated submonoid of $M \times M'$. Thus there exist an alphabet $Z$, and a morphism $\gamma: Z^* \to M \times M'$ such that $\gamma(Z^*) = N$. Since $A \in \mathrm{Rat}(N)$, $A = \gamma(K)$ for some regular language $K \subset Z^*$. Next define $\alpha: Z^* \to M$, $\beta: Z^* \to M'$ by $\gamma h = (\alpha h, \beta h)$, $(h \in Z^*)$. This yields the desired representation. The converse is clear. ■

## Exercises

**3.1** Let $M$ be a finitely generated, infinite monoid. Show that $A = \{(m, m): m \in M\}$ is a rational and not recognizable subset of $M \times M$.

**3.2** Let $X$ be an alphabet with at least two letters. Show that the relation $R = \{(f, \tilde{f}): f \in X^*\}$ is not rational.

**3.3** Give a counter example to the following version of the Iteration Lemma: For $A \in \mathrm{Rat}(X^* \times Y^*)$ there is an integer $N \geq 1$ such that for any $(g, g') \in A$, and for any factorization

$$(g, g') = (h_1, h_1')(f, f')(h_2, h_2') \qquad \text{with} \quad |f| + |f'| \geq N,$$

$(f, f')$ admits a factorization $(f, f') = (a, a')(u, u')(b, b')$ such that $0 < |u| + |u'| \leq N$ and

$$(h_1, h_1')(a, a')(u, u')^*(b, b')(h_2, h_2') \subset A.$$

**3.4** A right linear system of equations over $X^* \times Y^*$ is a system of equations of the form

$$\xi_i = \sum_{j=1}^{N} C_{ij}\xi_j + B_i \qquad i = 1, \ldots, N,$$

where $C_{ij}, B_i \subset X^* \times Y^*$. The system is strict iff $(1, 1) \notin C_{ij}$ for $j = 1, \ldots, N$. A vector $A = (A_1, \ldots, A_N)$ of subsets of $X^* \times Y^*$ is a solution of the system iff

$$A_i = \bigcup_{j=1}^{N} (C_{ij}A_j \cup B_j) \qquad i = 1, \ldots, N.$$

a) Show that a strict right linear system has a unique solution.
b) Show that $R \subset X^* \times Y^*$ is a rational relation iff $R$ is a component of the solution of a strict right linear system of equations with $C_{ij}$, $B_j$ finite. (Hint. Use the fact that a) and b) hold in free monoids and apply Nivat's Theorem.)

## III.4 Rational Transductions

The "static" notion of rational relation is now transformed into the "dynamic" notion of rational transduction.
A transduction $\tau$ from $X^*$ into $Y^*$ is a function from $X^*$ into the set $\mathfrak{P}(Y^*)$ of subsets of $Y^*$. For commodity, we write $\tau: X^* \to Y^*$. The domain $\mathrm{dom}(\tau)$ and the image $\mathrm{im}(\tau)$ are defined by

$$\mathrm{dom}(\tau) = \{f \in X^* : \tau(f) \neq \emptyset\};$$
$$\mathrm{im}(\tau) = \{g \in Y^* \mid \exists f \in X^* : g \in \tau(f)\}.$$

The transduction $\tau$ is extended to a mapping from $\mathfrak{P}(X^*)$ into $\mathfrak{P}(Y^*)$ by setting

$$\tau(A) = \bigcup_{f \in A} \tau(f) \qquad A \subset X^*.$$

The graph of $\tau$ is the relation $R$ defined by

$$R = \{(f, g) \in X^* \times Y^* \mid g \in \tau(f)\}.$$

Conversely, for any relation $R \subset X^* \times Y^*$, the transduction $\tau: X^* \to Y^*$ defined by $R$ is given by

$$\tau(f) = \{g \in Y^* \mid (f, g) \in R\}.$$

**Definition** A transduction $\tau: X^* \to Y^*$ is rational iff its graph $R$ is a rational relation over $X$ and $Y$.

Let $\tau: X^* \to Y^*$ be a rational transduction, and let $R \subset X^* \times Y^*$ be the graph of $\tau$. The monoids $X^* \times Y^*$ and $Y^* \times X^*$ are isomorphic. Thus the relation

$$R^{-1} = \{(g, f) \mid (f, g) \in R\}$$

is rational, and the transduction $\tau^{-1}: Y^* \to X^*$ defined by $R^{-1}$ is rational. $\tau^{-1}$ is the inverse transduction of $\tau$. Clearly

$$\tau^{-1}(B) = \{f \in X^* \mid \tau(f) \cap B \neq \emptyset\} \qquad B \subset Y^*.$$

In general, $\tau(\tau^{-1}(B)) \neq B$, and $\tau^{-1}(\tau(A)) \neq A$, $(A \subset X^*)$. The domain $\mathrm{dom}(\tau)$ and the image $\mathrm{im}(\tau)$ are homomorphic images of the rational relation $R$, and consequently are regular languages.
Let $\tau_1, \tau_2: X^* \to Y^*$ be rational transductions and let $R_1, R_2$ be the graphs of $\tau_1$ and $\tau_2$. We denote by $\tau_1 \cup \tau_2$, $\tau_1\tau_2$ and $\tau_1^+$ the transductions with graphs $R_1 \cup R_2$, $R_1R_2$, $R_1^+$. Obviously, these transductions are rational. They verify:

$$(\tau_1 \cup \tau_2)(f) = \tau_1(f) \cup \tau_2(f); \qquad (\tau_1\tau_2)(f) = \bigcup_{f_1 f_2 = f} \tau_1(f_1)\tau_2(f_2);$$
$$\tau_1^+(f) = \bigcup \{\tau_1(f_1) \cdots \tau_1(f_n) \mid n \geq 1, \quad f_1 \cdots f_n = f\}.$$

If $\mathrm{dom}(\tau_1) \cap \mathrm{dom}(\tau_2) = \emptyset$, we also write $\tau_1 + \tau_2$ instead of $\tau_1 \cup \tau_2$. Finally, we associate to $\tau: X^* \to Y^*$ a transduction $\tilde{\tau}: X^* \to Y^*$ by setting $\tilde{\tau}(f) = (\tau(\tilde{f}))^{\sim}$.

Let $R$ be the graph of $\tau$, and let $\tilde{R}$ be the graph of $\tilde{\tau}$. Then $\tilde{R} = \{(\tilde{f}, \tilde{g}) \mid (f, g) \in R\}$. The formulas

$$(A \cup B)^{\sim} = \tilde{A} \cup \tilde{B}; \qquad (AB)^{\sim} = \tilde{B}\tilde{A}; \qquad (A^+)^{\sim} = (\tilde{A})^+ \qquad (A, B \subset X^* \times Y^*)$$

show that $R$ is rational iff $\tilde{R}$ is rational. Thus the transduction $\tilde{\tau}$, the reversal of $\tau$ is rational iff $\tau$ is rational.

Nivat's Theorem proved in the preceding section can be formulated as follows for rational transductions.

**Theorem 4.1** (Nivat [1968]) *Let X and Y be alphabets. The following conditions are equivalent:*

(i) $\tau : X^* \to Y^*$ *is a rational transduction;*

(ii) *There exist an alphabet Z, two morphisms* $\varphi : Z^* \to X^*$, $\psi : Z^* \to Y^*$ *and a regular language* $K \subset Z^*$ *such that*

$$\tau(f) = \psi(\varphi^{-1}(f) \cap K) \qquad f \in X^*; \tag{4.1}$$

(iii) *There exist an alphabet Z, two alphabetic morphisms* $\alpha : Z^* \to X^*$, $\beta : Z^* \to Y^*$ *and a regular language* $K \subset Z^*$ *such that*

$$\tau(f) = \beta(\alpha^{-1}(f) \cap K) \qquad (f \in X^*);$$

(iv) *There exist an alphabet Z, two alphabetic morphisms* $\alpha : Z^* \to X^*$, $\beta : Z^* \to Y^*$ *and a local regular language* $K \subset Z^*$ *such that*

$$\tau(f) = \beta(\alpha^{-1}(f) \cap K) \qquad f \in X^*;$$

*further if* $X \cap Y = \emptyset$, *then* (i) *is equivalent to*

(v) *There exists a regular language* $K \subset (X \cup Y)^*$ *such that*

$$\tau(f) = \pi_Y(\pi_X^{-1}(f) \cap K) \qquad f \in X^*,$$

*where* $\pi_X$ *and* $\pi_Y$ *are the projections of* $(X \cup Y)^*$ *onto* $X^*$ *and* $Y^*$ *respectively.* ■

From (4.1), we deduce immediately that

$$\tau^{-1}(g) = \varphi(\psi^{-1}(g) \cap K) \qquad g \in Y^*.$$

It follows also from (4.1) that

$$\tau(A) = \psi(\varphi^{-1}(A) \cap K) \qquad A \subset X^*.$$

Thus:

**Corollary 4.2** *Each rational transduction preserves rational and algebraic languages. That is, for each rational transduction* $\tau$, $\tau(A)$ *is rational if A is rational, and* $\tau(A)$ *is algebraic if A is algebraic.* ■

**Example 4.1** Let $X = \{x, y\}$, $Y = \{a, b\}$, and consider the transduction $\tau : X^* \to Y^*$ defined by

$$\tau(f) = \begin{cases} \emptyset & \text{if } f \notin x^+y^*; \\ (a^+b)^n a^{2m} b & \text{if } f = x^n y^m, \quad n \geq 1, \quad m \geq 0. \end{cases}$$

Obviously $\mathrm{dom}(\tau)=x^+y^*$, $\mathrm{im}(\tau)=(a^+b)^+(a^2)^*b$. We claim that the transduction $\tau$ is rational. This can be shown in several ways. First, let $R$ be the graph of $\tau$. Then

$$R=(\{x\}\times a^+b)^+(y,a^2)^*(1,b)\in \mathrm{Rat}(X^*\times Y^*).$$

Next let $Z=\{r,s,t,u\}$ and define $\varphi: Z^*\to X^*$, $\psi: Z^*\to Y^*$ by

$$\varphi r=x,\quad \varphi s=1,\quad \varphi t=y,\quad \varphi u=1;$$

$$\psi r=b,\quad \psi s=a,\quad \psi t=a^2,\quad \psi u=b.$$

Let $K=(s^+r)^+t^*u$. Then $\varphi K=x^+y^*$, $\psi K=(a^+b)^+(a^2)^*b$. Further

$$\varphi^{-1}(x^ny^m)\cap K=(s^+r)^nt^mu \qquad n\geqslant 1,\quad m\geqslant 0,$$

thus $\tau(f)=\psi(\varphi^{-1}(f)\cap K)$ for all $f\in X^*$.
Finally, since $X\cap Y=\emptyset$, we can represent $\tau$ by projections. Consider indeed the regular language

$$K'=xa\{a,bxa\}^*b(ya^2)^*b\subset(X\cup Y)^*.$$

Then $\pi_X(K')=xx^*y^*$. Next if $f=x^ny^m$, $(n\geqslant 1, m\geqslant 0)$, then

$$\pi_X^{-1}(f)\cap K'=xa(a^*bxa)^{n-1}a^*b(ya^2)^mb,$$

$$\pi_Y(\pi_X^{-1}(f)\cap K')=a(a^*ba)^{n-1}a^*ba^{2m}b=(a^+b)^na^{2m}b.$$

As for rational relations, the definition of rational transductions can be extended to arbitrary monoids.

**Definition** Let $M$, $M'$ be monoids. A rational transduction $\tau: M\to M'$ is a function from $M$ into $\mathfrak{P}(M')$ such that the graph $R=\{(m,m')\mid m'\in\tau(m)\}$ of $\tau$ is a rational subset of $M\times M'$.

From Proposition 3.4, we immediately obtain:

**Proposition 4.3** *Let $M$, $M'$ be monoids. A transduction $\tau: M\to M'$ is rational iff there exist an alphabet $Z$, two morphisms $\alpha: Z^*\to M$, $\beta: Z^*\to M'$ and a rational language $K\subset Z^*$ such that*

$$\tau(m)=\beta(\alpha^{-1}(m)\cap K)\qquad m\in M. \qquad \blacksquare$$

If $A\subset M$, then

$$\tau(A)=\beta(\alpha^{-1}(A)\cap K). \tag{4.2}$$

Thus, if $A$ is a recognizable subset of $M$, then $\alpha^{-1}(A)$ is recognizable, hence regular, $\alpha^{-1}(A)\cap K$ is regular, hence rational, and finally $\tau(A)$ is a rational subset of $M'$. Note that $\tau(A)$ is not necessarily recognizable, and that $\tau(A)$ is not necessarily rational if $A$ is rational. This follows from Examples 1.5; 2.4, since morphisms and inverse morphisms are particular rational transductions.

We now consider composition of rational transductions. If $\tau: M \to M'$ and $\tau'$: $M' \to M''$ are transductions, then the composition $\tau' \circ \tau: M \to M''$ is defined by

$$(\tau' \circ \tau)(m) = \tau'(\tau(m)) = \bigcup_{m' \in \tau(m)} \tau'(m').$$

First, we settle the case of free monoids.

**Theorem 4.4** (Elgot and Mezei [1965]) *Let X, Y, Z be alphabets, and let $\tau: X^* \to Y^*$ and $\tau': Y^* \to Z^*$ be rational transductions. Then the transduction $\tau' \circ \tau: X^* \to Z^*$ is rational.*

We first prove the theorem in a special case. The general case follows then from this special case.

**Lemma 4.5** *Let X, Y, Z be three pairwise disjoint alphabets. Set $X' = X \cup Y$, $Z' = Y \cup Z$, $T = X \cup Y \cup Z$ and let*

$$\alpha: X'^* \to Y^*, \quad \beta: Z'^* \to Y^*, \quad \alpha': T^* \to X'^*, \quad \beta': T^* \to Z'^*$$

*be the projections. Then $\beta^{-1} \circ \alpha = \beta' \circ \alpha'^{-1}$.*

Lemma 4.5 is represented in Fig. III.1.

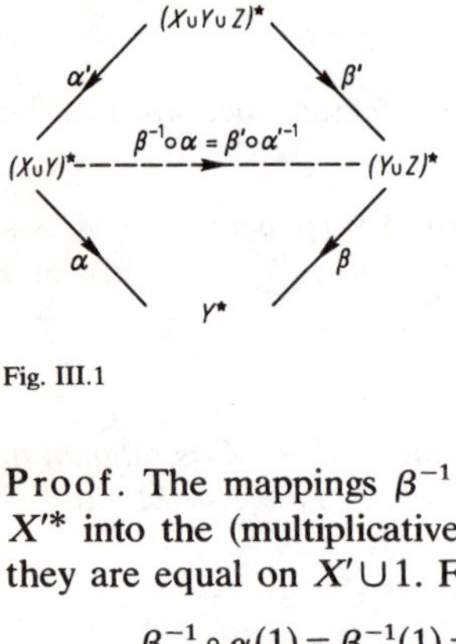


Fig. III.1

Proof. The mappings $\beta^{-1} \circ \alpha$ and $\beta' \circ \alpha'^{-1}$ are morphisms of the semigroup $X'^*$ into the (multiplicative) semigroup $\mathfrak{P}(Y^*)$. Thus it suffices to prove that they are equal on $X' \cup 1$. First

$$\beta^{-1} \circ \alpha(1) = \beta^{-1}(1) = Z^*; \qquad \beta' \circ \alpha'^{-1}(1) = \beta'(Z^*) = Z^*.$$

Then for $x \in X$, $\beta^{-1} \circ \alpha(x) = \beta^{-1}(1) = Z^*$ and $\beta' \circ \alpha'^{-1}(x) = \beta'(Z^* x Z^*) = Z^*$. Finally, if $y \in Y$, then $\beta^{-1} \circ \alpha(y) = \beta^{-1}(y) = Z^* y Z^*$ and $\beta' \circ \alpha'^{-1}(y) = \beta'(Z^* y Z^*) = Z^* y Z^*$. This proves the lemma. ■

Proof of Theorem 4.4. After a copy if necessary, we may assume that the alphabets $X$, $Y$, $Z$ are pairwise disjoint. Set $X' = X \cup Y$, $Z' = Y \cup Z$. In view of Nivat's Theorem, there exists a regular language $K \subset X'^*$ such that

$$\tau(f) = \alpha(\pi^{-1}(f) \cap K) \qquad f \in X^*, \tag{4.3}$$

where $\pi: X'^* \to X^*$ and $\alpha: X'^* \to Y^*$ are the projections. Next there is a regular language $M \subset Z'^*$ such that

$$\tau'(g) = \omega(\beta^{-1}(g) \cap M) \qquad g \in Y^*, \tag{4.4}$$

where $\beta: Z'^* \to Y^*$ and $\omega: Z'^* \to Z^*$ are the projections. Thus we have Fig. III.2. According to Lemma 4.5, $\beta^{-1} \circ \alpha = \beta' \circ \alpha'^{-1}$ where $T = X \cup Y \cup Z$, and

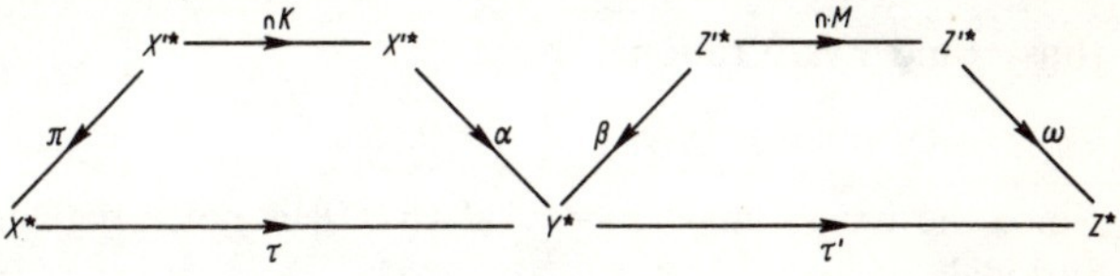


Fig. III.2

$\beta'$, $\alpha'$ are the projections of $T^*$ onto $Z'^*$ and $X'^*$ respectively. Thus the above diagram can be completed to Fig. III.3.

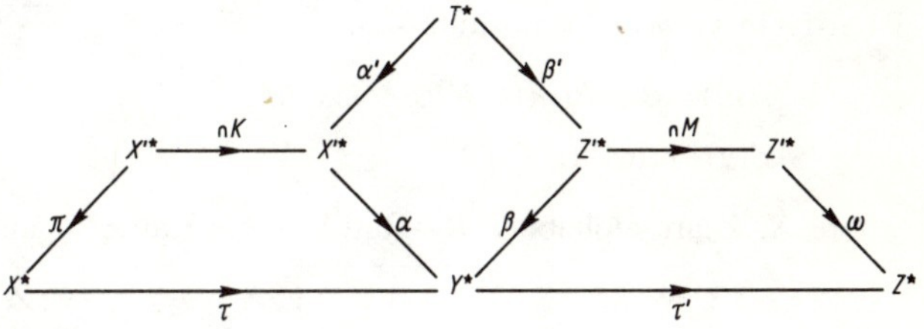


Fig. III.3

Next, setting $\tau'' = \tau' \circ \tau$, we have by (4.3), (4.4), for $f \in X^*$,

$$\tau''(f) = \omega[(\beta^{-1} \circ \alpha)(\pi^{-1}(f) \cap K) \cap M].$$

Since $\beta^{-1} \circ \alpha = \beta' \circ \alpha'^{-1}$,

$$\begin{aligned} \tau''(f) &= \omega[(\beta' \circ \alpha'^{-1})(\pi^{-1}(f) \cap K) \cap M] \\ &= \omega[\beta'((\pi \circ \alpha')^{-1}(f) \cap \alpha'^{-1}(K)) \cap M]. \end{aligned} \tag{4.5}$$

Define $\psi = \omega \circ \beta': T^* \to Z^*$, $\varphi = \pi \circ \alpha': T^* \to X^*$, and set $K' = \alpha'^{-1}(K)$, $M' = \beta'^{-1}(M)$, $N = K' \cap M'$. Then (4.5) implies

$$\begin{aligned} \tau''(f) &= \omega[\beta'(\varphi^{-1}(f) \cap K') \cap M] = \omega[\beta'(\varphi^{-1}(f) \cap K' \cap M')] \\ &= \psi(\varphi^{-1}(f) \cap N). \end{aligned} \tag{4.6}$$

Since $N \subset T^*$ is a regular language, the transduction $\tau''$ is rational by (4.6). ■

If $M$, $M'$, $M''$ are arbitrary monoids, then the composition of two rational transductions $\tau: M \to M'$ and $\tau' = M' \to M''$ is not necessarily rational.

**Example 4.2** Let $x$, $y$, $z$ be letters, $M=x^*$, $M'=y^*\times z^*$, $M''=\{y, z\}^*$. Define $\tau: M\to M'$ and $\tau': M'\to M''$ by

$$\tau(x^n)=(y^n, z^n), \qquad \tau'(y^n, z^k)=y^nz^k \qquad (n, k\geqslant 0).$$

The graphs $R$ and $R'$ of $\tau$ and $\tau'$ are:

$$R=(x, (y, z))^*, \qquad R'=((y, 1), y)^*((1, z), z)^*,$$

thus $\tau$ and $\tau'$ are rational. Next

$$(\tau'\circ\tau)(x^n)=y^nz^n \qquad n\geqslant 0.$$

Since the image $\operatorname{im}(\tau'\circ\tau)=\{y^nz^n : n\geqslant 0\}$ is not a regular language, $\tau'\circ\tau$ is not rational.

Despite this example, we have

**Proposition 4.6** *Let $M$, $M'$ be monoids, and let $Y$ be an alphabet. If $\tau: M\to Y^*$ and $\tau': Y^*\to M'$ are rational transductions, then $\tau'\circ\tau$ is rational.*

Proof. In view of Proposition 4.3,

$$\tau(m)=\beta(\alpha^{-1}(m)\cap K) \qquad m\in M$$
$$\tau'(g)=\delta(\gamma^{-1}(g)\cap L) \qquad g\in Y^*$$

where $X$, $Z$ are alphabets, $K\in\operatorname{Rat}(X^*)$, $L\in\operatorname{Rat}(Z^*)$, and

$$\alpha: X^*\to M, \quad \beta: X^*\to Y^*, \quad \gamma: Z^*\to Y^*, \quad \delta: Z^*\to M'$$

are morphisms. It follows that

$$(\tau'\circ\tau)(m)=\delta[(\gamma^{-1}\circ\beta)(\alpha^{-1}(m)\cap K)\cap L] \qquad m\in M.$$

Since $\gamma^{-1}\circ\beta: X^*\to Z^*$ is the composition of two rational transductions, it is a rational transduction by Elgot and Mezei's Theorem. Thus

$$(\gamma^{-1}\circ\beta)(f)=\psi(\varphi^{-1}(f)\cap N) \qquad f\in X^*$$

for some alphabet $T$, morphisms $\varphi: T^*\to X^*$, $\psi: T^*\to Z^*$ and some $N\in\operatorname{Rat}(T^*)$. Thus as in the proof of Theorem 4.4,

$$(\tau'\circ\tau)(m)=(\delta\circ\psi)[(\alpha\circ\varphi)^{-1}(m)\cap(\varphi^{-1}(K)\cap N\cap\psi^{-1}(L))] \qquad m\in M,$$

showing that $\tau'\circ\tau$ is rational. ■

It is natural to look for a generalization of rational transductions involving context-free languages. This can be done by developing a theory of algebraic sets in arbitrary monoids analogue to the theory of rational sets (see Eilenberg [1978], also Exercises 4.5, 4.6). This yields an analogue of Nivat's Theorem. We prefer in this context to take that analogue as a definition.

**Definition** A transduction $\tau: X^* \to Y^*$ is algebraic if there exist an alphabet $Z^*$, two morphisms $\alpha: Z^* \to X^*$, $\beta: Z^* \to Y^*$ and a context-free language $A \subset Z^*$ such that

$$\tau(f) = \beta(\alpha^{-1}(f) \cap A) \qquad f \in X^*.$$

It follows immediately that $\tau(L)$ is context-free if $L \subset X^*$ is regular, and it is easy to see that $\tau(L)$ is not necessarily context-free if $L$ is context-free. The following result is proved in the same way as Theorem 4.4.

**Proposition 4.7** *Let $\tau: X^* \to Y^*$ and $\tau': Y^* \to Z^*$ be transductions. If one of them is rational and the other is algebraic, then $\tau' \circ \tau$ is algebraic.* ∎

If both transductions are algebraic, then $\tau' \circ \tau$ is not necessarily algebraic.

## Exercises

**4.1** Give an example of a transduction $\tau: X^* \to Y^*$, and of subsets $A \subset X^*$, $B \subset Y^*$ such that $\tau^{-1}(\tau(A)) \neq A$ and $\tau(\tau^{-1}(B)) \neq B$.

**4.2** Prove Proposition 4.7.

**4.3** Give an example of two algebraic transductions $\tau$, $\tau'$ such that the composition $\tau' \circ \tau$ is not algebraic.

**4.4** Consider the Dyck reduction $\rho: Z_n^* \to Z_n^*$. Show that $\rho$ is an algebraic transduction. Show that $\rho$ is not a rational transduction.

**4.5** (Eilenberg [1978]) Let $M$ be a monoid and let $V$ be an alphabet disjoint from $M$. The set $M[V]$ of words

$$w = m_0 \xi_1 m_1 \cdots m_{k-1} \xi_k m_k$$

with $k \geq 0$, $m_0, \ldots, m_k \in M$, $\xi_1, \ldots, \xi_k \in V$ is a monoid when multiplication of $w$ with $w' = n_0 \zeta_1 \cdots \zeta_l n_l$ is defined by

$$ww' = m_0 \xi_1 \cdots \xi_k (m_k n_0) \zeta_1 \cdots \zeta_l n_l.$$

An algebraic grammar $G = \langle V, M, P \rangle$ over $M$ is given by a finite subset $P$ of $V \times M[V]$. Derivations are defined as in free monoids. The language $L_G(\xi)$ generated by $\xi$ is the sets of all $m \in M$ derived from $\xi$. Languages generated by algebraic grammars over $M$ are called algebraic subsets of $M$.

a) Show that for any algebraic grammar $G = \langle V, M, P \rangle$, there is an algebraic grammar $G_1 = \langle V, M_1, P \rangle$, where $M_1$ is a finitely generated submonoid of $M$, such that $L_G(\xi) = L_{G_1}(\xi)$ for all $\xi \in V$.

b) Show that $A$ is an algebraic subset of $M$ iff there exist an alphabet $X$, a morphism $\varphi: X^* \to M$, and a context-free language $L \subset X^*$ such that $\varphi(L) = A$.

c) Show that any rational subset of $M$ is algebraic.

d) Show that a transduction $\tau: X^* \to Y^*$ is algebraic in the sense given in the text iff its graph is an algebraic subset of $X^* \times Y^*$.

**4.6** (continuation of 4.5) Show that in a free commutative monoid $X^{\oplus}$, any algebraic subset is rational. (This is Parikh's Theorem. For a proof, see Conway [1971], Ginsburg [1966].) Show that the same result holds in any commutative monoid.

**4.7** Nivat's Theorem implies that morphisms and inverse morphisms can be represented by means of projections, inverse projections and intersection with regular sets. Give such representations explicitely.

**4.8** (Elgot and Mezei [1965]) Let $\tau: X^* \to Y^*$ be a rational transduction. Then $\tau = \tau_0 + \tau_\infty$, where

$$\tau_0(f) = \tau(f), \quad \tau_\infty(f) = \emptyset, \qquad \text{if} \quad \mathrm{Card}(\tau f) < \infty;$$

$$\tau_0(f) = \emptyset, \quad \tau_\infty(f) = \tau(f), \qquad \text{if} \quad \mathrm{Card}(\tau f) = \infty.$$

Show that $\tau_0$, $\tau_\infty$ are rational transductions.

**4.9** Let $M$, $M'$, $M''$ be finitely generated monoids, and let $\tau: M \to M'$, $\tau': M' \to M''$ be transductions. Show that if one of them is recognizable (i.e. its graph is recognizable) and the other is rational, then $\tau' \circ \tau: M \to M''$ is rational, and even recognizable, provided $M$, $M'$, $M''$ are free monoids. Show that if $\tau$ and $\tau'$ are recognizable, then $\tau' \circ \tau$ is always recognizable.

## III.5 Examples

The explicit description of a rational transduction is a simple method to prove that certain transformations preserve regular and context-free languages. Rational transductions can also be used to prove that a given language is context-free, by representing it as the image of a language "known" to be context-free. One of the most important applications of rational transductions will be shown in later chapters: They are used there as a tool of comparison of subfamilies of the family of context-free languages. The proof of the rationality of a given transduction is frequently through one of the versions of Nivat's Theorem, or else through a rational expression for the graph of the transduction.

**5.1** The identity mapping $f \mapsto f$ from $X^*$ into itself is a rational transduction. This is straightforward by Nivat's Theorem. The graph of this mapping is

$$\Delta = \{(f, f) : f \in X^*\} = \Big(\bigcup_{x \in X} (x, x)\Big)^*.$$

**5.2** The rational constants $\tau_K: X^* \to Y^*$ defined for a fixed regular language $K \subset Y^*$ by $\tau_K(f) = K$ are rational. The graph of $\tau_K$ is $X^* \times K$.

**5.3** Any morphism, any inverse morphism is a rational transduction.

**5.4** A *rational substitution* is a substitution $\sigma: X^* \to Y^*$ such that $\sigma(x) \in \mathrm{Rat}(Y^*)$ for $x \in X$. The graph of $\sigma$ is

$$\left(\bigcup_{x \in X} \{x\} \times \sigma(x)\right)^* ;$$

thus $\sigma$ is a rational transduction.

**5.5** The *union* (and of course the *intersection*) with a regular language is performed by a rational transduction. Let $K \subset X^*$ be a regular language, and consider the transduction $f \mapsto f \cup K$ from $X^*$ into $X^*$. Then its graph is $\Delta \cup (X^* \times K)$.

**5.6** The *product* with a rational language: let $K \in \mathrm{Rat}(X^*)$ and consider the transduction $f \mapsto Kf$ $(f \in X^*)$. Its graph is $(\{1\} \times K)\Delta$.

**5.7** The (left or right) *quotient* by a rational language. Let $K \subset X^*$ be a rational language. The transduction $X^* \to X^*$ defined by

$$f \mapsto K^{-1}f = \{g \in X^* \mid \exists u \in K : ug = f\}$$

is the inverse of the transduction of Example 5.6, and consequently is rational. (This proves that $K^{-1}L$ is context-free if $L$ is a context-free language.)

**5.8** The transduction $\tau: X^* \to X^*$ with $\tau(f) = f^{-1}K = \{g \in X^* : fg \in K\}$ is rational if $K \subset X^*$ is a regular language. Consider indeed its graph

$$R = \{(f, g) : fg \in K\}.$$

Let $\bar{X} = \{\bar{x} \mid x \in X\}$ be a copy of $X$, disjoint from $X$, set $Z = X \cup \bar{X}$, and define morphisms $\alpha$, $\beta$, $\gamma: Z^* \to X^*$ by

$$\alpha(x) = x, \quad \beta(x) = 1, \quad \gamma(x) = x \qquad x \in X;$$

$$\alpha(\bar{x}) = 1, \quad \beta(\bar{x}) = x, \quad \gamma(\bar{x}) = x \qquad \bar{x} \in \bar{X}.$$

Consider the regular language $K' = \gamma^{-1}(K) \cap X^*\bar{X}^* \subset Z^*$. Any word $h \in K'$ factorizes in a unique way into $h = fg$, $(f \in X^*, g \in \bar{X}^*)$. It follows that

$$R = \{(\alpha h, \beta h) : h \in K'\}.$$

**5.9** The transduction $X^* \to X^*$ which associates to any word $f \in X^*$ the set of its subwords (resp. factors, left factors, right factors) is a rational relation. Let indeed $\bar{X}$, $Z$, $\alpha$, $\gamma$ be as above. Then the set of subwords of $f$ is $\alpha(\gamma^{-1}(f))$, the set of factors of $f$ is $\alpha(\gamma^{-1}(f) \cap \bar{X}^*X^*\bar{X}^*)$, etc. (Note that the set of left factors of $f$ is $f(X^*)^{-1}$. Thus the rationality of this transduction follows from Example 5.7.)

**5.10** Let $\tau: X^* \to X^*$ be defined for $f = x_1x_2 \cdots x_n$, $(n \geq 1, x_i \in X)$ by $\tau(f) = x_1x_3x_5 \ldots$, and $\tau(1) = 1$. With the notations of Example 5.8,

$$\tau(f) = \alpha(\gamma^{-1}(f) \cap [(X\bar{X})^* \cup (X\bar{X})^*X]).$$

**5.11** The transduction $\tau: x^* \to \{a, b\}^*$ defined by

$$\tau(x^n) = \begin{cases} a^n & n \text{ even}, \\ b^n & n \text{ odd}, \end{cases}$$

is rational. Indeed, its graph is $(x^2, a^2)^* \cup (x, b)(x^2, b^2)^*$.

**5.12** According to Greibach [1973], the "hardest" context-free language is the language $L_0 \subset X^*$, with $X = \{x_1, x_2, \bar{x}_1, \bar{x}_2, \not{c}, |, \#\}$ defined in the following way. Set $T = X \backslash \#$. Then $f \in L_0$ iff either

$$f = 1$$

or $$f = u_1 \, |v_1| w_1 \# u_2 \, |v_2| \, w_2 \# \cdots \# u_n \, |v_n| \, w_n,$$

with $u_1, w_1, \ldots, u_n, w_n \in T^*$, $v_1, \ldots, v_n \in \{x_1, x_2, \bar{x}_1, \bar{x}_2\}^*$ and $v_1 v_2 \cdots v_n \in \not{c} D_2'^*$. In order to show that $L_0$ is indeed context-free, consider the two transductions $\tau, \tau_1: X^* \to X^*$:

$$\tau(h) = \# T^* \, |h| \, T^* \qquad \tau_1(h) = T^* \, |\not{c} h| \, T^* \qquad (h \in X^*)$$

In view of Example 5.6, these transductions are rational. Consequently, the transduction $\tau' = \tau_1 \tau^*$ is rational. Since $L_0 = 1 \cup \tau'(D_2'^*)$, $L_0$ is context-free.

**5.13** The transduction $X^* \to X^*$ which associates to any $f \in X^*$ the reversal $\tilde{f}$ is not rational if $\operatorname{Card}(X) \geqslant 2$ since its graph $\{(f, \tilde{f}) : f \in X^*\}$ is not a rational relation (Exercise 3.2). (This is an example of an irrational relation that preserves regular and context-free languages.)

All the transductions above are unary operations. Some of the examples, like product, union, etc. are binary operations. Thus we consider them now as binary transductions.

**5.14** The transduction $(f, g) \mapsto fg$ from $X^* \times X^*$ into $X^*$ is rational. Its graph is indeed

$$A = \{(x, 1, x) : x \in X\}^* \{(1, x, x) : x \in X\}^*.$$

**5.15** The *shuffle* $f \amalg g$ of two words $f, g \in X^*$ is defined as

$$f \amalg g = \{f_1 g_1 \cdots f_n g_n \mid f_1, \ldots, f_n, g_1, \ldots, g_n \in X^*, \; f_1 \cdots f_n = f, \; g_1 \cdots g_n = g\}.$$

The transduction $(f, g) \mapsto f \amalg g$ is rational, since its graph is $A^*$, with $A$ given as in Example 5.14.

**5.16** Finally we show that *addition* of nonnegative integers in some fixed base $k \geqslant 2$ can be performed by a rational transduction. (For $k = 1$, this is done by Example 5.14.)

Let $k \geq 2$, and let $\Bbbk = \{0, 1, \ldots, k-1\}$. The empty word of $\Bbbk^*$ is denoted by $\varepsilon$. For each $f = x_0 x_1 \cdots x_n$, $(x_i \in \Bbbk)$, let

$$\langle f \rangle = \sum_{i=0}^{n} x_i k^{n-i}$$

be the integer represented by $f$ in base $\Bbbk$. Then $\langle \varepsilon \rangle = 0$, and for any $m \in \mathbb{N}$, there is a unique $f \in \Bbbk^* \backslash 0\Bbbk^*$ such that $\langle f \rangle = n$. The transduction

$$\oplus : \Bbbk^* \times \Bbbk^* \to \Bbbk^*$$

which associates to any $(f, g) \in \Bbbk^* \times \Bbbk^*$ the unique word $h \in \Bbbk^* \backslash 0\Bbbk^*$ such that $\langle h \rangle = \langle f \rangle + \langle g \rangle$ is rational. The construction is in three steps. The first step just adds leading zeros in order to make the two arguments of the same length. Consider a transduction

$$\tau_1 : \Bbbk^* \times \Bbbk^* \to (\Bbbk \times \Bbbk)^*.$$

In order to avoid confusion, elements of $\Bbbk \times \Bbbk$ are noted $[x, y]$. If $f = x_1 x_2 \cdots x_n$, $g = y_1 y_2 \cdots y_m$, $(x_i, y_i \in \Bbbk)$, $\tau_1(f, g)$ is defined to be equal either to

$$[0, 0]^+ [x_1, 0] \cdots [x_{n-m}, 0][x_{n-m+1}, y_1] \cdots [x_n, y_m]$$

or $$[0, 0]^+ [0, y_1] \cdots [0, y_{m-n}][x_1, y_{m-n+1}] \cdots [x_n, y_m]$$

according to $n \geq m$ or $m \geq n$. Define $R, S, T \subset \Bbbk^* \times \Bbbk^* \times (\Bbbk \times \Bbbk)^*$ by

$$S = \{(\varepsilon, y, [0, y]) : y \in \Bbbk\}^* \cup \{(x, \varepsilon, [x, 0]) : x \in \Bbbk\}^*,$$
$$T = \{(x, y, [x, y]) : x, y \in \Bbbk\}^*, \qquad R = (\varepsilon, \varepsilon, [0, 0])^+ ST.$$

Then $R$ is rational and is the graph of $\tau_1$. Next, we define

$$\tau_2 : (\Bbbk \times \Bbbk)^* \to \Bbbk^*$$

to perform the addition step: For

$$w = [x_0, y_0] \cdots [x_n, y_n], \qquad x_0 = y_0 = 0, \tag{5.1}$$
$$f = x_0 \cdots x_n, \qquad g = y_0 \cdots y_n,$$

$\tau_2(w)$ will be the word $h = z_0 \cdots z_n$ such that $\langle h \rangle = \langle f \rangle + \langle g \rangle$. For this, we introduce an auxiliary alphabet $\{0, 1\} \times \Bbbk \times \Bbbk \times \Bbbk$ composed of quadruples $[r, z, x, y]$. During the computation, $z + r \cdot k$ represents the number $x + y + 1$ or $x + y + 0$, according to the existence or not of a "carry" from a previous computation. Formally, we define morphisms

$$\varphi : Y^* \to (\Bbbk \times \Bbbk)^*, \qquad \psi : Y^* \to \Bbbk^*$$

by $$\varphi[r, z, x, y] = [x, y], \qquad \psi[r, z, x, y] = z \qquad [r, z, x, y] \in Y.$$

Next, we define a local regular language

$$K=(UY^*\cap Y^*V)\setminus Y^*WY^* \qquad U=\{[0,1,0,0],[0,0,0,0]\}$$
$$V=\{[r,z,x,y]: z+rk=x+y\}$$
$$Y^2\setminus W=\{[r,z,x,y][r',z',x',y']: z+rk=x+y+r'\}.$$

Then for $w$ given by (5.1),

$$\varphi^{-1}(w)\cap K=[r_0,z_0,x_0,y_0]\cdots[r_n,z_n,x_n,y_n]$$

with $x_n+y_n=z_n+kr_n$, $x_i+y_i+r_{i+1}=z_i+kr_i$, $(i=n-1,\ldots,0,\ r_0=0)$. Hence $\psi(\varphi^{-1}(w)\cap K)=\tau_2(w)$ and $\tau_2$ is rational. The final step just deletes initial zeros from the result. It is performed by the transduction

$$\tau_3:\mathbb{k}^*\to\mathbb{k}^*, \qquad \tau_3(h)=(0^*)^{-1}h\cap\mathbb{k}^*\setminus 0\mathbb{k}^*$$

which is clearly rational. Thus, by Proposition 4.6, the transduction $\oplus=\tau_3\circ\tau_2\circ\tau_1$ is rational.

(For further properties of arithmetic operations considered as rational transductions, see Eilenberg [1974], and Exercises 5.3, 5.4.)

## Exercises

**5.1** Let $X=\{x_1,x_2,\ldots,x_k\}$. Define an order on $X^*$ by

$$f\underset{l}{<}g \iff \begin{cases} g=fu & \text{for some } u\in X^+ \text{ or}\\ f=ux_iv,\quad g=ux_jv' & \text{and } i<j;\end{cases}$$

this is the lexicographical order. The "radix" order is defined by

$$f\underset{r}{<}g \iff \begin{cases} |f|<|g| & \\ |f|=|g| & \text{and } f\underset{l}{<}g.\end{cases}$$

Show that the four transductions from $X^*$ into itself which associate to $f$ the sets

$$\{g\mid g>f\} \qquad (\text{resp. } \{g\mid g<f\})$$

are rational for both orders.

**5.2** Show that a transduction $\tau: X^*\times Y^*\to Z^*$ is rational iff there are an alphabet $T$, three morphisms

$$\alpha_1: T^*\to X^*, \qquad \alpha_2: T^*\to Y^*, \qquad \beta: T^*\to Z^*$$

and a regular language $K\subset T^*$ such that

$$\tau(f,g)=\beta(\alpha_1^{-1}(f)\cap\alpha_2^{-1}(g)\cap K) \qquad (f,g)\in X^*\times Y^*.$$

Show that for $R\subset X^*$, the transduction $\tau': Y^*\to Z^*$ given by $\tau'(g)=\tau(R\times\{g\})$ $(g\in Y^*)$ is rational if $R$ is regular, and is algebraic if $R$ is context-free. Use this to deduce Example 5.5 from Example 5.14.

**5.3** Show that multiplication $\otimes$ from $\mathbb{k}^*\times\mathbb{k}^*$ into $\mathbb{k}^*$ is not a rational transduction. (Hint (Messerschmidt). Compute for $k=2$ the language $\otimes(10^*1\times1^*)$.

**5.4** Show that for fixed $q \geq 1$, the multiplication by $q: \mathbb{k}^* \to \mathbb{k}^*$ which associates to $f \in \mathbb{k}^*$ the word $f' \in \mathbb{k}^* \setminus 0\mathbb{k}^*$ such that $\langle f' \rangle = q \cdot \langle f \rangle$ is rational.

**5.5** Let $X = \{x, y\}$, and let $\sigma$ be the congruence generated by $yxx \sim xyy$. Show that the transduction $\tau: X^* \to X^*$ which associates to $f$ the class $[f]_\sigma = \{g \mid g \equiv f \pmod{\sigma}\}$ is rational. (Note that this is not true for all congruences: thus the result does not hold for the Lukasiewicz congruence $\lambda$ of Section II.4.)

## III.6 Transducers

The machines realizing rational transductions are called transducers. As for transductions, transducers can be regarded either in a static or in a dynamic way. In the first case, a transducer is a finite nondeterministic acceptor reading on two tapes. It then recognizes the pairs of words of a rational relation. In the second case, the automaton reads input words on one tape, and prints output words on a second tape. The automaton thus realizes a rational transduction. Both aspects clearly are equivalent. In the following presentation, we adopt the second point of view which corresponds to the use of transductions as a tool for transformation of languages.

**Definition** A transducer $\boldsymbol{T} = \langle X, Y, Q, q_-, Q_+, E \rangle$ is composed of an input alphabet $X$, an output alphabet $Y$, a finite set of states $Q$, an initial state $q_-$, a set of final states $Q_+$, and a finite set of transitions or edges $E$ satisfying

$$E \subset Q \times X^* \times Y^* \times Q. \tag{6.1}$$

The terminology stems from Elgot and Mezei [1965] and Eilenberg [1974]. Ginsburg [1975] uses the term "*a*-transducer", the letter "*a*" emphasizing the presence of accepting states. Transductions are defined by means of transducers in the paper of Elgot and Mezei [1965]. They prove then that a transduction is realized by a transducer iff its graph is a rational relation.

Transducers have a graphical representation very similar to the usual representation of finite automata. Each state $q$ is represented by a circle, labelled $q$ and to each transition $e = (q, u, v, q')$ is associated an arrow directed from $q$ to $q'$ and labelled $u/v$. The initial state has an arrow entering in it, and final states are doubly circled.

**Example 6.1** Consider the transducer given by $X = Y = \{x, y\}$, $Q = \{a, b\}$, $q_- = a$, $Q_+ = \{b\}$, $E = \{(a, x^2y, 1, a), (a, 1, 1, b), (b, x, x, b), (b, y, y, b)\}$. Its representation is shown in Fig. III.4.

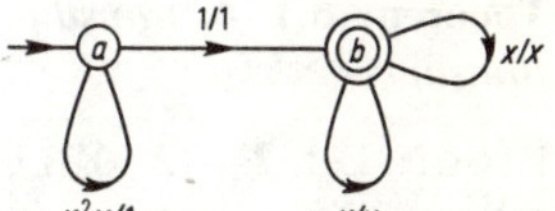


Fig. III.4

We now introduce some supplementary definitions. Consider the free monoid $E^*$ generated by the set $E$ of transitions. The empty word of $E^*$ is denoted by $\varepsilon$. Given a word

$$e=(p_1, u_1, v_1, q_1)\cdots(p_n, u_n, v_n, q_n), \tag{6.2}$$

the label of $e$ is the pair of words $|e|=(f, g)$ defined by $f=u_1\cdots u_n$, $g=v_1\cdots v_n$. By convention, $|\varepsilon|=(1, 1)$. Clearly the function $e\mapsto|e|$ is a morphism from $E^*$ into $X^*\times Y^*$ which can be decomposed into two morphisms $\alpha: E^*\to X^*$, $\beta: E^*\to Y^*$ defined by $|e|=(\alpha e, \beta e)$. $\alpha e$ and $\beta e$ are the input label and the output label of $e$. The word $e$ given by (6.2) is a path or a computation in $\boldsymbol{T}$ from $p_1$ to $q_n$ iff $q_i=p_{i+1}$, $(i=1,\ldots,n-1)$. For $p, q\in Q$, $\Lambda(p, q)$ is the set of all paths from $p$ to $q$. By convention, $\varepsilon\in\Lambda(p, p)$ for all $p\in Q$. We extend this notation by setting

$$\Lambda(p, Q')=\bigcup_{q\in Q'}\Lambda(p, q) \qquad Q'\subset Q.$$

Finally, define

$$T(p, q)=\{|e|: e\in\Lambda(p, q)\} \qquad T(p, Q')=\{|e|: e\in\Lambda(p, Q')\}.$$

A path $e$ from $p$ to $q$ is successful iff $p=q_-$ and $q\in Q_+$. Thus, the set of all successful paths is $\Lambda(q_-, Q_+)$.

**Definition** The transduction $|\boldsymbol{T}|: X^*\to Y^*$ realized by $\boldsymbol{T}$ is defined by

$$|\boldsymbol{T}|(f)=\{g\in Y^* \mid (f, g)\in T(q_-, Q_+)\}. \tag{6.3}$$

Thus $g\in|\boldsymbol{T}|(f)$ iff there exists a successful path in $\boldsymbol{T}$ with label $(f, g)$. With the morphisms $\alpha$ and $\beta$, (6.3) can be reformulated as

$$|\boldsymbol{T}|(f)=\beta(\alpha^{-1}(f)\cap\Lambda(q_-, Q_+)). \tag{6.4}$$

**Example 6.1** (continued). The set of successful paths is

$$\Lambda(a, b)=(a, x^2y, 1, a)^*(a, 1, 1, b)\{(b, x, x, b), (b, y, y, b)\}^*.$$

The set of labels of successful paths is

$$T(a, b)=(x^2y, 1)^*\{(x, x), (y, y)\}^*.$$

The transduction $\tau$ realized by the transducer is

$$\tau(f)=K^{-1}f, \qquad \text{with} \quad K=(x^2y)^*.$$

**Theorem 6.1** *A transduction $\tau: X^*\to Y^*$ is rational iff $\tau$ is realized by a transducer.*

Proof. Let $\boldsymbol{T}=\langle X, Y, Q, q_-, Q_+, E\rangle$ be a transducer. The rationality of $|\boldsymbol{T}|$ follows immediately from (6.4), provided $\Lambda(q_-, Q_+)$ is a regular language. To show this, it suffices to show that the set $\Lambda(p, q)$ of all paths from $p$ to $q$ is regular. This follows from the fact that

$$\Lambda(p, q) = \Omega \cup (U_p E^* \cap E^* V_q) \setminus E^* W E^*,$$

where $U_p = \{(q_1, u, v, q_2) \in E \mid q_1 = p\}$, $\quad V_q = \{(q_1, u, v, q_2) \in E \mid q_2 = q\}$,

$W = \{(q_1, u_1, v_1, q_1')(q_2, u_2, v_2, q_2') \in E^2 \mid q_1' \neq q_2\}$, $\quad \Omega = \Lambda(p, q) \cap \{\varepsilon\}$.

Conversely, let $\tau: X^* \to Y^*$ be a rational transduction. After a copy, we may assume $X \cap Y = \emptyset$. Thus

$$\tau(f) = \pi_Y(\pi_X^{-1}(f) \cap K) \qquad f \in X^*,$$

with $K \subset (X \cup Y)^*$ a regular language and $\pi_X$, $\pi_Y$ the projections of $(X \cup Y)^*$ onto $X^*$ and $Y^*$. Let $\mathbf{A} = \langle X \cup Y, Q, q_-, Q_+ \rangle$ be a finite automaton recognizing $K$, and define a transducer

$$\boldsymbol{T} = \langle X, Y, Q, q_-, Q_+, E \rangle$$

$$E = \{(q, \pi_X(z), \pi_Y(z), q \cdot z) \mid q \in Q, z \in X \cup Y\}. \tag{6.5}$$

Then $\boldsymbol{T}$ realizes $\tau$. ■

We easily obtain the following corollary

**Corollary 6.2** *Any rational transduction $\tau: X^* \to Y^*$ can be realized by a transducer $\boldsymbol{T} = \langle X, Y, Q, q_-, Q_+, E \rangle$ such that*

$$E \subset Q \times (X \cup \{1\}) \times (Y \cup \{1\}) \times Q \tag{6.6}$$

*and further $Q_+$ consists of a single state $q_+ \neq q_-$, and $(p, u, v, q) \in E$ implies $p \neq q_+$ and $q \neq q_-$.*

Proof. The condition (6.6) is fullfilled with $E$ satisfying (6.5). Next, add to $Q$ two new states $q^0$ and $q^1$, and to $\boldsymbol{T}$ the new transitions

$\{(q^0, u, v, q') \mid (q_-, u, v, q') \in E\}$,

$\{(q, u, v, q^1) \mid (q, u, v, q') \in E \text{ and } q' \in Q_+\}$

and $\quad (q^0, 1, 1, q^1) \;$ if $\; q_- \in Q_+$.

Let $\boldsymbol{T}'$ be the transducer obtained in this way with initial state $q^0$ and unique final state $q^1$. Then obviously $\tau = |\boldsymbol{T}'|$. ■

Remark. If $\tau(1) = \emptyset$, then (6.6) can be replaced by

$$E \subset (Q \times X \times \{1\} \times Q) \cup (Q \times \{1\} \times Y \times Q).$$

Note that the proof of Theorem 6.1, and equation (6.4) give an effective procedure to construct a transducer from a bimorphism and conversely. For the construction of a bimorphism, it is frequently easier to take as alphabet the set of labels of transitions instead of the set of transitions itself.

**Example 6.1** (continued). Consider the alphabet $Z$ composed of the three "letters" $x^2y/1$, $x/x$, $y/y$. Define morphisms $\varphi$, $\psi$ from $Z^*$ into $X^*$ by $\varphi(u/v)=u$, $\psi(u/v)=v$. Then

$$T(a, b)=\{(\varphi h, \psi h) \mid h \in R\} \qquad \text{with} \quad R=(x^2y/1)^*\{x/x, y/y\}^*.$$

**Example 6.2** Consider the transduction $\tau: X^* \to Y^*$ with $X=\{x, y\}$, $Y=\{a, b\}$ of Example 4.1 defined by

$$\tau(f)=\begin{cases}\emptyset & \text{if } f \notin x^+y^*;\\ (a^+b)^n a^{2m} b & \text{if } f=x^ny^m \in x^+y^*.\end{cases}$$

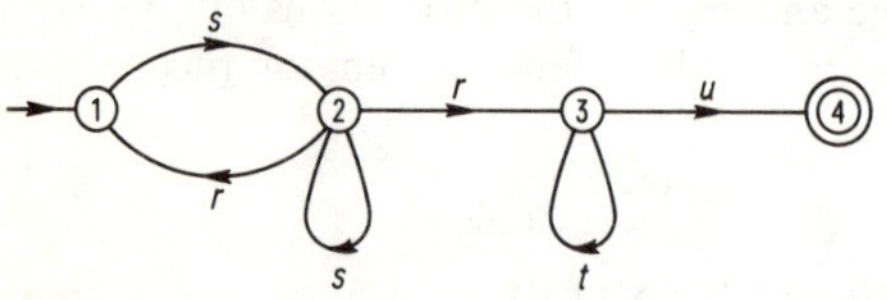


Fig. III.5

With the notations of that example, a finite nondeterministic automaton for $K=(s^+r)^+t^*u$ is given in Fig. III.5. Thus we obtain the following transducer realizing $\tau$ (Fig. III.6).

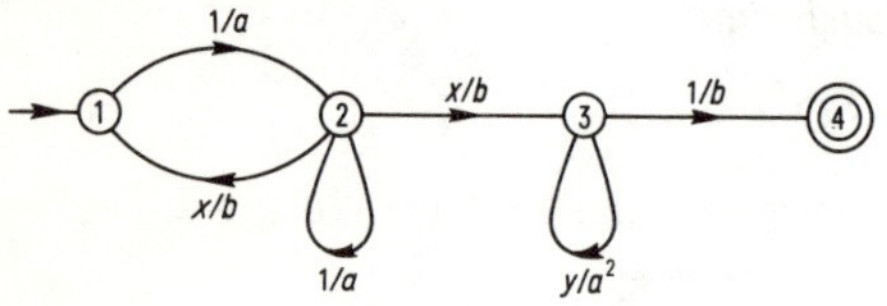


Fig. III.6

**Exercise**

**6.1** Let $\boldsymbol{T}$ be a transducer realizing a transduction $\tau$. Show how finite automata recognizing $\mathrm{dom}(\tau)$ and $\mathrm{im}(\tau)$ can be obtained from $\boldsymbol{T}$.

## III.7 Matrix Representations

Matrix representations are another equivalent definition of rational transductions. They constitute a compact formulation of transducers, obtained by grouping in one matrix all output words corresponding to a fixed input word by considering all pairs of states. The multiplication of matrices corresponds then to the concatenation of paths in the transducer, and to the union of the sets of output words of these paths.

Let $S$ be a semiring, and let $Q$ be a finite set. Then the set $S^{Q\times Q}$ of all $Q\times Q$-matrices with entries in $S$ is again a semiring for addition and multiplication of matrices induced by the operations in $S$ (see Section I.2). The identity matrix is denoted by $I$.

Let $X$ be an alphabet. A **morphism** $\mu: X^* \to S^{Q\times Q}$ is a monoid morphism from $X^*$ into the multiplicative monoid $S^{Q\times Q}$. Thus

$$\mu(fg)=\mu f\cdot\mu g \qquad f, g\in X^* \tag{7.1}$$

$$\mu 1 = I. \tag{7.2}$$

If only (7.1) is verified, then $\mu$ is a **semigroup morphism**. In this case, $\mu(X^*)=\{\mu f: f\in X^*\}$ is a monoid of $Q\times Q$-matrices with neutral element $\mu 1$, and $\mu 1$ is idempotent ($\mu 1\cdot\mu 1=\mu 1$) by (7.1). We are interested here in matrices whose entries are regular languages over an alphabet $Y$. Thus the semiring $S$ is $\mathrm{Rat}(Y^*)$. Consequently, the identity matrix $I$ is given by

$$I_{pq}=\begin{cases}1 & \text{if } \ p=q;\\ \varnothing & \text{if } \ p\neq q.\end{cases}$$

For simplicity, we frequently write 0 instead of $\varnothing$.

**Definition** A **matrix representation** $\boldsymbol{M}=\langle\mu, Q, q_-, Q_+\rangle$ **from** $X^*$ **into** $Y^*$ is composed of a finite set of states $Q$, an **initial state** $q_-$, a set of **final states** $Q_+$, and a **semigroup morphism** $\mu: X^*\to\mathrm{Rat}(Y^*)^{Q\times Q}$. The transduction $|\boldsymbol{M}|: X^*\to Y^*$ **realized** by $\boldsymbol{M}$ is defined by

$$|\boldsymbol{M}|(f)=\bigcup_{q\in Q_+}\mu f_{q_-q}. \tag{7.3}$$

For $p, q\in Q$, note $\mu_{pq}$ the transduction $f\mapsto\mu f_{pq}$. Then (7.3) can be written as

$$|\boldsymbol{M}|=\bigcup_{q\in Q_+}\mu_{q_-q}.$$

**Example 7.1** Let $X$ be an alphabet and $x_0\in X$. Set $Q=\{1, 2\}$, and define a monoid morphism $\mu: X^*\to\mathrm{Rat}(Y^*)^{2\times 2}$ by

$$\mu x=\begin{bmatrix}0 & 0\\ 0 & x\end{bmatrix} \quad x\in X\backslash x_0, \qquad \mu x_0=\begin{bmatrix}0 & 1\\ 0 & x_0\end{bmatrix}.$$

Then $\mu f=\begin{bmatrix}0 & 0\\ 0 & f\end{bmatrix}$ if $f\notin x_0X^*$, $f\neq 1$,

and $\mu f=\begin{bmatrix}0 & g\\ 0 & f\end{bmatrix}$ if $f=x_0 g$.

Thus for $\boldsymbol{M}=\langle\mu, Q, 1, \{2\}\rangle$, $|\boldsymbol{M}|(f)=\mu f_{12}=x_0^{-1}f$, ($f\in X^*$).

**Theorem 7.1** *A transduction $\tau: X^* \to Y^*$ is rational iff there exists a matrix representation $\boldsymbol{M} = \langle \mu, Q, q_-, Q_+ \rangle$ realizing $\tau$. Then the following hold:*
(i) *if $\tau(1) = 0$ or $\tau(1) = 1$, then $\mu$ can be chosen to be a monoid morphism;*
(ii) *$Q_+$ can be assumed to consist of a single state $q_+ \neq q_-$, and $\mu f_{q,q_-} = \mu f_{q_+q} = 0$ for all $f \in X^+$, $q \in Q$:*
(iii) *if $\tau(1) = 0$, both* (i) *and* (ii) *can be satisfied simultaneously.*

Note that $\mu$ cannot always be chosen to be a monoid morphism. Indeed, in this case $\tau(1) = \bigcup_{q \in Q_+} \mu 1_{q_-q}$ is equal to 1 or 0 according to $q_- \in Q_+$ or $q_- \notin Q_+$. The fact that semigroup morphisms are necessary is equivalent to the possibility for transducers to have transitions with the empty word as input label. This complicates the proof of the theorem.

Proof. We first prove the existence of a matrix representation. Let $\boldsymbol{T} = \langle X, Y, Q, q_-, Q_+, E \rangle$ be a transducer realizing $\tau$. In view of Corollary 6.2, we may suppose that $Q_+ = \{q_+\}$, $q_+ \neq q_-$,

$$E \subset Q \times (X \cup \{1\}) \times Y^* \times Q,$$

and moreover $(q, u, v, q') \in E$ implies $q' \neq q_-$, $q \neq q_+$. Let $\alpha: E^* \to X^*$ and $\beta: E^* \to X^*$ be the input and output morphisms as defined in the preceding section. Then

$$\tau(f) = \beta(\alpha^{-1}(f) \cap \Lambda(q_-, q_+)). \tag{7.4}$$

Next note that for $p, q, r \in Q$,

$$\Lambda(p, q) = \bigcup_{r \in Q} \Lambda(p, r) \Lambda(r, q).$$

Since $\alpha$ is an alphabetic morphism by the assumption on $E$, we have $\alpha^{-1}(fg) = \alpha^{-1}(f)\alpha^{-1}(g)$ for $f, g \in X^*$. Consequently

$$\alpha^{-1}(fg) \cap \Lambda(p, q) = \bigcup_{r \in Q} (\alpha^{-1}(f) \cap \Lambda(p, r))(\alpha^{-1}(g) \cap \Lambda(r, q)). \tag{7.5}$$

Define a mapping $\mu: X^* \to \mathfrak{P}(Y^*)^{Q \times Q}$ by

$$\mu f_{pq} = \beta(\alpha^{-1}(f) \cap \Lambda(p, q)) \qquad p, q \in Q, f \in X^*. \tag{7.6}$$

In view of (7.5), $\mu$ is a semigroup morphism, and by (7.6), $\mu f_{pq}$ is a regular language. Let $\boldsymbol{M} = \langle \mu, Q, q_-, q_+ \rangle$. Then by (7.4)

$$|\boldsymbol{M}|(f) = \mu f_{q_-q_+} = \tau(f).$$

This proves the existence of a matrix representation. Further, since $\Lambda(q, q_-) \cap E^+ = \Lambda(q_+, q) \cap E^+ = \emptyset$ for $q \in Q$ by the assumptions on $\boldsymbol{T}$, condition (ii) holds. Next, define the (monoid) morphism $\bar{\mu}$ by

$$\bar{\mu}f = \mu f \quad (f \in X^+), \qquad \bar{\mu}1 = I,$$

and $\bar{\boldsymbol{M}} = \langle \bar{\mu}, Q, q_-, q_+ \rangle$. Since $q_- \neq q_+$, $|\bar{\boldsymbol{M}}| = \tau$ in the case where $\tau(1) = 0$. This proves (i) in that case and proves also (iii). It remains to prove (i) in the case where $\tau(1) = 1$. For this, consider $\mu$ given by (7.6), let $q_0 \notin Q$, set $P = q_0 \cup Q$ and define

$$\mu' : X^* \to \mathrm{Rat}(Y^*)^{P \times P}$$

by $\mu' 1 = I$ and

$$\mu' f_{pq} = \begin{cases} \mu f_{pq} & \text{if} \quad p, q \in Q; \\ \mu f_{q_- q} & \text{if} \quad p = q_0, \quad q \in Q; \quad (f \in X^+) \\ 0 & \text{otherwise.} \end{cases}$$

Then $\mu'$ is easily seen to be a morphism, and

$$\tau(f) = \mu' f_{q_0 q_+} \cup \mu' f_{q_0 q_0} \qquad (f \in X^*).$$

Thus $\tau$ is realized by the matrix representation $\langle \mu', P, q_0, \{q_0, q_+\} \rangle$.

Conversely, let $\tau : X^* \to Y^*$ be the transduction realized by a matrix representation $\boldsymbol{M} = \langle \mu, Q, q_-, Q_+ \rangle$. In order to prove the rationality of $\tau$, we proceed in several steps.

First we show that $Q_+$ can be assumed to consist of a single state $q_+ \neq q_-$. Let indeed $s \notin Q$, set $P = s \cup Q$ and define a semigroup morphism $\upsilon : X^* \to \mathrm{Rat}(Y^*)^{P \times P}$ for $u \in 1 \cup X$ by

$$\begin{aligned} \upsilon u_{pq} &= \mu u_{pq} && (p, q \in Q) \\ \upsilon u_{sq} &= \upsilon u_{ss} = 0 && (q \in Q) \\ \upsilon u_{qs} &= \bigcup_{p \in Q_+} \mu u_{qp} && (q \in Q). \end{aligned}$$

Then these formulas hold for any word $f \in X^*$. This is obvious for the two first formulas; the third follows by induction from:

$$\upsilon(fg)_{qs} = \bigcup_{r \in Q} \mu f_{qr} \upsilon g_{rs} = \bigcup_{p \in Q_+} \bigcup_{r \in Q} \mu f_{qr} \mu g_{rp} = \bigcup_{p \in Q_+} \mu(fg)_{qp}.$$

Thus $\quad \tau(f) = \bigcup_{p \in Q_+} \mu f_{q_- p} = \upsilon f_{q_- s}$.

Consequently, we assume $Q_+ = \{q_+\}$ and $q_+ \neq q_-$. Next, define the monoid morphism $\bar{\mu} : X^* \to \mathrm{Rat}(Y^*)^{Q \times Q}$ by $\bar{\mu}x = \mu x \, (x \in X)$. Then $\bar{\mu} f_{q_- q_+} = \mu f_{q_- q_+} = \tau(f)$ for $f \in X^+$, and $\bar{\mu} 1_{q_- q_+} = \emptyset$. Thus

$$\tau = \tau_1 \cup \bar{\tau},$$

where $\bar{\tau}$ is the transduction realized by $\langle\bar{\mu}, Q, q_-, q_+\rangle$, and where $\tau_1: X^* \to Y^*$ is defined by $\tau_1(1) = \mu 1_{q_-q_+} \in \mathrm{rat}(Y^*)$ and $\tau_1(f) = \emptyset$ for $f \in X^+$. Since $\tau_1$ is obviously a rational transduction, it suffices to show that $\bar{\tau}$ is rational.

Thus we may assume that $\mu$ is a monoid morphism and $\tau(1) = 0$. Let $Z = Q \times X \times Q$ and define the strictly alphabetic morphism $\varphi: Z^* \to X^*$ by

$$\varphi((p, x, q)) = x.$$

Let $\quad K = [(q_- \times X \times Q)Z^* \cap Z^*(Q \times X \times q_+)] \setminus Z^*WZ^*$

with $\quad W = \{(q, x, q')(p, y, p') \in Z^2 \mid q' \neq p\}$.

Then $K$ is a local regular language, $\varphi^{-1}(1) \cap K = \emptyset$, and for $f = x_1x_2 \cdots x_n$, $(x_i \in X)$,

$$\varphi^{-1}(f) \cap K = \{(q_-, x_1, q_1)(q_1, x_2, q_2) \cdots (q_{n-1}, x_n, q_+) \mid q_1, \ldots, q_{n-1} \in Q\}. \tag{7.7}$$

Define a rational substitution $\sigma: Z^* \to X^*$ by

$$\sigma((p, x, q)) = \mu x_{pq}.$$

Then $\quad \sigma(\varphi^{-1}(1) \cap K) = \emptyset = \tau(1) \qquad (7.8)$

and, in view of (7.7),

$$\sigma(\varphi^{-1}(f) \cap K) = \mu f_{q_-q_+} = \tau(f) \qquad (f \in X^+) \tag{7.9}$$

Consequently, $\tau$ is a composition of rational transductions and therefore is rational. ■

The proof of Theorem 7.1 yields the following corollary which is another variation of Nivat's Theorem.

**Corollary 7.2** *Let $\tau: X^* \to Y^*$ be a transduction with $\tau(1) = 0$ or $\tau(1) = 1$. Then $\tau$ is rational iff there exist an alphabet $Z$, a strictly alphabetic morphism $\varphi: Z^* \to X^*$, a rational substitution $\sigma: Z^* \to Y^*$ and a local regular language $K \subset Z^*$ such that*

$$\tau(f) = \sigma(\varphi^{-1}(f) \cap K) \qquad (f \in X^*). \tag{7.10}$$

Proof. Let $\tau$ be given by (7.10). Then $\tau$ is clearly rational. Conversely, the conclusion holds if $\tau(1) = 0$ in view of (7.8) and (7.9). If $\tau(1) = 1$, then it suffices to replace the language $K$ of the preceding proof by $K \cup 1$. ■

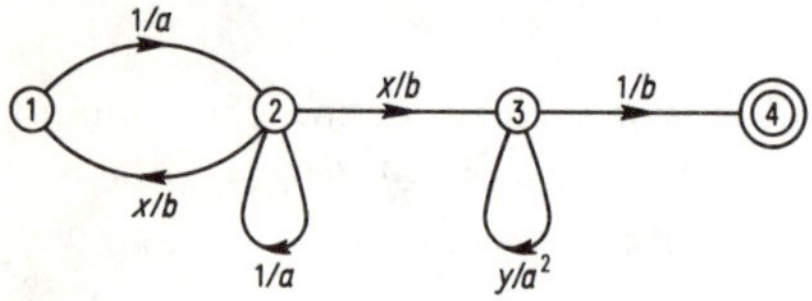


Fig. III.7

**Example 7.2** Consider for $X=\{x, y\}$, $Y=\{a, b\}$, the transducer of Example 6.2 (Fig. III.7). Formula (7.6) gives the following semigroup morphism $\mu$:

$$\mu 1=\begin{bmatrix} 1 & a^+ & 0 & 0 \\ 0 & a^* & 0 & 0 \\ 0 & 0 & 1 & b \\ 0 & 0 & 0 & 1 \end{bmatrix} \qquad \mu x=\begin{bmatrix} a^+b & a^+ba^+ & 0 & 0 \\ a^*b & a^*ba^+ & a^*b & a^*b^2 \\ 0 & 0 & 0 & 0 \\ 0 & 0 & 0 & 0 \end{bmatrix}$$

$$\mu y=\begin{bmatrix} 0 & 0 & 0 & 0 \\ 0 & 0 & 0 & 0 \\ 0 & 0 & a^2 & a^2b \\ 0 & 0 & 0 & 0 \end{bmatrix}$$

Theorem 7.1 shows a relationship between rational transductions and formal power series: In the terminology of Eilenberg [1974] or Salomaa and Soittola [1978], a transduction $\tau: X^* \to Y^*$ is rational iff the formal power series $\sum_{f\in X^*} \tau(f)\cdot f$ with coefficients in the semiring $\mathrm{Rat}(Y^*)$ is recognizable. The following theorem is the analogue to a well-known characterization of recognizable formal power series.

**Proposition 7.3** *A transduction $\tau: X^* \to Y^*$ is rational iff there exist a finite set $Q$, a monoid morphism $\mu: X^* \to \mathrm{Rat}(Y^*)^{Q\times Q}$, a row $Q$-vector $\lambda$, a column $Q$-vector $\rho$ with entries in* $\mathrm{Rat}(Y^*)$ *such that*

$$\tau(f)=\lambda\mu f\rho \qquad (f\in X^*). \tag{7.11}$$

Proof. Let $\tau$ be given by (7.11). Then

$$\tau(f)=\bigcup_{p,q\in Q} \lambda_p \mu f_{pq} \rho_q.$$

By Theorem 7.1, the transductions $\mu_{pq}: f\mapsto \mu f_{pq}$ are rational. Since $\lambda_p$, $\rho_q$ are regular languages, the transductions $f\mapsto \lambda_p\mu f_{pq}\rho_q$ are rational. Thus $\tau$ is rational.

Conversely, let $\tau$ be realized by the matrix representation $\boldsymbol{M}=\langle \upsilon, P, q_-, \{q_+\}\rangle$ with $q_-\neq q_+$. Let $s\notin P$, set $Q=s\cup P$ and define a monoid morphism $\mu: X^* \to \mathrm{Rat}(Y^*)^{Q\times Q}$ by

$$\begin{aligned} \mu x_{pq} &= \upsilon x_{pq} && p, q\in P \\ \mu x_{pq} &= \emptyset && p=s \quad \text{or} \quad q=s \end{aligned} \qquad (x\in X)$$

Then clearly

$$\mu f_{pq}=\begin{cases} \upsilon f_{pq} & p, q\in P \\ \emptyset & p=s \quad \text{or} \quad q=s \end{cases} \qquad (f\in X^+)$$

Next define the $Q$-vectors $\lambda$ and $\rho$ by

$$\lambda_s = (\upsilon 1)_{q_-q_+}; \quad \rho_s = \{1\}, \quad \lambda_{q_-} = \{1\}, \quad \rho_{q_+} = \{1\},$$
$$\lambda_q = \rho_q = \emptyset \text{ otherwise.}$$

Then $\lambda\mu 1\rho = \lambda\rho = \lambda_s\rho_s = \tau(1), \qquad \lambda\mu f\rho = \lambda_{q_-}\mu f_{q_-q_+}\rho_{q_+} = \tau(f) \quad f \in X^+.$ ■

**Example 7.3** Let $X = \{x\}$, $Y = \{a, b\}$, $Q = \{1, 2, 3, 4\}$, let $\mu$ be the monoid morphism defined by

$$\mu x = \begin{bmatrix} 0 & a & 0 & 0 \\ a & 0 & 0 & 0 \\ 0 & 0 & 0 & b \\ 0 & 0 & b & 0 \end{bmatrix} \qquad \lambda = [1, 0, 1, 0] \qquad \rho = \begin{bmatrix} 1 \\ 0 \\ 0 \\ 1 \end{bmatrix}.$$

A simple computation shows that

$$\lambda\mu x^n = \begin{cases} [a^n, 0, b^n, 0] & \text{if } n \text{ is even;} \\ [0, a^n, 0, b^n] & \text{if } n \text{ is odd.} \end{cases}$$

Thus $\lambda\mu x^n\rho = \begin{cases} a^n & n \text{ even;} \\ b^n & n \text{ odd.} \end{cases}$

We conclude this section by considering a useful technical notion.

**Definition** Let $\boldsymbol{M} = \langle \mu, Q, q_-, Q_+ \rangle$ be a matrix representation from $X^*$ into $Y^*$. Then $\boldsymbol{M}$ is trim if the following condition is satisfied. For any $q \in Q$, there exist $f, g \in X^*$, $q_+ \in Q_+$, such that

$$\mu f_{q_-q} \neq \emptyset \quad \text{and} \quad \mu g_{qq_+} \neq \emptyset. \tag{7.12}$$

**Proposition 7.4** *Let $\tau: X^* \to Y^*$ be a transduction with* $\mathrm{dom}(\tau) \neq \emptyset$, *and let $\boldsymbol{M} = \langle \mu, Q, q_-, Q_+ \rangle$ with $\mu$ a monoid morphism be a matrix representation realizing $\tau$. Then there exists a trim matrix representation $\boldsymbol{M}' = \langle \upsilon, P, q_-, P_+ \rangle$ realizing $\tau$ with $P \subset Q$ and $P_+ = Q_+ \cap P$.*

Proof. Let $P \subset Q$ be the set of states such that (7.12) holds. Since $\mathrm{dom}(\tau) \neq \emptyset$ and $\mu$ is a monoid morphism, we have $q_- \in P$ and $P_+ = Q_+ \cap P \neq \emptyset$. Moreover, for any $q \in Q_+$, $q \in P_+$ iff $\mu h_{q_-q} \neq \emptyset$ for at least one word $h \in X^*$. Consequently

$$|\boldsymbol{M}| = \bigcup_{q \in Q_+} \mu_{q_-q} = \bigcup_{q \in P_+} \mu_{q_-q} \tag{7.13}$$

Define a monoid morphism $\upsilon: X^* \to \mathrm{Rat}(Y^*)^{P \times P}$ by

$$\upsilon x_{pq} = \mu x_{pq} \qquad p, q \in P, \quad x \in X.$$

In order to prove the desired result, it suffices to show that

$$\upsilon f_{pq} = \mu f_{pq} \qquad p, q \in P, \quad f \in X^* \tag{7.14}$$

since then in view of (7.13)

$$|\boldsymbol{M}'| = \bigcup_{q\in P_+} \upsilon_{q_-q} = \bigcup_{q\in P_+} \mu_{q_-q} = |\boldsymbol{M}|.$$

To show (7.14), we first verify that for $p, q \in P$, $r \in Q \backslash P$,

$$\mu f_{pr} = \emptyset \quad \text{or} \quad \mu g_{rq} = \emptyset \tag{7.15}$$

for any pair of words $f$, $g$. Assume the contrary. Then there exist words $f$, $g$ such that both $\mu f_{pr} \neq \emptyset$ and $\mu g_{rq} \neq \emptyset$. Since $p \in P$, there is a word $f'$ such that $\mu f'_{q_-p} \neq \emptyset$, and similarly $\mu g'_{qq_+} \neq \emptyset$ for some word $g'$ and some $q_+ \in Q_+$. But then $\emptyset \neq \mu f'_{q_-p} \mu f_{pr} \subset \mu f' f_{q_-r}$ and $\emptyset \neq \mu g_{rq} \mu g'_{qq_+} \subset \mu g g'_{rq_+}$ and by (7.12), $r \in P$ contrary to the assumption. This proves (7.15).

Now (7.14) is true if $|f| \leq 1$. Arguing by induction, let $h \in X^*$, $x \in X$. Then for $p, q \in P$,

$$(\mu hx)_{pq} = \bigcup_{r\in Q} \mu h_{pr} \mu x_{rq} = \bigcup_{r\in P} \mu h_{pr} \mu x_{rq}$$

by (7.15). Consequently

$$(\mu hx)_{pq} = \bigcup_{r\in P} \upsilon h_{pr} \upsilon x_{rq} = (\upsilon hx)_{pq}.$$

■

**Exercises**

**7.1** Show that $\boldsymbol{M} = \langle \mu, Q, q_-, Q_+ \rangle$ is trim iff for any $q \in Q$, there exist $q_+ \in Q_+$, and $f, g \in X^*$ with $|f|, |g| \leq \operatorname{card} Q$ such that $\mu f_{q_-q} \neq \emptyset$ and $\mu g_{qq_+} \neq 0$.

**7.2** A transduction $\tau: X^* \to Y^*$ is faithful if $\tau^{-1}(g)$ is finite for all $g \in Y^*$, and is continuous if $\tau(f) \subset Y^+$ for $f \in X^+$. Show that a rational transduction $\tau: X^* \to Y^*$ is both faithful and continuous iff

$$\tau(f) = \psi(\varphi^{-1}(f) \cap K) \qquad f \in X^*$$

for some alphabet $Z$, $K \in \operatorname{Rat}(Z^*)$, $\varphi: Z^* \to X^*$ a morphism and $\psi: Z^* \to Y^*$ a strictly alphabetic morphism. (Hint. Apply Corollary 7.2 to $\tau^{-1}$.) Show that the composition of two faithful (continuous) transductions is still faithful (continuous).

**7.3** Let $\mathscr{R}$ be the least family of subsets of $X^* \times Y^*$ closed under union, product and the plus operation, and containing $\emptyset$, $\{(1, 1)\}$, and the relations $\{(u, y)\}$ for $u \in 1 \cup X$, $y \in Y$.

a) Show that $RS, SR \in \mathscr{R}$ for all $R \in \mathscr{R}$ such that $(1, 1) \notin R$, and $S \in \operatorname{Rat}(X^* \times Y^*)$.

b) Show that $R \in \mathscr{R}$ iff the transduction with graph $R$ is rational, faithful and continuous.

## III.8 Decision Problems

We show in this section that most of the usual questions are undecidable for rational transductions. We shall see in the next chapter that some of these questions become decidable for rational functions. The results of this section are mainly from Fischer and Rosenberg [1968].

The proofs of undecidability are "relative" in the following sense: We give (without proof) a particular undecidable problem (Post's Correspondence Problem) and we prove that some property is undecidable by showing that the existence of a decision procedure for this property would imply a decision procedure for the Correspondence Problem.

**Post's Correspondence Problem** *Given an alphabet X with at least two letters, and given two sequences*

$$u_1, u_2, \ldots, u_p \quad \text{and} \quad v_1, v_2, \ldots, v_p \tag{8.1}$$

*of words of $X^*$, decide whether there exist indices $i_1, i_2, \ldots, i_k$, $(k>0)$ such that*

$$u_{i_1}u_{i_2}\cdots u_{i_k} = v_{i_1}v_{i_2}\cdots v_{i_k}. \tag{8.2}$$

**Theorem 8.1** (Post's Correspondence Theorem) *Post's Correspondence Problem is undecidable.*

For a proof, see for instance Davis [1958] or Schnorr [1974]. The theorem means that there exists no algorithm that has as input the two sequences (8.1), and yields as output "yes" or "no" according to the existence or the non-existence of a sequence $i_1, \ldots, i_k$ such that (8.2) holds.

First we give two decidable properties. As usual for decision problems, the word "given" in the statement should be interpreted to mean that an explicit description of the object, here of the rational relation $R$ is provided. This can be done in the present context by a rational expression, by a matrix representation, by a transducer or by a bimorphism. From the constructions of the previous sections, it should be clear that any of the above descriptions of a rational relation can be obtained effectively from another one.

**Proposition 8.2** *Given a rational relation $R \subset X^* \times Y^*$ it is decidable whether $R$ is empty and whether $R$ is finite.*

Proof. $R$ is empty iff one of the two projections $\pi_X(R)$ and $\pi_Y(R)$ on $X^*$ and $Y^*$ are empty, and $R$ is finite iff both projections are finite. Each projection is a regular language, and an explicit description of these languages is readily obtained from any effective description of $R$. Since emptiness and finiteness are decidable for regular languages, the conclusion follows. ■

We now prove a lemma which will be of use later. Let $X = \{a, b\}$, let $Y$ be an alphabet, and let $u_1, u_2, \ldots, u_p \in Y^*$. Define

$$U = \{(ab, u_1), (ab^2, u_2), \ldots, (ab^p, u_p)\}.$$

Clearly $U$, hence $U^+$ is a rational relation over $X$ and $Y$.

**Lemma 8.3** *The relation* $(X^* \times Y^*)\backslash U^+$ *is rational.*

Usually, $\text{Rat}(X^* \times Y^*)$ is not closed under complementation, thus Lemma 8.3 has to be proved.

Proof. We show that $W = (X^* \times Y^*)\backslash U^+$ is a rational relation by writing $W$ as a union of four rational relations. First the relation $H$ composed of all $(f, g) \in X^* \times Y^*$ such that

$$f \notin \{ab, ab^2, \ldots, ab^p\}^+$$

is rational and even recognizable since

$$H = (X^* \backslash \{ab, ab^2, \ldots, ab^p\}^+) \times Y^*.$$

Next $\quad (f, g) \in W \quad$ and $\quad (f, g) \notin H$

if and only if

$$f = ab^{i_1}ab^{i_2} \cdots ab^{i_r} \quad \text{for some} \quad r > 0, \quad 1 \leq i_1, \ldots, i_r \leq p \tag{8.3}$$

and $\quad g \neq u_{i_1}u_{i_2} \cdots u_{i_r}$. $\qquad$ (8.4)

Now (8.4) holds iff one of the three following conditions hold

$$|g| < |u_{i_1}u_{i_2} \cdots u_{i_r}| \tag{8.5}$$

$$|g| > |u_{i_1}u_{i_2} \cdots u_{i_r}| \tag{8.6}$$

$$|g| = |u_{i_1}u_{i_2} \cdots u_{i_r}|, \tag{8.7}$$

and there is a factorization $g = g'hg''$ and $k \in \{1, \ldots, p\}$ with

$$|g'| = |u_{i_1} \cdots u_{i_{k-1}}|, \qquad |h| = |u_{i_k}|, \qquad h \neq u_{i_k}, \qquad |g''| = |u_{i_{k+1}} \cdots u_{i_r}|.$$

Define the following relations which are clearly rational:

$$F = \bigcup_{i=1}^{p} ab^i \times Y^{|u_i|}; \qquad G = \bigcup_{i=1}^{p} ab^i \times (Y^{|u_i|} \backslash u_i) = F \backslash U;$$

$$D = \bigcup_{i=1}^{p} ab^i \times Y^{|u_i|}Y^+ = F \cdot (1 \times Y^+); \qquad C = \bigcup_{i=1}^{p} ab^i \times Y_i,$$

with $\quad Y_i = \{u \in Y^* : |u| < |u_i|\}$.

Then $CF = FC$, $DF = FD$, and

$$\{(f, g) \mid (f, g) \text{ verifies (8.3) and (8.5)}\} = C^+F^*;$$

$$\{(f, g) \mid (f, g) \text{ verifies (8.3) and (8.6)}\} = D^+F^*;$$

$$\{(f, g) \mid (f, g) \text{ verifies (8.3) and (8.7)}\} = F^*GF^*.$$

Thus $\quad W = H \cup C^+F^* \cup D^+F^* \cup F^*GF^* \in \text{Rat}(X^* \times Y^*)$. ■

**Theorem 8.4** *Let X, Y be alphabets with at least two letters. Given rational relations* $A, B \subset X^* \times Y^*$, *it is undecidable to determine whether*

(i) $A \cap B = \emptyset$;

(ii) $A \subset B$;

(iii) $A = B$;

(iv) $A = X^* \times Y^*$;

(v) $(X^* \times Y^*) \backslash A$ *is finite;*

(vi) *A is recognizable.*

Proof. We may assume that $X$ contains exactly two letters, and set $X = \{a, b\}$. Consider two sequences

$$u_1, u_2, \ldots, u_p \quad \text{and} \quad v_1, v_2, \ldots, v_p \tag{8.8}$$

of words of $Y^*$ and define

$$U = \{(ab, u_1), \ldots, (ab^p, u_p)\}, \qquad V = \{(ab, v_1), \ldots, (ab^p, v_p)\}.$$

Then $U^+$, $V^+$ are rational relations, and by the preceding lemma, $\bar{U} = (X^* \times Y^*) \backslash U^+$ and $\bar{V} = (X^* \times Y^*) \backslash V^+$ are rational relations.

(i) Let $A = U^+$, $B = V^+$. Then $A \cap B \neq \emptyset$ iff there exist integers $i_1, \ldots, i_k$ such that $u_{i_1} \cdots u_{i_k} = v_{i_1} \cdots v_{i_k}$, thus iff the Correspondence Problem (8.8) has a solution. Thus if the emptiness $A \cap B = \emptyset$ would be decidable, Post's Correspondence Problem would be decidable. This proves (i).
(ii) Let $A = U^+$, $B = \bar{V}$. Then $A \subset B$ iff $U^+ \cap V^+ = \emptyset$. Thus (ii) follows from (i).
(iii) is a consequence of (iv) since $X^* \times Y^*$ is rational.
(iv) Let $A = \bar{U} \cup \bar{V}$. Then $A = X^* \times Y^*$ iff $X^* \times Y^* \backslash A = U^+ \cap V^+ = \emptyset$. Thus (iv) is undecidable by (i).
(v) Let again $A = \bar{U} \cup \bar{V}$. Then $(m, u) \in X^* \times Y^* \backslash A$ iff there exist $i_1, \ldots, i_r$ such that $m = ab^{i_1} \cdots ab^{i_r}$ and $u = u_{i_1} \cdots u_{i_r} = v_{i_1} \cdots v_{i_r}$. Thus $(m, u) \in X^* \times Y^* \backslash A$ implies $(m^k, u^k) \in X^* \times Y^* \backslash A$ for any $k \geq 1$. Consequently, $X^* \times Y^* \backslash A$ is finite iff $X^* \times Y^* \backslash A$ is empty, and the last property is undecidable by (iv).
(vi) Let again $A = \bar{U} \cup \bar{V}$. Then $A$ is recognizable iff $X^* \times Y^* \backslash A = U^+ \cap V^+$ is recognizable since $\text{Rec}(X^* \times Y^*)$ is closed under complementation. We shall see that $U^+ \cap V^+$ is recognizable iff $U^+ \cap V^+ = \emptyset$. Assume $U^+ \cap V^+$ recognizable. Then by Mezei's Theorem,

$$U^+ \cap V^+ = P_1 \times Q_1 \cup \cdots \cup P_l \times Q_l$$

for $P_1, \ldots, P_l \in \text{Rat}(X^*)$, $Q_1, \ldots, Q_l \in \text{Rat}(Y^*)$. Next assume $(m, u) \in U^+ \cap V^+$. Then $(m^k, u^k) \in U^+ \cap V^+$ for $k \geq 1$, thus there exist integers $r$, $s$,

$(r > s \geq 1)$ such that

$$(m^r, u^r), (m^s, u^s) \in P_j \times Q_j$$

for some $j$, $(1 \leq j \leq l)$. Then $(m^s, u^r) \in P_j \times Q_j$ but $(m^s, u^r) \notin U^+ \cap V^+$ since $s \neq r$. Thus $U^+ \cap V^+ = \emptyset$, and (vi) follows from (i). ■

## Exercises

**8.1** Show that all properties of Theorem 8.4 are decidable for recognizable relations.

**8.2** Show that for a rational relation $A \subset X^* \times Y^*$, the word problem: $(u, v) \in A$ can be solved in $O(n)$ steps, where $n = |u| + |v|$. (For connections with the same problem for linear languages, see Proposition V.6.5.)

**8.3** Let $\boldsymbol{M} = \langle \mu, Q, q_-, Q_+ \rangle : X^* \to Y^*$ be matrix representation. Show that a trim matrix representation realizing $|\boldsymbol{M}|$ can effectively be constructed.

**8.4** (continuation of Exercise 4.8) Assume that a rational transduction $\tau : X^* \to Y^*$ is effectively given. Show that the transductions $\tau_0$ and $\tau_\infty$ can be computed effectively.

# IV Rational Functions

The present chapter deals with rational functions, i.e. rational transductions which are partial functions. Rational functions have remarkable properties. First, several decision problems become solvable. This is shown in Section 1. Then there exist special representations, called unambiguous representations for rational functions. They are defined by the property that there is at most one successful path for each input word. Two different methods for constructing unambiguous representations are given in Section 3 and 4, the first by means of a cross-section theorem due to Eilenberg, the second through so-called semimonomial representations and due to Schützenberger. Section 2 is concerned with sequential functions which are a particular case of rational functions. In Section 5, bimachines are defined and are used to show that any rational function can be obtained as a composition of a left sequential followed by a right sequential function. In Section 6, we prove that it is decidable whether a rational function is sequential.

## IV.1 Rational Functions

In this section, rational functions are defined and some examples are given. Further a decidability result is proved. A more detailed description of rational functions will be given in Section 4 and 5.

**Definition** A rational function $\alpha : X^* \to Y^*$ is a rational transduction which is a partial function, i.e. such that $\mathrm{Card}(\alpha f) \leq 1$ for all $f \in X^*$.

In order to simplify statements and proofs, we first make a general observation. Given any transduction

$$\tau : X^* \to Y^*$$

define two transductions $\tau_1, \tau_+ : X^* \to Y^*$ by

$$\tau_1(1) = \tau(1); \quad \tau_+(1) = \emptyset;$$

$$\tau_1(f) = \emptyset; \qquad \tau_+(f) = \tau(f) \qquad f \in X^+.$$

Then $\tau = \tau_1 \cup \tau_+$ (and even $\tau = \tau_1 + \tau_+$), and $\tau$ is rational iff $\tau_1$ and $\tau_+$ are rational. Further, any transduction $\tau' : X^* \to Y^*$ with $\tau'_+ = \tau_+$ is rational iff $\tau$ is rational and $\tau'(1)$ is a rational language. Thus, rational transductions can always be considered "up to the value $\tau(1)$". Therefore we stipulate that in this chapter, $\tau(1)$ is always equal to $\emptyset$ or $\{1\}$. Then, according to Theorem III.7.1, the morphism of a matrix representation $\boldsymbol{M}=\langle\mu, Q, q_-, Q_+\rangle$ realizing $\tau$ can always be chosen to be a monoid morphism.

Further, we recall that if $\tau(1)=0$, then we may assume that $Q_+=\{q_+\}$ and $q_+\neq q_-$, and that $\mu f_{qq_-}=\mu f_{q_+q}=0$ for $f\in X^+$, $q\in Q$. As a result of the above discussion we thus may assume that this relation also holds if $\tau(1)=1$, and that $Q_+=\{q_-, q_+\}$. Then indeed $\tau(1)=\mu 1_{q_-q_-}=1$, and $\tau(f)=\mu f_{q_-q_+}$ if $f\in X^+$.

A matrix representation which satisfies the above conditions and which is trim is called normalized. Normalization clearly is effective.

**Proposition 1.1** *Let $\tau: X^*\rightarrow Y^*$ be the transduction realized by a normalized matrix representation $\boldsymbol{M}=\langle\mu, Q, q_-, Q_+\rangle$. Then $\tau$ is a partial function iff* $\mathrm{Card}(\mu f_{pq})\leqslant 1$ *for any $f\in X^*$, $(p, q\in Q)$.*

Proof. If the conclusion holds, then $\mathrm{Card}(\mu f_{q_-q_+})\leqslant 1$ for any $f\in X^+$, thus $\mathrm{Card}(\tau(f))\leqslant 1$ for $f\in X^*$. Conversely, assume that $\mathrm{Card}(\mu f_{pq})\geqslant 2$ for some $f\in X^*$, $(p, q\in Q)$. Then $f\in X^+$ since $\mu 1$ is the identity matrix. Since $M$ is trim, $\mu h_{q_-p}\neq\emptyset$ and $\mu h'_{qq_+}\neq\emptyset$ for some $h, h'\in X^*$. Then

$$\tau(hfh')=\mu(hfh')_{q_-q_+}\supset\mu h_{q_-p}\mu f_{pq}\mu h'_{qq_+},$$

and thus $\mathrm{Card}\tau(hfh')\geqslant 2$. ■

Let $\alpha: X^*\rightarrow Y^*$ be a rational function realized by a normalized matrix representation $\boldsymbol{M}=\langle\mu, Q, q_-, Q_+\rangle$. Then we associate to $\boldsymbol{M}$ the transducer $\boldsymbol{T}=\langle X, Y, Q, q_-, Q_+, E\rangle$ with

$$E=\{(p, x, \mu x_{pq}, q)\mid p, q\in Q, x\in X, \mu x_{pq}\neq 0\}.$$

Thus $E\subset Q\times X\times Y^*\times Q$, and for any $(p, x, q)\in Q\times X\times Q$, there is at most one $h\in Y^*$ such that $(p, x, h, q)\in E$. Conversely, if $E$ satisfies these conditions then the formula

$$\mu x_{pq}=\begin{cases} h & \text{if } (p, x, h, q)\in E;\\ 0 & \text{otherwise}\end{cases}$$

defines a matrix representation $\boldsymbol{M}=\langle\mu, Q, q_-, Q_+\rangle$. The transducer $\boldsymbol{T}$ and the matrix representation $\boldsymbol{M}$ are called associated, and we sometimes identify them. Thus we speak of a normalized transducer, of a path in a matrix representation and so on.

**Example 1.1** Let $\alpha: x^*\rightarrow\{a, b\}^*$ be given by

$$\alpha(x^n)=\begin{cases} a^n & n \text{ even};\\ b^n & n \text{ odd}.\end{cases}$$

Then $\alpha$ is a rational transduction (Example III.7.4), hence a rational function.

**Example 1.2** Let $X=\{x\}$, $Y=\{b\}$ and consider the transducer in Fig. IV.1

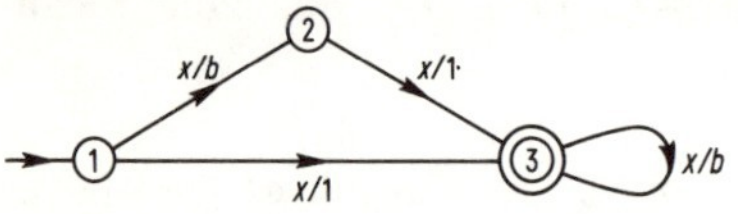


Fig. IV.1

corresponding to the matrix

$$\mu x = \begin{bmatrix} 0 & b & 1 \\ 0 & 0 & 1 \\ 0 & 0 & b \end{bmatrix}.$$

The transduction $\alpha$ realized by this transducer is given by $\alpha(1)=0$, $\alpha(x^n)=b^{n-1}$, $(n \geqslant 1)$; hence $\alpha$ is a rational function.

**Example 1.3** (Choffrut) Let again $X=\{x\}$, $Y=\{b\}$, and consider the transducer in Fig. IV.2. Let $\alpha$ be the transduction realized. It is easy to see that

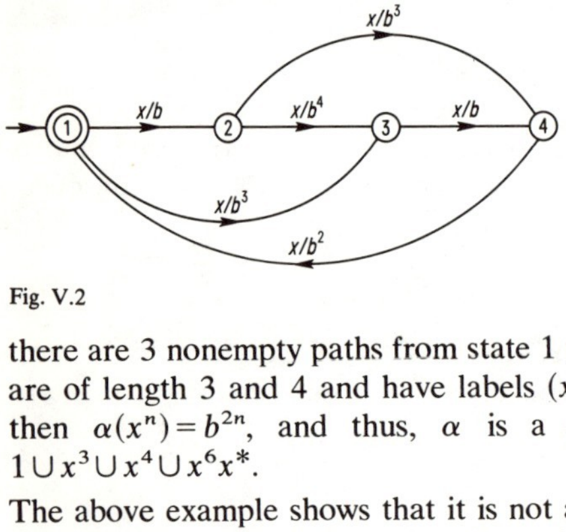


Fig. V.2

there are 3 nonempty paths from state 1 to itself without internal node 1. They are of length 3 and 4 and have labels $(x^3, b^6)$ and $(x^4, b^8)$. Thus if $\alpha(x^n) \neq 0$, then $\alpha(x^n)=b^{2n}$, and thus, $\alpha$ is a partial function. Further $\mathrm{dom}(\alpha)= 1 \cup x^3 \cup x^4 \cup x^6 x^*$.

The above example shows that it is not always easy to determine whether the transduction realized by a transducer is a (partial) function. However, this property has been shown to be decidable by Schützenberger [1975] (see also Blattner and Head [1977]):

**Theorem 1.2** *Let $\tau: X^* \rightarrow Y^*$ be a transduction realized by a normalized matrix representation $\boldsymbol{M}=\langle \mu, Q, q_-, q_+ \rangle$, and let $m=\mathrm{Card}(Q)$. Then $\tau$ is a rational function iff $\mathrm{Card}(\mu f_{pq}) \leqslant 1$ for all $p, q \in Q$ and all $f \in X^*$ with $|f| \leqslant 1+2m(m-1)$.*

Proof. By Proposition 1.1, the condition is nccessary. Assume the converse is false. Then, still by Proposition 1.1, $\mathrm{Card}(\mu f_{pq}) \geqslant 2$ for some $f \in X^+$ and $p, q \in Q$. Choose a word $f$ of minimal length such that $\mathrm{Card}(\mu f_{pq}) \geqslant 2$ for some $p, q \in Q$. Then $|f| > 1+2m(m-1)$. Set $f = x_1 \cdots x_n$ with $x_1, \ldots, x_n \in X$. There is a sequence of $n+1$ pairs of states $(q_j, q'_j)$, $(j=0,\ldots,n)$ such that $q_0=q'_0=p$, $q_n=q'_n=q$, and such that with

$$u_j=(\mu x_j)_{q_{j-1}q_j}, \qquad v_j=(\mu x_j)_{q'_{j-1}q'_j},$$

the words $u=u_1\cdots u_n$, $v=v_1\cdots v_n$ are two distinct elements of $\mu f_{pq}$. Since $f$ has been chosen of minimal length, $q_j\neq q'_j$ for $j=1,\ldots,n-1$. Indeed, assume $q_j=q'_j$ for some $j$. Then either $u_1\cdots u_j\neq v_1\cdots v_j$ or $u_{j+1}\cdots u_n\neq v_{j+1}\cdots v_n$. Next, since $n-1>2m(m-1)$, there are three indices $1\leqslant i<j<k\leqslant n-1$ such that

$$(q_i, q'_i)=(q_j, q'_j)=(q_k, q'_k). \tag{1.1}$$

Define

$$\begin{aligned}
&f_1=x_1\cdots x_i, && f_2=x_{i+1}\cdots x_j, && f_3=x_{j+1}\cdots x_k, && f_4=x_{k+1}\cdots x_n;\\
&a_1=u_1\cdots u_i, && a_2=u_{i+1}\cdots u_j, && a_3=u_{j+1}\cdots u_k, && a_4=u_{k+1}\cdots u_n;\\
&b_1=v_1\cdots v_i, && b_2=v_{i+1}\cdots v_j, && b_3=v_{j+1}\cdots v_k, && b_4=v_{k+1}\cdots v_n.
\end{aligned}$$

Then by (1.1)

$$a_1a_4,\ b_1b_4\in\mu(f_1f_4)_{pq};\quad a_1a_2a_4,\ b_1b_2b_4\in\mu(f_1f_2f_4)_{pq};$$
$$a_1a_3a_4,\ b_1b_3b_4\in\mu(f_1f_3f_4)_{pq}.$$

By the minimality of $f$, we have

$$a_1a_4=b_1b_4,\quad a_1a_2a_4=b_1b_2b_4,\quad a_1a_3a_4=b_1b_3b_4. \tag{1.2}$$

We shall deduce from (1.2) that $u=v$, in contradiction with the assumption. By symmetry, we may suppose $|a_1|\leqslant|b_1|$, hence $b_1=a_1w$ for some $w\in Y^*$. Then the first of the equations (1.2) implies $a_4=wb_4$. Reporting this in the two other equations (1.2) yields: $a_2w=wb_2$ and $a_3w=wb_3$. It follows that:

$$u=a_1a_2a_3a_4=a_1a_2a_3wb_4=a_1a_2wb_3b_4=a_1wb_2b_3b_4=v. \qquad \blacksquare$$

Given two rational functions $\alpha,\beta: X^*\to Y^*$, we write $\alpha\subset\beta$ if $\alpha(f)\neq 0\Rightarrow \alpha(f)=\beta(f)$, $(f\in X^*)$.

**Corollary 1.3** *Given two rational functions $\alpha,\beta: X^*\to Y^*$, it is decidable whether $\alpha\subset\beta$, and whether $\alpha=\beta$.*

Proof. Clearly $\alpha\subset\beta$ iff the two following conditions hold:

(i) $\mathrm{dom}(\alpha)\subset\mathrm{dom}(\beta)$;
(ii) $\alpha\cup\beta$ is a rational function.

Condition (i) is decidable since $\mathrm{dom}(\alpha)$ and $\mathrm{dom}(\beta)$ are regular languages. Condition (ii) is decidable by the previous theorem. Next $\alpha=\beta$ iff $\alpha\subset\beta$ and $\beta\subset\alpha$, thus equality of functions is decidable. $\blacksquare$

**Exercises**

**1.1** Show that it is decidable whether a rational function $\alpha$ is recognizable (i.e. its graph is a recognizable relation).

**1.2** Show that it is undecidable, for rational functions $\alpha, \beta : X^* \to Y^*$, whether there exists a word $f \in X^*$ such that $\alpha(f) = \beta(f)$.

## IV.2 Sequential Transductions

For practical purposes, a rational transduction is required not only to be a partial function, but also to be computable in some sequential way. Such a model is provided by sequential transductions. In fact, the transducers which are used for instance in compilation are more general, since there is usually an output after the lecture of the last letter of the input word. In order to fit into the model of sequential transducers, the input word is frequently considered to be followed by some "endmarker". Another way to describe this situation is to add a supplementary output function to a sequential transducer. This is the definition of the subsequential transducers.

In this section, we define sequential and subsequential transductions and give a "machine independent" characterization of these particular rational functions. Sequential transductions are among the oldest concepts in formal language theory. For a complete exposition, see Eilenberg [1974]. Subsequential transductions are defined in Schützenberger [1977a]. A systematic exposition can be found in Choffrut [1978].

**Definition** A left sequential transducer (or sequential transducer for short) $\boldsymbol{L}$ consists of an input alphabet $X$, an output alphabet $Y$, a finite set of states $Q$, an initial state $q_- \in Q$, and of two partial functions

$$\delta : Q \times X \to Q, \qquad \lambda : Q \times X \to Y^*$$

having the same domain and called the next state function and the output function respectively.

We usually denote $\delta$ by a dot, and $\lambda$ by a star. Thus we write $q \cdot x$ for $\delta(q, x)$ and $q * x$ for $\lambda(q, x)$. Then $\boldsymbol{L}$ is specified by

$$\boldsymbol{L} = \langle X, Y, Q, q_- \rangle.$$

With the conventions of Section I.1, $Q$ can be considered as a subset of $\mathfrak{P}(Q)$, and $q \cdot x$ is undefined iff $q \cdot x = \emptyset$ (or $q \cdot x = 0$, by writing 0 for $\emptyset$). Further $0 \cdot x = 0$ for all $x \in X$. Thus, the next state function can also be viewed as a total function from $Q \cup \{0\} \times X$ into $Q \cup \{0\}$, and 0 can be considered as a new, "sink" state.

A sequential transducer is called a generalized sequential machine (gsm) by Eilenberg [1974] and Ginsburg [1966].

The next state and the output function are extended to $Q \times X^*$ by setting, for $f \in X^*$, $x \in X$

$$\begin{aligned} q \cdot 1 &= q; \qquad q \cdot (fx) = (q \cdot f) \cdot x; \\ q * 1 &= 1; \qquad q * (fx) = (q * f)((q \cdot f) * x). \end{aligned} \tag{2.1}$$

The parentheses in (2.1) can be omitted without ambiguity. We agree that concatenation has higher priority than the dot, and that the dot has higher priority than the star. For $f, g \in X^*$, $(q \in Q)$, the following formula hold

$$q \cdot fg = (q \cdot f) \cdot g; \tag{2.2}$$

$$q * fg = (q * f)(q \cdot f * g). \tag{2.3}$$

Indeed (2.2) is clear, and (2.3) is proved by induction on $|g|$: the formula is obvious for $|g| = 0$. If $g = hx$ with $h \in X^*$, $x \in X$, then

$$\begin{aligned} q * fg &= q * fhx = (q * fh)(q \cdot fh * x) \\ &= (q * f)(q \cdot f * h)((q \cdot f) \cdot h * x) = (q * f)(q \cdot f * hx) = (q * f)(q \cdot f * g). \end{aligned}$$

The partial function $|\boldsymbol{L}| : X^* \to Y^*$ realized by $\boldsymbol{L}$ is defined by

$$|\boldsymbol{L}|(f) = q_- * f \qquad (f \in X^*).$$

**Definition** A partial function $\alpha : X^* \to Y^*$ is a (left) sequential transduction or (left) sequential function if $\alpha = |\boldsymbol{L}|$ for some sequential transducer $\boldsymbol{L}$.

If $\alpha = |\boldsymbol{L}|$ with $\boldsymbol{L}$ as above, then

$$\alpha(1) = 1 \tag{2.4}$$

$$\alpha(fg) = \alpha(f)(q_- \cdot f * g). \tag{2.5}$$

By (2.4), dom($\alpha$) is nonempty. Say that a partial function $\alpha : X^* \to Y^*$ preserves left factors if (2.4) holds and if further

$$\alpha(fg) \neq 0 \Rightarrow \alpha(fg) \in \alpha(f) Y^*.$$

Then by (2.5), a sequential function preserves left factors. Note that this is a rather strong constraint. In particular, the domain of such a function is prefix-closed, i.e. it contains the left factors of its elements. Of course, this is due to the lack of final states.

To each sequential transducer $\boldsymbol{L} = \langle X, Y, Q, q_- \rangle$ we associate a transducer $\boldsymbol{T} = \langle X, Y, Q, q_-, Q_+, E \rangle$ by setting $Q_+ = Q$ and

$$E = \{(q, x, q * x, q \cdot x) \mid q \in Q, x \in X, q \cdot x \neq 0\}.$$

Then clearly $|\boldsymbol{L}| = |\boldsymbol{T}|$. Thus

**Proposition 2.1** *Any sequential function is rational.* ∎

**Example 2.1** Any morphism is a sequential function.

**Example 2.2** Let $X=\{x, y\}$, $Y=\{a, b\}$, and define $\alpha: X^* \to Y^*$ by

$$\alpha(f)=\begin{cases} a^{|f|} & \text{if } f\in xX^*, \\ b^{|f|} & \text{otherwise.} \end{cases}$$

Then $\alpha$ is a sequential function realized by the following transducer (Fig. IV.3).

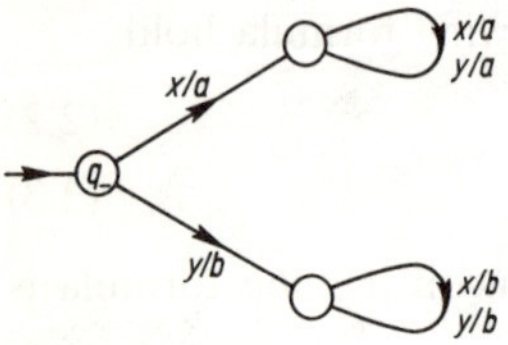


Fig. IV.3

**Example 2.3** The function $\tau: x^* \to \{a, b\}^*$ defined by

$$\tau(x^n)=\begin{cases} a^n & n \text{ even;} \\ b^n & n \text{ odd,} \end{cases}$$

is rational, but not sequential, since it does not preserve left factors. Sometimes, it is useful to have some "reversal" of a left sequential transducer.

**Definition** A right sequential transducer $\boldsymbol{R}=\langle X, Y, Q, q_-\rangle$ is given by objects $X, Y, Q, q_-$ which have the same meaning as for left sequential transducers, and by two partial functions

$$X\times Q\to Q; \qquad X\times Q\to Y^*$$

with same domain, called next-state and output function and denoted by a dot and by a star respectively.
As above, these functions are extended to $X^*\times Q$ by setting

$$1\cdot q=q; \qquad xf\cdot q=x\cdot(f\cdot q);$$
$$1*q=1; \qquad xf*q=(x*f\cdot q)(f*q).$$

Then the "reversal" of formulas (2.3), (2.4) hold:

$$fg\cdot q=f\cdot(g\cdot q); \qquad fg*q=(f*g\cdot q)(g*q). \qquad (f, g\in X^*)$$

The partial function $|\boldsymbol{R}|$ realized by $\boldsymbol{R}$ is defined by

$$|\boldsymbol{R}|(f)=f*q_- \qquad (f\in X^*),$$

and a partial function realized by a right sequential transducer is called a right sequential transduction or right sequential function.

**Proposition 2.2** *Let $\alpha: X^*\to Y^*$ be a partial function, and define $\beta: X^*\to Y^*$ by $\beta(f)=[\alpha(\tilde{f})]^\sim$ $(f\in X^*)$. Then $\alpha$ is left sequential iff $\beta$ is right sequential.*

Proof. Note first that $\alpha(f)=[\beta(\tilde{f})]^{\sim}$, $(f\in X^*)$. Thus it suffices to show that if $\alpha$ is right or left sequential, then $\beta$ is left or right sequential. Assume that $\alpha$ is realized by some right sequential transducer $\boldsymbol{R}=\langle X, Y, Q, q_-\rangle$, and define a left sequential transducer $\boldsymbol{L}=\langle X, Y, Q, q_-\rangle$ by setting

$$q\cdot x = x\cdot q; \qquad q*x=(x*q)^{\sim}.$$

Then $\quad q\cdot f=\tilde{f}\cdot q \quad$ and $\quad q*f=(\tilde{f}*q)^{\sim}$,

since by induction, for $h\in X^*$, $x\in X$,

$$q\cdot hx=(q\cdot h)\cdot x = x\cdot(\tilde{h}\cdot q)=(hx)^{\sim}\cdot q,$$

$$q*hx=(q*h)(q\cdot h*x)=(\tilde{h}*q)^{\sim}(x*\tilde{h}\cdot q)^{\sim}$$
$$=[(x*\tilde{h}\cdot q)(\tilde{h}*q)]^{\sim}=[(hx)^{\sim}*q]^{\sim}.$$

Thus $|\boldsymbol{L}|(f)=[|\boldsymbol{R}|(\tilde{f})]^{\sim}$ for all $f\in X^*$, and $\beta=|\boldsymbol{L}|$. ■

**Corollary 2.3** *A right sequential function is rational.*

Proof. Let $\alpha: X^*\to Y^*$ be a right sequential function, and let $\beta$ be defined by $\beta(f)=[\alpha(\tilde{f})]^{\sim}$. Then $\beta$ is sequential, hence rational, and its graph $B$ is a rational relation. Let $A$ be the graph of $\alpha$. Then $A=\{(\tilde{f},\tilde{g})\mid(f,g)\in B\}$, and $A$ is rational (see Section III.4). ■

**Example 2.2** (continued). The function $\alpha$ is not right sequential since it does not preserve right factors.

**Example 2.3** (continued). For the same reason, the function $\tau$ is not right sequential.

**Example 2.4** The basic step for addition in some base $k$ is realized (see Example III.5.16) by a function $\alpha$ which associates, to two words $u, v\in\Bbbk^*$ of same length, the shortest word $w$ such that $\langle u\rangle+\langle v\rangle=\langle w\rangle$. The number $\langle u\rangle$ can be defined, for $u=x_0x_1\cdots x_n$, $(x_i\in\Bbbk)$ either as in Example III.5.16, or by

$$\langle u\rangle = x_0+x_1k+\cdots+x_nk^n.$$

This is the "reversal interpretation" which is more convenient when the input is read from left to right, as will be done here. Since $u$ and $v$ have the same length, $\alpha$ can be considered as a function $\alpha:(\Bbbk\times\Bbbk)^*\to\Bbbk^*$. If $v=y_0y_1\cdots y_n$, then the argument of $\alpha$ is $f=(x_0, y_0)(x_1, y_1)\cdots(x_n, y_n)$. For simplicity, we write indistinctly $\alpha(u, v)$ or $\alpha(f)$. By Example III.5.16, $\alpha$ is known to be rational, but $\alpha$ is neither left nor right sequential. Consider for instance $k=2$. Then

$$\alpha(11,10)=w_1=001 \qquad \alpha(11111,10010)=w_2=000101.$$

The word $f_1=(1,1)(1,0)$ is both a left and a right factor of $f_2=(1,1)(1,0)(1,0)(1,1)(1,0)$, but $w_1$ is neither a left nor a right factor of $w_2$. Now consider the following (left) sequential transducer (Fig. IV.4) and let $\beta$ be the

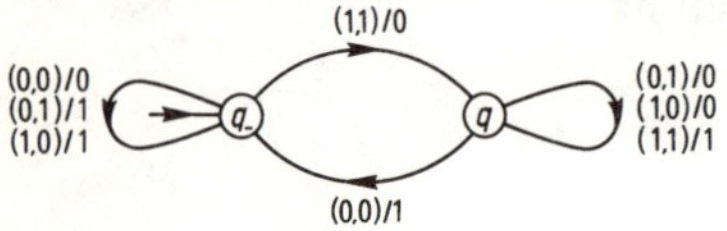


Fig. IV.4

sequential function realized. Then

$$\alpha(f)=\begin{cases}\beta(f) & \text{if } q_-\cdot f=q_-;\\ \beta(f)1 & \text{if } q_-\cdot f=q.\end{cases}$$

Thus $\alpha$ is "almost" a sequential function. This leads to the following definition.

**Definition** A (left) subsequential transducer $S=\langle X, Y, Q, q_-, \rho\rangle$ is composed of a sequential transducer $\langle X, Y, Q, q_-\rangle$ and of a partial function $\rho: Q\to Y^*$. The partial function $|S|: X^*\to Y^*$ realized by $S$ is defined by

$$|S|(f)=(q_-*f)\rho(q_-\cdot f). \tag{2.6}$$

A subsequential function is a partial function realized by some subsequential transducer.

According to the discussion at the beginning of this section, $\rho(q_-\cdot f)$ has the value 0 in (2.6) whenever $q_-\cdot f=0$.

**Example 2.4** (continued). The function $\alpha$ is subsequential with $p(q_-)=\varepsilon$, $\rho(q)=1$.

**Example 2.5** Any sequential function is subsequential: it suffices to define $\rho(q)$ to be the empty word for all $q\in Q$.

**Example 2.6** Any partial function with finite domain is subsequential (this is not true for sequential functions). Consider indeed $\alpha: X^*\to Y^*$ and suppose $\text{dom}(\alpha)$ is finite. We define a subsequential transducer $S=\langle X, Y, Q, q_-, \rho\rangle$ as follows: $Q=\text{dom}(\alpha)(X^*)^{-1}$ is the set of left factors of words in $\text{dom}(\alpha)$; $q_-=1$. The next state and the output function are defined by

$$u\cdot x=\begin{cases}ux & \text{if } ux\in Q;\\ 0 & \text{otherwise.}\end{cases}\qquad u*x=\begin{cases}1 & \text{if } ux\in Q;\\ 0 & \text{otherwise.}\end{cases}$$

$u\in Q,\ x\in X.$

Finally,

$$\rho(u)=\begin{cases}\alpha(u) & \text{if } u\in\text{dom}(\alpha)\\ 0 & \text{otherwise.}\end{cases}\qquad u\in Q.$$

Then clearly $\alpha=|S|$.

**Example 2.7** The function $\tau$ of Example 2.3 is not subsequential. Assume indeed that $\tau=|\boldsymbol{S}|$ for $\boldsymbol{S}$ as in the definition and set $K=\max\{|\rho(q)|:\rho(q)\neq 0, q\in Q\}$. Let $n$ be even. Then

$$|\boldsymbol{S}|(x^n)=(q_-*x^n)\rho(q_-\cdot x^n)=a^n$$

$$|\boldsymbol{S}|(x^{n+1})=(q_-*x^n)(q_-\cdot x^n*x)\rho(q_-\cdot x^{n+1})=b^{n+1}.$$

If $n>K$, then $w=q_-*x^n$ is not the empty word, and $w\in a^+\cap b^+$, which is impossible.

**Proposition 2.4** *A subsequential function is rational.*

Proof. Consider a subsequential transducer $\boldsymbol{S}=\langle X, Y, Q, q_-, \rho\rangle$ and define a morphism $\mu: X^*\to \mathrm{Rat}(Y^*)^{Q\times Q}$ by

$$\mu x_{p,q}=\begin{cases} p*x & \text{if } p\cdot x=q;\\ 0 & \text{otherwise.}\end{cases} \qquad (x\in X) \tag{2.7}$$

Then an obvious induction shows that (2.7) still holds if $x$ is replaced by a word $f\in X^*$. Next consider $\rho$ as a column $Q$-vector, and define a row vector $\lambda$ by

$$\lambda_q=\begin{cases}1 & \text{if } q=q_-;\\ 0 & \text{otherwise.}\end{cases}$$

Then: $\lambda\mu f\rho=\bigcup_{q\in Q}\mu f_{q_-q}\rho(q)=(q_-*f)\rho(q_-\cdot f)=|\boldsymbol{S}|(f).$

Thus $|\boldsymbol{S}|$ is rational by Proposition III.7.3. ■

Note that the matrices $\mu f$ of the preceding proof are row monomial, i.e. for each $p\in Q$, there is at most one $q\in Q$ such that $\mu f_{pq}\neq 0$. Thus the transductions $f\mapsto \mu f_{q_-q}$, $(q\in Q)$ have disjoint domains, and the same holds for the transductions $f\mapsto\mu f_{q_-q}\rho(q)$.

**Proposition 2.5** *Let $\alpha: X^*\to Y^*$ and $\beta: Y^*\to Z^*$ be subsequential functions. Then $\beta\circ\alpha: X^*\to Z^*$ is subsequential. If further $\alpha$ and $\beta$ are sequential (right sequential), then $\beta\circ\alpha$ is sequential (right sequential).*

Proof. Consider two subsequential transducers

$$\boldsymbol{S}=\langle X, Y, Q, q_-, \rho\rangle, \qquad \boldsymbol{T}=\langle Y, Z, P, p_-, \sigma\rangle$$

realizing $\alpha$ and $\beta$ respectively. Elements of the product $P\times Q$ are noted $[p, q]$ for easier checking. Define

$$\boldsymbol{T}\circ\boldsymbol{S}=\langle X, Z, P\times Q, [p_-, q_-], \omega\rangle$$

$$[p,q]\cdot x=[p\cdot(q*x), q\cdot x] \tag{2.8}$$

$$[p,q]*x=p*(q*x) \qquad p\in P,\quad q\in Q,\quad x\in X \tag{2.9}$$

$$\omega([p,q])=(p*\rho(q))\sigma(p\cdot\rho(q)). \tag{2.10}$$

We prove that (2.8) and (2.9) remain true if $x$ is replaced by a word $f \in X^*$. This is clear for $f=1$. Arguing by induction, consider $f=hx$, with $h \in X^*$, $x \in X$, and set

$$w = q * h, \qquad w' = q \cdot h * x.$$

Then $ww' = q * hx = q * f$ by (2.3). Next

$$\begin{aligned}[p, q] \cdot f &= [p \cdot (q * h), q \cdot h] \cdot x = [p \cdot w, q \cdot h] \cdot x \\ &= [p \cdot w \cdot (q \cdot h * x), q \cdot h \cdot x] \\ &= [p \cdot ww', q \cdot hx] = [p \cdot (q * f), q \cdot f].\end{aligned}$$

$$\begin{aligned}[p, q] * f &= ([p, q] * h)([p, q] \cdot h * x) = (p * (q * h))([p \cdot (q * h), q \cdot h] * x) \\ &= (p * w)([p \cdot w, q \cdot h] * x) \\ &= (p * w)(p \cdot w * (q \cdot h * x)) = (p * w)(p \cdot w * w') = p * ww' \\ &= p * (q * f).\end{aligned}$$

Finally

$$\begin{aligned}\omega([p_-, q_-] \cdot f) &= \omega([p_- \cdot (q_- * f), q_- \cdot f]) \\ &= (p_- \cdot (q_- * f) * \rho(q_- \cdot f))\sigma(p_- \cdot (q_- * f) \cdot \rho(q_- \cdot f)) \\ &= (p_- \cdot (q_- * f) * \rho(q_- \cdot f))\sigma(p_- \cdot \alpha(f)).\end{aligned}$$

Consequently

$$\begin{aligned}|\boldsymbol{T} \circ \boldsymbol{S}|(f) &= ([p_-, q_-] * f)\omega([p_-, q_-] \cdot f) \\ &= (p_- * (q_- * f))(p_- \cdot (q_- * f) * \rho(q_- \cdot f))\sigma(p_- \cdot \alpha(f)) \\ &= (p_- * (q_- * f)\rho(q_- \cdot f))\sigma(p_- \cdot \alpha(f)) \\ &= (p_- * \alpha(f))\sigma(p_- \cdot \alpha(f)) = \beta(\alpha(f)).\end{aligned}$$

Thus $|\boldsymbol{T} \circ \boldsymbol{S}| = \beta \circ \alpha$. If both $\alpha$ and $\beta$ are sequential, then $\rho$ and $\sigma$ can be assumed to have always the value 1. Then by (2.10), $\omega([p, q]) = (p * 1)\sigma(p \cdot 1) = 1$ and $\beta \circ \alpha$ is sequential. For right sequential functions, the result follows from Proposition 2.2. ■

If one of the two partial functions $\alpha$ and $\beta$ is left sequential and the other is right sequential, then $\beta \circ \alpha$ is a rational function. It is quite remarkable that conversely any rational function can be factorized as a composition of a left and a right sequential function. This will be proved in Section 5.

A sequential function preserves left factors. We show now that a subsequential function which preserves left factors is sequential.

**Proposition 2.6** *Let $\alpha : X^* \to Y^*$ be a partial function. Then $\alpha$ is sequential iff the following conditions hold:*

(i) *$\alpha$ is subsequential;*

(ii) *$\alpha$ preserves left factors.*

Proof. Clearly the conditions are necessary. Conversely, assume that $\alpha$ satisfies (i) and (ii), and consider a subsequential transducer $\boldsymbol{S} = \langle X, Y, Q, q_-, \rho\rangle$ realizing $\alpha$. We first put $\boldsymbol{S}$ into some standard form. Consider a state $q \in Q$. If $q$ is not accessible, i.e. if there is no word $u$ such that $q_- \cdot u = q$, then the state $q$ can clearly be deleted. Thus we may assume that all states are accessible. Next, if $\rho(q) = 0$, then $\alpha(u) = 0$ for all $u \in X^*$ such that $q_- \cdot u = q$, and further $\alpha(uv) = 0$ for all $v \in V^*$ since $\alpha$ preserves left factors. Consequently, if the next-state function and the output function are modified by setting $q' \cdot x = 0$, $q' * x = 0$ for all $(q', x)$ such that $q' \cdot x = q$, then the new subsequential transducer realizes the same function. Thus $q$ can be deleted (since it is no longer accessible), and consequently we can assume that $\rho(q) \neq 0$ for all $q \in Q$.

Next we claim that for all $q \in Q$, $x \in X$, there exists $\lambda(q, x) \in Y^* \cup \{\emptyset\}$ such that

$$\rho(q)\lambda(q, x) = (q * x)\rho(q \cdot x) \tag{2.11}$$

$$\lambda(q, x) \neq 0 \quad \Leftrightarrow \quad q * x \neq 0. \tag{2.12}$$

Indeed (2.12) follows from (2.11) since $\rho(q) \neq 0$ for all $q \in Q$. Next in order to prove (2.11), let $u$ be a word such that $q_- \cdot u = q$. If $q * x \neq 0$, then

$$\alpha(ux) = (q_- * u)(q * x)\rho(q \cdot x) \neq 0,$$

and since $\alpha$ preserves left factors,

$$\alpha(ux) = \alpha(u)g = (q_- * u)\rho(q)g$$

for some word $g \in Y^*$. Thus

$$\rho(q)g = (q * x)\rho(q \cdot x),$$

showing that $g$ is independent of $u$. We define

$$\lambda(q, x) = \begin{cases} g & \text{if } q * x \neq 0; \\ 0 & \text{otherwise.} \end{cases}$$

Then $\lambda$ has the same domain as the output function of $\boldsymbol{S}$.

Consider now the sequential transducer $\boldsymbol{L} = \langle X, Y, Q, q_-\rangle$ with the same next state function as $\boldsymbol{S}$, and with output function $\lambda$. We claim that $\alpha = |\boldsymbol{L}|$, i.e. that $\alpha(f) = \lambda(q_-, f)$ for $f \in X^*$. By (ii), this holds for $f = 1$. If $f = hx$, with $h \in X^*$, $x \in X$, then

$$\begin{aligned}\lambda(q_-, f) &= \lambda(q_-, h)\lambda(q_- \cdot h, x) = \alpha(h)\lambda(q_- \cdot h, x)\\ &= (q_- * h)\rho(q_- \cdot h)\lambda(q_- \cdot h, x)\\ &= (q_- * h)(q_- \cdot h * x)\rho(q_- \cdot hx) = \alpha(hx).\end{aligned}$$

■

Subsequential functions preserve left factors only if they are sequential. However, they satisfy a property which is closely related to the preservation of left factors. Consider indeed a subsequential transducer $\boldsymbol{S} = \langle X, Y, Q, q_-, \rho\rangle$, and let $\alpha = |\mathbf{S}|$. If $u_1, u_2 \in \operatorname{dom}(\alpha)$ are "near" in the sense that $u_1 = uv_1$, $u_2 = uv_2$ and $|v_1| + |v_2|$ is "small", then $\alpha(u_1)$, $\alpha(u_2)$ are also near, since

$$\alpha(uv_i) = (q_- * u)(q_- \cdot u * v_i)\rho(q_- \cdot uv_i) \qquad i = 1, 2,$$

and the length of the words $(q_- \cdot u * v_i)\rho(q_- \cdot uv_i)$ are bounded by some function of $|v_1|$ and $|v_2|$. This observation expresses some topological property. In order to explain it, we introduce some definitions.

For words $u, v \in X^*$, we define

$$u \wedge v = \text{the greatest common left factor of } u \text{ and } v.$$

More generally, if $A$ is a nonempty language, define

$$\bigwedge A = \text{the longest word which is a left factor of all words in } A.$$

The notation is justified by the following remark: Define a relation $\leqslant$ by: $u \leqslant v$ iff $u$ is a left factor of $v$. Then $\leqslant$ is a partial order, sometimes called the "prefix order". Since $u \leqslant v$ iff $u \wedge v = u$, $X^*$ is a semi-lattice, and $u \wedge v$ is the greatest lower bound of $u$ and $v$.

**Definition** The (left) distance of $u$ and $v$ is the number

$$\|u, v\| = |u| + |v| - 2\,|u \wedge v|.$$

Thus $\|u, v\|$ is the sum of the length of those words which remain when the greatest common left factor of $u$ and $v$ is erased. In order to verify that we get a distance, we first observe that $\|u, v\| = 0$ iff $|u| + |v| = 2\,|u \wedge v|$. Since $|u \wedge v| \leqslant |u|, |v|$, this is equivalent to $|u \wedge v| = |u| = |v|$, i.e. to $u = v$. Next we verify that

$$\|u, v\| \leqslant \|u, w\| + \|w, v\| \qquad u, v, w \in X^*.$$

A straightforward computation shows that this inequality is equivalent to

$$|u \wedge w| + |w \wedge v| \leqslant |w| + |u \wedge v|.$$

Since $u \wedge w$ and $v \wedge w$ are left factors of $w$, either $u \wedge w$ is a left factor of $v \wedge w$, thus of $u$ and of $v$, and $|u \wedge w| \leqslant |u \wedge v|$, or $v \wedge w$ is a left factor of $u$ and of $v$, and $|v \wedge w| \leqslant |u \wedge v|$. Both cases give the desired inequality.

From $|u \wedge v| \leqslant |u|, |v|$, we obtain immediately

$$\big|\,|u| - |v|\,\big| \leqslant \|u, v\| \qquad u, v \in X^*. \tag{2.13}$$

Another useful inequality is the following: if $A \subset X^*$, $A \neq \emptyset$, and $w = \bigwedge A$, then

$$\|w, u\| \leqslant \max_{f, g \in A} \|f, g\| \qquad u \in A. \tag{2.14}$$

Indeed, for any $u \in A$, there is some $v \in A$ such that $u \wedge v = w$, (since otherwise $w$ would be a proper left factor of all $u' \wedge v$, $(v \in A)$ for some $u' \in A$, thus of all $v \in A$). Consequently $\|w, u\| = \|u \wedge v, u\| \leqslant \|v, u\| \leqslant \max_{f, g \in A} \|f, g\|$.

**Definition** A partial function $\alpha: X^* \to Y^*$ has bounded variation iff for all $k \geq 0$, there exists $K \geq 0$ such that

$$u, v \in \mathrm{dom}(\alpha), \|u, v\| \leq k \Rightarrow \|\alpha(u), \alpha(v)\| \leq K.$$

**Example 2.8** A subsequential function has bounded variation. Let indeed $\alpha = |\boldsymbol{S}|$, with $\boldsymbol{S} = \langle X, Y, Q, q_-, \rho\rangle$, and set

$$M = \max\{|q * x| : q \in Q,\ x \in X, q * x \neq 0\},$$

$$N = \max\{|\rho(q)| : q \in Q,\ \rho(q) \neq 0\}.$$

If $uv \in \mathrm{dom}(\alpha)$, then $\alpha(uv) = (q_- * u)(q_- \cdot u * v)\rho(q_- \cdot uv)$. Thus $|\alpha(uv)| \leq |q_- * u| + |v| \cdot M + N$.

Let $k \geq 0$, and define $K = k \cdot M + 2N$. If $u_1, u_2 \in \mathrm{dom}(\alpha)$ and $\|u_1, u_2\| \leq k$, then $u_1 = uv_1$, $u_2 = uv_2$ with $|v_1| + |v_2| \leq k$. Consequently $\alpha(uv_1) = (q_- * u)w_1$, $\alpha(uv_2) = (q_- * u)w_2$ and

$$\|\alpha(uv_1), \alpha(uv_2)\| \leq |w_1| + |w_2| \leq (|v_1| + |v_2|)M + 2N \leq K.$$

Note that for $M' = \max(M, 2N)$, we have the stronger inequality:

$$\|\alpha(u_1), \alpha(u_2)\| \leq M'(1 + \|u_1, u_2\|).$$

The following result gives a characterization of subsequential functions.

**Theorem 2.7** (Choffrut [1978]) *Let $\alpha: X^* \to Y^*$ be a partial function. Then $\alpha$ is subsequential iff*
(i) *$\alpha$ has bounded variation;*
(ii) *for all $L \in \mathrm{Rat}(Y^*)$, $\alpha^{-1}(L) \in \mathrm{Rat}(X^*)$.*

This theorem is an extension of a characterization of sequential functions:

**Theorem 2.8** (Ginsburg, Rose [1966]) *Let $\alpha: X^* \to Y^*$ be a partial function. Then $\alpha$ is sequential iff*
(i) *$\alpha$ preserves left factors;*
(ii) *there exists an integer $M$ such that, for all $u \in X^*$, $x \in X$:*

$$ux \in \mathrm{dom}(\alpha),\ \alpha(ux) = \alpha(u)g \quad \textit{imply} \quad |g| \leq M;$$

(iii) *for all rational languages $L \subset Y^*$, $\alpha^{-1}(L)$ is a rational language.*

Proof. In order to deduce Theorem 2.8 from Theorem 2.7, it suffices to show that $\alpha$ has bounded variation. The desired conclusion then follows by Proposition 2.6. Let $k \geq 0$, and let $uv_1, uv_2 \in \mathrm{dom}(\alpha)$ be such that $v_1 \wedge v_2 = 1$, and $\|uv_1, uv_2\| = |v_1| + |v_2| \leq k$. Then $\alpha(uv_1) = \alpha(u)g_1$, $\alpha(uv_2) = \alpha(u)g_2$ and, by (ii), $|g_1| < |v_1|\, M$, $|g_2| < |v_2|\, M$. Consequently, $\|\alpha(uv_1), \alpha(uv_2)\| \leq |g_1| + |g_2| \leq k \cdot M$. ■

Proof of Theorem 2.7. The proof is an adaptation of the proof of the Ginsburg–Rose Theorem, as given for instance in Ginsburg [1966] or in

Eilenberg [1974]. Since a subsequential function verifies (i) and (ii), we have to prove that these conditions are sufficient.

The proof is in four parts. We first associate to $\alpha$ a finite set $R$ of partial functions from $X^*$ into $Y^*$. We then prove that a certain family $A_r$, $(r \in R)$ of subsets of $X^*$ is composed of regular languages. This enables us to construct a machine realizing $\alpha$, which works like a subsequential transducer excepted that the output function has its values in the free group $Y^{(*)}$ generated by $Y$. The last step consists in replacing this transducer by a sequential transducer satisfying our definition. We use the following abbreviation:

If $Z$ is an alphabet and $n \geq 0$ is an integer, then

$$Z^{(n)} = 1 \cup Z \cup \cdots \cup Z^n = Z^* \backslash Z^n Z^+.$$

Let $\alpha : X^* \to Y^*$ satisfy (i) and (ii). Then $\mathrm{dom}(\alpha) = \alpha^{-1}(Y^*)$ is a regular language by (ii). Let $N$ be the number of states of a finite automaton recognizing $\mathrm{dom}(\alpha)$. Then we note, for later reference, that

$$uX^* \cap \mathrm{dom}(\alpha) \neq \emptyset \quad \text{iff} \quad uX^{(N-1)} \cap \mathrm{dom}(\alpha) \neq \emptyset. \tag{2.15}$$

Indeed, if $uv \in \mathrm{dom}(\alpha)$ and $|v| \geq N$, then there exists, by the Iteration Lemma for Regular Languages, a word $v'$ such that $|v'| < |v|$ and $uv' \in \mathrm{dom}(\alpha)$.

For $u \in X^*$ define

$$J(u) = \{v \in X^{(N)} \mid \alpha(uv) \neq 0\},$$

and define a partial function $\beta : X^* \to Y^*$ by

$$\beta(u) = \begin{cases} 0 & \text{if } J(u) = \emptyset; \\ \bigwedge\{\alpha(uv) : v \in J(u)\} & \text{otherwise.} \end{cases}$$

Thus $\beta(u) \neq 0$ iff $J(u) \neq \emptyset$ iff $uX^{(N)} \cap \mathrm{dom}(\alpha) \neq \emptyset$. In this case, there exists, for each $v \in J(u)$, a word $r_u(v) \in Y^*$ such that

$$\alpha(uv) = \beta(u) r_u(v) \tag{2.16}$$

and further there are $v_1, v_2 \in J(u)$ such that

$$r_u(v_1) \wedge r_u(v_2) = 1. \tag{2.17}$$

(Note that (2.17) holds even if $J(u)$ is a singleton $v$, since then $r_u(v) = 1$.) We complete the definition by setting $r_u(v) = 0$ for $v \in X^{(N)} \backslash J(u)$. Thus for any $u \in X^*$, there is a partial function $r_u$ from $X^{(N)}$ into $Y^*$ satisfying (2.16), with $\mathrm{dom}(r_u) = J(u)$. Further $\alpha(u) \neq 0$ iff $1 \in J(u)$ iff $r_u(1) \neq 0$.

a) We prove that there exists an integer $M$ such that $\max\{|r_u(v)| : v \in \mathrm{dom}(r_u)\} \leq M$ for all $u \in X^*$. For this, consider $v_1, v_2 \in J(u)$. Then $\|uv_1, uv_2\| \leq |v_1| + |v_2| \leq 2N$. In view of condition (i), there exists an integer $M$ such that

$$\|\alpha(uv_1), \alpha(uv_2)\| \leq M \qquad v_1, v_2 \in J(u).$$

Consequently by (2.14),

$$|r_u(v)|=\|\beta(u),\alpha(uv)\|\leq \max_{v_1,v_2\in J(u)} \|\alpha(uv_1),\alpha(uv_2)\|\leq M$$

for all $v\in J(u)$. Thus each $r_u$ is a partial function $X^{(N)}\to Y^{(M)}$ and the set

$$R=\{r_u : u\in X^*\},$$

as a subset of the finite set of all partial functions from $X^{(N)}$ into $Y^{(M)}$ is itself finite. We note $\boldsymbol{O}$ the partial function $X^{(N)}\to Y^{(M)}$ with empty domain.
b) For $r\in R$, define $A_r=\{u\in X^* \mid r_u=r\}$. We claim: the languages $A_r$ are rational. To prove this, define for $i=0,\ldots,2M$ and $h\in Y^M$:

$$D_{i,h}=\{g\in Y^* : |g|\equiv i(\text{mod } 2M+1) \text{ and } (g\in Y^*h \text{ or } h\in Y^*g)\}.$$

(Note that there is at most one $g\in D_{i,h}$ such that $h\in Y^*g$.) Clearly the languages $D_{i,h}$ are rational. Consider a fixed $r\in R$ and define

$$L_{v,i,h}=[\alpha^{-1}(D_{i,h}r(v))]v^{-1} \qquad \text{for } v\in\text{dom}(r)$$
$$K_v=[X^*\backslash\text{dom}(\alpha)]v^{-1} \qquad \text{for } v\in X^{(N)}\backslash\text{dom}(r).$$

By (ii), these languages are rational. Set

$$K=\bigcap_{v\in X^{(N)}\backslash\text{dom}(r)} K_v;\quad L_{i,h}=\bigcap_{v\in\text{dom}(r)} L_{v,i,h};\quad B_r=K\cap\left(\bigcup_{i=0}^{2M}\bigcup_{h\in Y^M} L_{i,h}\right).$$

Then $B_r$ is a rational language. We show that $A_r=B_r$, which proves the claim. Consider first $u\in B_r$. We must show that $r_u=r$. There exist $i\in\{0,\ldots,2M\}$, $h\in Y^M$ such that $u\in K\cap L_{i,h}$. Thus if $v\in\text{dom}(r)$, $\alpha(uv)\in D_{i,h}r(v)$, and if $v\in X^{(N)}\backslash\text{dom}(r)$, then $\alpha(uv)=0$. Thus $\text{dom}(r)=\text{dom}(r_u)$. If $r=\boldsymbol{O}$, then $r=r_u$. Next if dom$(r)$ is a singleton $v$, then since $r=r_{u'}$ for some $u'\in X^*$, $r(v)=1$ by the remark following (2.17), and $r=r_u$. Thus assume Card(dom$(r)$)$\geq 2$. Then for each $v\in\text{dom}(r)$, there is a word $g_v\in D_{i,h}$ such that

$$\alpha(uv)=\beta(u)r_u(v)=g_v r(v), \tag{2.18}$$

and it suffices to show that $g_v=\beta(u)$ for $v\in\text{dom}(r)$. Let $v$, $v'\in\text{dom}(r)$. By (2.13), we have

$$|g_v|-|g_{v'}|=|\alpha(uv)|-|\alpha(uv')|+|r(v')|-|r(v)|\leq\|\alpha(uv),\alpha(uv')\|+M\leq 2M,$$

thus $||g_v|-|g_{v'}||\leq 2M$. Since further $|g_v|\equiv|g_{v'}|\ (\text{mod } 2M+1)$, it follows that $|g_v|=|g_{v'}|$ for all $v$, $v'\in\text{dom}(r)$. Let $n$ be the common length of the words $g_v\,(v\in\text{dom}(r))$.
Since $r=r_{u'}$ for some $u'\in X^*$, there exist, by (2.17), words $v_1$, $v_2\in\text{dom}(r)$ such that $r(v_1)\wedge r(v_2)=1$. Consider (2.18) for these words. This shows that $|\beta(u)|\leq n$. Next let $v_1, v_2$ be two words in dom$(r)$ such that $r_u(v_1)\wedge r_u(v_2)=1$. By

(2.18) there are two words $g_1$, $g_2$ of the same length such that

$$r_u(v_1)=g_1r(v_1), \qquad r_u(v_2)=g_2r(v_2).$$

Therefore $|g_1|=|g_2|\leqslant M$, and $g_1$, $g_2$ are both right factors of $h$. Thus $g_1=g_2$. Since $r_u(v_1)\wedge r_u(v_2)=1$, $g_1=g_2=1$ and $|\beta(u)|=n$. Thus $u\in A_r$.

Conversely, let $u\in A_r$. If $r=r_u=\boldsymbol{O}$, then $L_{i,h}=X^*$ for all $i$ and all $h$, and clearly $u\in K$. Thus $u\in B_r$ in this case. Thus assume $\mathrm{dom}(r)\neq\emptyset$, and let $i$ be the integer such that $0\leqslant i\leqslant 2M$ and $|\beta(u)|\equiv i(\mathrm{mod}\,2M+1)$. Next let $h$ be either the unique right factor of length $M$ of $\beta(u)$ if $|\beta(u)|\geqslant M$, or any word in $Y^M\cap Y^*\beta(u)$ otherwise. Then $\alpha(uv)\in D_{i,h}r(v)$ for all $v\in\mathrm{dom}(r)$, and consequently $u\in L_{i,h}$. Since clearly $u\in K$, we have $u\in B_r$.

c) Let $\boldsymbol{S}'=\langle X, Q', q_-\rangle$ be an accessible semiautomaton recognizing simultaneously all $A_r (r\in R)$ (for the construction, see Section I.4). Then for each $r\in R$, there is a subset $Q_r$ of $Q'$ such that

$$A_r=|\boldsymbol{S}'(Q_r)|=\{u\in X^* \mid q_-\cdot u\in Q_r\}.$$

Clearly the $Q_r$ are pairwise disjoint, and $Q'=\bigcup_{r\in R} Q_r$. Next set

$$R_+=\{r\in R \mid r(1)\neq 0\}, \qquad Q_+=\bigcup_{r\in R_+} Q_r.$$

Then $\boldsymbol{O}\notin R_+$ and $\mathrm{dom}(\alpha)=|\boldsymbol{S}'(Q_+)|$. Observe that

$$u\in A_{\boldsymbol{O}} \quad\Leftrightarrow\quad q_-\cdot u\in Q_{\boldsymbol{O}} \quad\Leftrightarrow\quad \beta(u)=0 \tag{2.19}$$

and that further, in view of (2.15)

$$u\in A_{\boldsymbol{O}} \quad\Leftrightarrow\quad uX^*\cap\mathrm{dom}(\alpha)=\emptyset \quad\Leftrightarrow\quad uX^{(N-1)}\cap\mathrm{dom}(\alpha)=\emptyset. \tag{2.20}$$

Thus $u\in A_{\boldsymbol{O}}$ implies $uX^*\subset A_{\boldsymbol{O}}$ (in other terms, $A_{\boldsymbol{O}}$ is a right ideal). Thus $q\in Q_{\boldsymbol{O}}$ implies $q\cdot f\in Q_{\boldsymbol{O}}$ for all $f\in X^*$, and further $q_-\notin Q_{\boldsymbol{O}}$, since otherwise $\alpha=0$ and there is nothing to prove.
Define $\boldsymbol{S}=\langle X, Q, q_-\rangle$ by setting $Q=Q'\backslash Q_{\boldsymbol{O}}$, and by defining the next state function of $\boldsymbol{S}$ to be the partial function obtained by restriction of the next state function of $\boldsymbol{S}'$ to $Q$. Thus $q\cdot x=0$ in $\boldsymbol{S}$ iff $q\cdot x\in Q_{\boldsymbol{O}}$ in $\boldsymbol{S}'$. Then for $q\in Q_r\subset Q$, $x\in X$:

$$q\cdot x=0 \quad\text{iff}\quad \mathrm{dom}(r)\cap xX^{(N-1)}=\emptyset. \tag{2.21}$$

Indeed, let $u\in X^*$ be such that $q_-\cdot u=q$. Then by (2.20), $q\cdot x=0$ iff $uxX^{(N-1)}\cap\mathrm{dom}(\alpha)=\emptyset$, and this holds iff $\mathrm{dom}(r)\cap xX^{(N-1)}=\emptyset$. After these preliminaries, we now construct a subsequential transducer realizing $\alpha$, but with output function into the free group $Y^{(*)}$ generated by $Y$. (Since each word of $Y^*$ is reduced, $Y^*$ can be identified with its image in $Y^{(*)}$, and hence $Y^*$ can be viewed as a submonoid of $Y^{(*)}$. In particular, if $u$, $v\in Y^*$ then $u^{-1}v$ is always a welldefined element of $Y^{(*)}$, and $u^{-1}v$ is in $Y^*$ iff $u$ is a left factor of $v$. See also II.3.) Consider a new state $q_0$, and extend the next state function of $\boldsymbol{S}$ by setting

$$q_0 \cdot x = q_- \cdot x \qquad (x \in X)$$

Next define

$$q_0 * x = \beta(x) \qquad (x \in X),$$

and for $q \in Q_r$, $x \in X$,

$$q * x = \begin{cases} 0 & \text{if} \quad q \cdot x = 0 \\ r(xv)r'(v)^{-1} & \text{if} \quad q \cdot x \in Q_{r'}, \qquad v \in \mathrm{dom}(r) \cap xX^{(N-1)}. \end{cases}$$

First we verify that the definition is correct. If $q \cdot x = q' \in Q_{r'}$, then $\mathrm{dom}(r) \cap xX^{(N-1)} \neq \emptyset$ by (2.20). Thus if $u \in X^*$ is such that $q_- \cdot u = q$, then for $v \in \mathrm{dom}(r) \cap xX^{(N-1)}$,

$$0 \neq \alpha(uxv) = \beta(u)r(xv) = \beta(ux)r'(v), \tag{2.22}$$

showing first that $r'(v) \neq 0$, and next that $r(xv)r'(v)^{-1} = \beta(u)^{-1}\beta(ux)$ is an element of $Y^{(*)}$ which is independent of the choice of $v$ in $\mathrm{dom}(r) \cap xX^{(N-1)}$. Thus the next state function and the output function have the same domain. We claim:

$$q_0 * u = \beta(u) \qquad u \in X^+. \tag{2.23}$$

This holds for $|u| = 1$. Arguing by induction, consider $u \in X^+$, $x \in X$. If $\beta(u) = 0$, then $u \in A_{\boldsymbol{O}}$, and $ux \in A_{\boldsymbol{O}}$. Consequently $q_0 * ux = 0 = \beta(ux)$ by (2.19). If $\beta(u) \neq 0$, then $q_0 \cdot u = q$ for some $q \in Q_r$, $(r \in R \backslash \boldsymbol{O})$. Then by (2.22),

$$q_0 * ux = \beta(u)(q * x) = \begin{cases} 0 & \text{if} \quad q \cdot x = 0 \\ \beta(u)(\beta(u))^{-1}\beta(ux) & \text{otherwise.} \end{cases}$$

Since by (2.19), $q \cdot x = 0$ iff $\beta(ux) = 0$, (2.23) is proved. Finally define $\rho : q_0 \cup Q \to Y^*$ by $\rho(q_0) = \alpha(1)$, and

$$\rho(q) = \begin{cases} 0 & \text{if} \quad q \notin Q_+ \\ r(1) & \text{if} \quad q \in Q_r \quad \text{and} \quad r \in R_+. \end{cases} \qquad q \in Q$$

Then $(q_0 * u)\rho(q_0 \cdot u) = \alpha(u)$ for all $u \in X^*$.

d) It remains to transform the above transducer into a subsequential transducer which agrees with our definition. For $q \in Q \cup q_0$, set

$$U_q = \{u \in X^* \mid q_0 \cdot u = q\}$$

$$\sigma(q) = \text{the longest right factor common to the words } q_0 * u, \ (u \in U_q).$$

Then for all $u \in U_q$, there is a word $\theta(u)$ such that $\beta(u) = q_0 * u = \theta(u)\sigma(q)$. Extend the definition by setting $\theta(u) = 0$ whenever $q_0 \cdot u = 0$. Then

$\theta: X^* \to Y^*$ is a partial function with same domain as $\beta$. Let $q \in Q \cup q_0$, and let $u \in U_q$, and suppose there is a letter $x$ such that $q \cdot x \neq 0$. Then

$$q_0 * ux = \theta(u)\sigma(q)(q * x) = \theta(ux)\sigma(q \cdot x).$$

Since $q_0 * u$, $q_0 * ux \in Y^*$ and $q * x \in Y^{(*)}$, there are words $f$, $g$, $h \in Y^*$ such that $\theta(u)\sigma(q) = fh$, $h^{-1}g = q * x$. The last relation shows that $h$, $g$ are independent of the choice of $u$ in $U_q$. Thus $h$ is a common right factor to all $q_0 * u (u \in U_q)$. Consequently $\sigma(q) = th$ for some $t \in Y^*$, and $\sigma(q)(q * x) = w \in Y^*$ with $w = thh^{-1}g = tg$. Thus

$$\theta(u)w = \theta(ux)\sigma(q \cdot x) \qquad u \in U_q.$$

Assume $|w| < |\sigma(q \cdot x)|$. Then the words $\theta(u)$ have a nonempty common right factor, in contradiction with the definition of $\sigma(q)$. Thus $|w| \geqslant |\sigma(q \cdot x)|$, and there is a word $\lambda(q, x) \in Y^*$ such that

$$\theta(u)\lambda(q, x) = \theta(ux) \qquad (u \in U_q). \tag{2.24}$$

Define $\lambda(q, x) = 0$ whenever $q \cdot x = 0$, and consider $\boldsymbol{S}$ equipped with the output function $\lambda$. $\lambda$ and the next state function have the same domain, further $\lambda(q_0, 1) = 1$ and by (2.24), $\theta(u) = \lambda(q_0, u)$ for all $u \in X^*$. Define $\tau: Q \cup q_0 \to Y^*$ by $\tau(q) = \sigma(q)\rho(q)$. Then

$$\alpha(u) = \beta(u)\rho(q_0 \cdot u) = \lambda(q_0, u)\tau(q_0 \cdot u) \qquad u \in X^*,$$

and $\alpha$ is realized by the subsequential transducer $\boldsymbol{S} = \langle X, Y, Q \cup q_0, q_0, \tau \rangle$ with output function $\lambda$. ■

Remark. Consider a partial function $\alpha: X^* \to Y^*$ realized by a "generalized" sequential transducer defined as a sequential transducer, but with an output function from $X^*$ into the free group $Y^{(*)}$. Such a transducer can erase a right factor of an already computed output word, and can replace it by another word. The last part of the preceding proof shows that such a transducer can be simulated by a subsequential transducer working without erasing, i.e. $\alpha$ is subsequential.

## Exercises

**2.1** Let $\alpha: X^* \to Y^*$ be a partial function, and let # be a new symbol. Show that $\alpha$ is subsequential iff there exists a sequential transduction $\beta: (X \cup \#)^* \to Y^*$ such that $\alpha(u) = \beta(u\#)$ for all $u \in X^*$.

**2.2** Let $\alpha_1$, $\alpha_2: X^* \to Y^*$ be sequential functions. Show that if $\alpha_1 \cup \alpha_2$ is a partial function, then $\alpha_1 \cup \alpha_2$ is sequential. Show that $\alpha_1 \cup \alpha_2$ is not necessarily subsequential, if $\alpha_1$, $\alpha_2$ are subsequential.

**2.3** Let $\alpha: X^* \to Y^*$ be a subsequential function, and let $R \subset X^*$ be a rational language. Show that the restriction $\alpha|_R$ is subsequential.

## IV.3 The Cross-Section Theorem

The following theorem is due to Eilenberg [1974]. It will be used in the next section in order to construct special representations for rational functions.

**Theorem 3.1** (Cross-Section Theorem) *Let $\alpha : X^* \to Y^*$ be a morphism. For any rational language $A \subset X^*$ there exists a rational language $B \subset A$ such that $\alpha$ maps bijectively $B$ onto $\alpha(A)$.*

Set $C = \alpha(A)$. The theorem asserts that in each class $A \cap \alpha^{-1}(f)$, $(f \in C)$ a unique word $u_f$ can be chosen in such a way that the language $B = \{u_f \mid f \in C\}$ is rational. The language $B$ is called a cross-section of $\alpha$ on $A$. We shall see that the proof is effective. Thus given $\alpha$ and $A$, a cross-section of $\alpha$ on $A$ can be constructed effectively.

Proof. We shall factorize $\alpha$ in morphisms of special form. Therefore we first verify that if $\beta : Y^* \to Z^*$ is a second morphism, and if the conclusion holds for $\alpha$ and $\beta$, it also holds for $\beta \circ \alpha : X^* \to Z^*$. Indeed, let $A \subset X^*$ be rational, and let $B$ be a rational cross-section of $\alpha$ on $A$. Set $C = \alpha(A) = \alpha(B)$, and let $D \subset Y^*$ be a rational cross-section of $\beta$ on $C$. Define $E = B \cap \alpha^{-1}(D)$. Then $E$ is rational, $\alpha$ is injective on $E$, and $\alpha(E) = D$. Since $\beta$ is injective on $D$, it follows that $\beta \circ \alpha$ is injective on $E$. Further $\beta \circ \alpha(E) = \beta(D) = \beta(C) = \beta \circ \alpha(A)$. Thus $E$ is a cross-section of $\beta \circ \alpha$ on $A$.

Next note that if $\alpha$ is injective, the conclusion holds trivially by taking $B = A$. Since any morphism $\alpha : X^* \to Y^*$ can be factorized into $\alpha = \beta \circ \gamma$, where $\gamma : X^* \to Z^*$ is injective and $\beta : Z^* \to Y^*$ is alphabetic, it suffices to consider the case where $\alpha$ is alphabetic. Further, any alphabetic morphism factorizes into projections and strictly alphabetic morphisms. Thus if suffices to consider the following two cases:

$$X = \{x_1, \ldots, x_n\} \qquad Y = \{x_1, \ldots, x_{n-1}\} \qquad (n \geqslant 2)$$
$$\alpha(x_i) = x_i \qquad i = 1, \ldots, n-1$$
$$\alpha(x_n) = x_{n-1} \quad \text{or} \quad \alpha(x_n) = 1.$$

Define the lexicographical order on $X^*$ by setting $u < v$ if either of the following cases holds

$$v = uw \qquad \text{with} \quad w \neq 1$$
$$u = fx_i g, \quad v = fx_j g' \qquad \text{with} \quad i < j.$$

Next define a transduction $\tau : X^* \to X^*$ by setting

$$\tau(u) = \{v \mid v > u \text{ and } \alpha(v) = \alpha(u)\},$$

and set

$$B = A \backslash \tau(A).$$

Thus for each $u \in A$, the smallest element of $\alpha^{-1}\alpha(u) \cap A$ is selected, and $B$ is the set of all elements so selected. thus $B$ is a cross-section of $\alpha$ on $A$.
To prove that $B$ is rational, it suffices to show that the transduction $\tau$ is rational. This will be done by constructing transducers realizing $\tau$. If $\alpha(x_n) = x_{n-1}$, we define a transducer by Fig. IV.5 with $V = \{x_i/x_i : i = 1, \ldots, n\}$. If

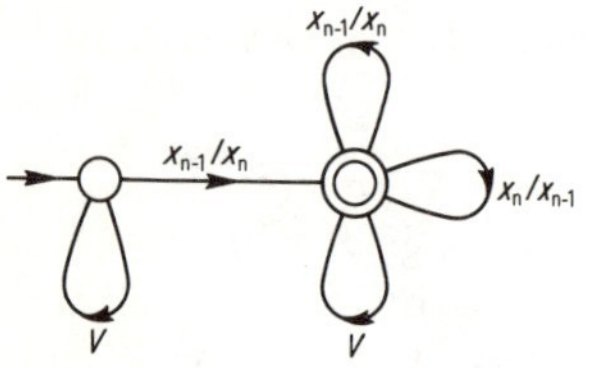


Fig. IV.5

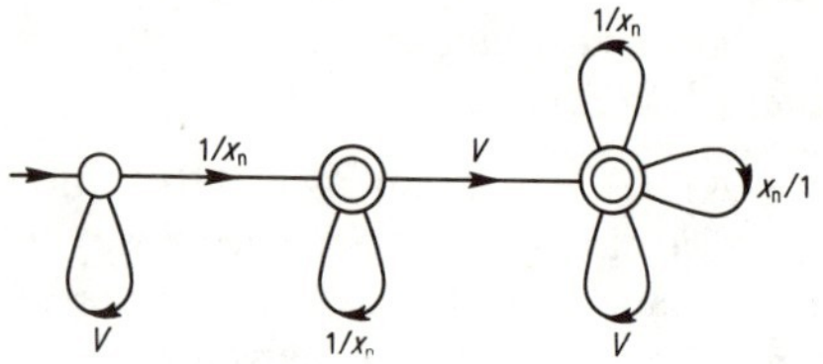


Fig. IV.6

$\alpha(x_n) = 1$, then we consider the transducer in Fig. IV.6. Then it is easily seen that these transducers realize $\tau$. ∎

Note that any morphism can be factorized into an injective morphism followed by a projection (Exercise I.3.3). Thus in the above proof, the case $\alpha(x_n) = x_{n-1}$ can be skipped. We conserved it since in the construction of unambiguous representations for rational functions, precisely that case appears, and it is easier to handle directly than through an additional decomposition.
We emphasize the fact that the cross-section $B$ can be obtained effectively from $A$. Assume $\alpha(x_n) = x_{n-1}$. Then we can proceed as follows. Let $y$ and $z$ be new letters, define $Z = X \cup \{y, z\}$, and let $\varphi, \psi : Z^* \to X^*$ be the morphisms

$$\begin{aligned} &\varphi(x_i) = \psi(x_i) = x_i && i = 1, \ldots, n, \\ &\varphi(y) = \psi(z) = x_{n-1}, && \varphi(z) = \psi(y) = x_n. \end{aligned} \tag{3.1}$$

Then for $A \subset X^*$,

$$\tau(A) = \psi(\varphi^{-1}(A) \cap K) \tag{3.2}$$

where $K$ is the rational language over $Z$ recognized by the automaton in Fig. IV.7. Thus if $A$ is given by a finite automaton, finite automata recognizing $\varphi^{-1}(A)$, $\varphi^{-1}(A) \cap K$, $\tau(A)$ and $A \backslash \tau(A)$ can be effectively constructed.

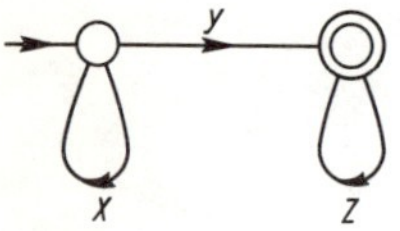


Fig. IV.7

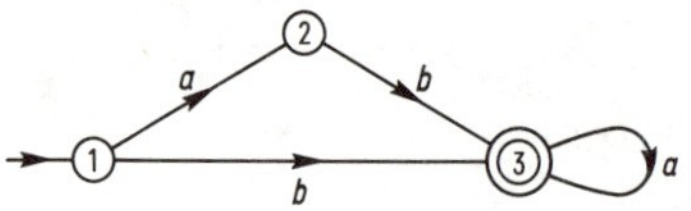


Fig. IV.8

**Example 3.1** Let $X = \{a, b\}$ and let $\alpha : X^* \to a^*$ be the morphism given by $\alpha(a) = \alpha(b) = a$. Further, let $A \subset X^*$ be given by Fig. IV.8. The lexicographical order is given here by $a<b$. Then

$$\tau(A)=bX^+$$

and the desired cross-section is

$$B=A\backslash\tau(A)=aba^*\cup b.$$

**Example 3.2** Let $X=\{a, b, c, d\}$, let $\alpha: X^*\to a^*$ be the morphism given by $\alpha(X)=a$, and let $A\subset X^*$ be recognized by Fig. IV.9 Thus

$$A=[(bdb\cup bc\cup cb)a]^*.$$

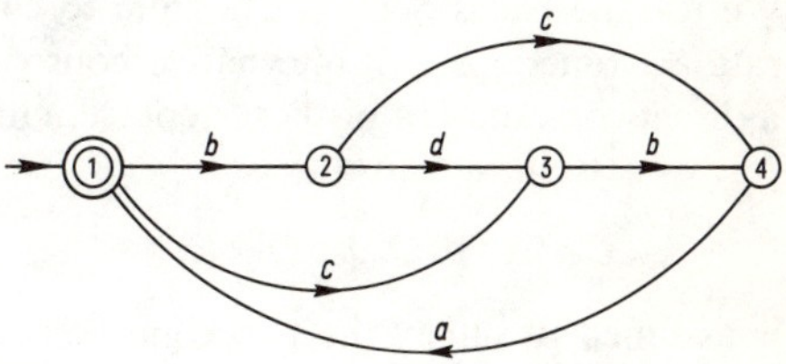


Fig. IV.9

Define a factorization $\alpha=\alpha_3\circ\alpha_2\circ\alpha_1$:

$$X^*\xrightarrow{\alpha_1}\{a, b, c\}^*\xrightarrow{\alpha_2}\{a, b\}^*\xrightarrow{\alpha_3}a^*$$

by $\alpha_1(d)=c$, $\alpha_2(c)=b$, $\alpha_3(b)=a$, the other letters being unchanged. We compute a cross-section of $\alpha$ on $A$.

First $\quad A_1=\alpha_1(A)=[(bcb\cup bc\cup cb)a]^*$,

and $\alpha_1$ is injective on $A$. Next

$$A_2=\alpha_2(A_1)=[(b^3\cup b^2)a]^*.$$

We have $b<c$, and

$$A_1'=[(bcb\cup bc)a]^*\subset A_1$$

is a rational cross-section of $\alpha_2$ on $A_1$. Then

$$A_3=\alpha_2(A_2)=[a^4\cup a^3]^*.$$

Since $a<b$, the construction of the proof yields

$$A_2'=(b^2a)^*(1\cup b^3a\cup(b^3a)^2)$$

as rational cross-section of $\alpha_3$ on $A_2$. By backward computation

$$A_1''=A_1'\cap\alpha_2^{-1}(A_2')=(bca)^*(1\cup bcba\cup(bcba)^2)$$

is a rational cross-section of $\alpha_3\circ\alpha_2$ on $A_1$, and

$$B=A\cap\alpha_1^{-1}(A_1'')=(bca)^*(1\cup bdba\cup(bdba)^2)$$

is a rational cross-section of $\alpha$ on $A$.

**Exercise**

**3.1** Replace $\alpha$ in Example 3.1 by: $\alpha(a)=\alpha(b)=b$, and compute a cross-section of $\alpha$ on $A$.

## IV.4 Unambiguous Transducers

We use the Cross-Section Theorem to construct particular representations for rational functions. An alternative construction is also presented which allows a direct computation of these representations.
In this section, a transducer

$$\boldsymbol{T}=\langle X, Y, Q, q_-, Q_+, E\rangle$$

is assumed to satisfy the two conditions

$$E\subset Q\times X\times Y^*\times Q, \tag{4.1}$$

$$(p, x, h, q), (p, x, h', q)\in E \Rightarrow h=h'. \tag{4.2}$$

**Definition** The transducer $\boldsymbol{T}$ is called unambiguous if any word $f\in X^*$ is the input label of at most one successful path $e$ in $\boldsymbol{T}$.

Let $\tau$ be the transduction realized by $\boldsymbol{T}$. If $\boldsymbol{T}$ is unambiguous and if $f\in \mathrm{dom}(\tau)$, then there exists a successful path $e$ in $\boldsymbol{T}$ with input label $f$. Let $g$ be the output label of $e$. Then $\tau(f) = g$. Thus

**Proposition 4.1** *The transduction realized by an unambiguous transducer is a partial function.* ■

Conversely, we have

**Theorem 4.2** (Eilenberg [1974]) *Let* $\tau: X^*\to Y^*$ *be a rational function* (*with* $\tau(1)=0$ *or* $\tau(1)=1$). *Then there exists an unambiguous transducer realizing* $\tau$.

Proof. By Corollary III.7.2, there are an alphabet $Z$, a strictly alphabetic morphism $\alpha: Z^*\to X^*$, a rational substitution $\sigma: Z^*\to Y^*$, and a regular language $K\subset Z^*$ such that

$$\tau(f)=\sigma(\alpha^{-1}(f)\cap K) \qquad f\in X^*.$$

Clearly we may assume $Z$ minimal, i.e. each letter $z\in Z$ has at least one occurrence in a word in $\alpha^{-1}(X^*)\cap K$. Then $\sigma$ is a morphism, since $\tau$ is a partial function.
Since $\mathrm{dom}(\tau)=\alpha(K)$, there exists, by the Cross-Section Theorem, a rational language $R\subset K$ that maps bijectively $R$ onto $\mathrm{dom}(\tau)$. Let

$$\mathbf{A}=\langle Z, Q, q_-, Q_+\rangle$$

be a finite automaton recognizing $R$, and define

$$\boldsymbol{T} = \langle X, Y, Q, q_-, Q_+, E\rangle$$
$$E = \{(q, \alpha z, \sigma z, q \cdot z) \mid q \in Q, z \in Z\}.$$

Note that $\boldsymbol{T}$ satisfies (4.1) and (4.2). Indeed, since $\alpha$ is strictly alphabetic, $\alpha z \in X$ for $z \in Z$. Next consider two successful paths

$$e = (q_-, u_1, v_1, q_1) \cdots (q_{n-1}, u_n, v_n, q)$$
$$e' = (q_-, u_1', v_1', q_1') \cdots (q_{m-1}, u_m', v_m', q')$$

and let $z_i, z_j' \in Z$ such that $\alpha z_i = u_i$, $(1 \leq i \leq n)$, $\alpha z_j' = u_j'$, $(1 \leq j \leq m)$. Assume that $e$ and $e'$ have the same input label $f = \alpha h = \alpha h'$ with $h = z_1 \cdots z_n$, $h' = z_1' \cdots z_m'$. Since $h, h' \in R$ and $\alpha$ is injective on $R$, it follows that $h = h'$ and $e = e'$. This shows that $\boldsymbol{T}$ is unambiguous. ∎

**Corollary 4.3** *Let $\tau: X^* \to Y^*$ be a rational function. Then there exists a normalized unambiguous transducer realizing $\tau$.*

Proof. Let $\boldsymbol{T} = \langle X, Y, Q, q_-, Q_+, E\rangle$ be an unambiguous transducer realizing $\tau$. Add two new states $q_0$, $q_1$ to $Q$, and the transitions

$$\{(q_0, x, h, q) \mid (q_-, x, h, q) \in E\} \cup \{q, x, h, q_1) \mid (q, x, h, q_+) \in E, q_+ \in Q_+\}$$

to $E$. Take $q_0$ as new initial state and $\{q_1\}$ or $\{q_0, q_1\}$ as final states, according to $\tau(1) = 0$ or $= 1$. Next the resulting transducer can be made trim by deleting unnecessary states. Clearly it is unambiguous. ∎

**Example 4.1** Consider a left sequential transducer. Any path starting at the initial state is successful, and two distinct successful paths have distinct input labels. Thus any left sequential transducer is unambiguous.

**Example 4.2** Let $X = \{x\}$, $Y = \{a, b\}$, and consider the transducer in Fig. IV.10

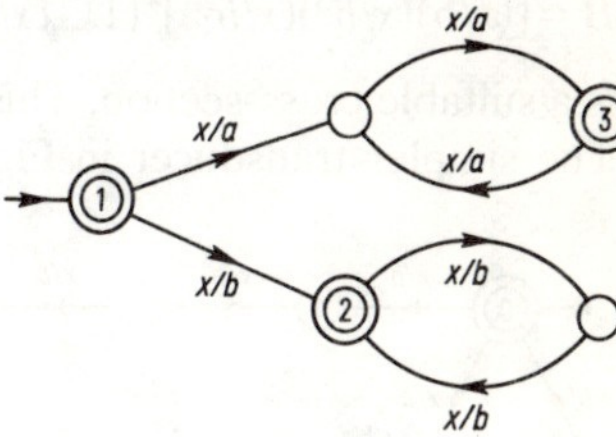


Fig. IV.10

realizing the function

$$\tau(x^n) = \begin{cases} a^n & n \text{ even;} \\ b^n & n \text{ odd.} \end{cases}$$

The transducer is unambiguous since if $n$ is even the only successful path leads to state 3, and if $n$ is odd the only successful path leads to state 2.

**Example 4.3** Consider the transducer of Example 1.2 (Fig. IV.11). This transducer is ambiguous. Take as alphabet $Z$ the labels of the transitions: $Z=\{(x/b),(x/1)\}$, and consider the morphism $\alpha:Z^*\to x^*$ defined by

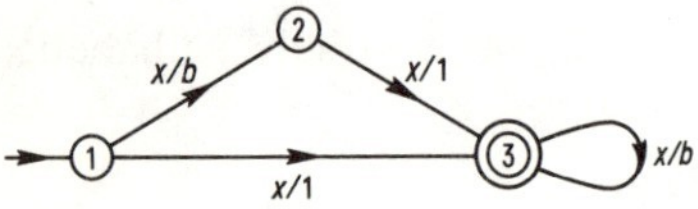


Fig. IV.11

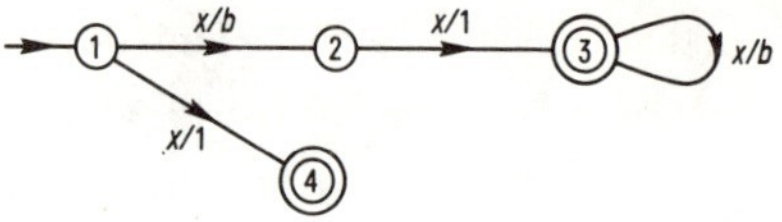


Fig. IV.12

$\alpha((x/b))=\alpha((x/1))=x$. Then, up to a renaming, we are in the situation of Example 3.1. Thus $B=(x/b)(x/1)(x/b)^*\cup(x/1)$ is a suitable cross-section, giving the unambiguous transducer in Fig. IV.12.

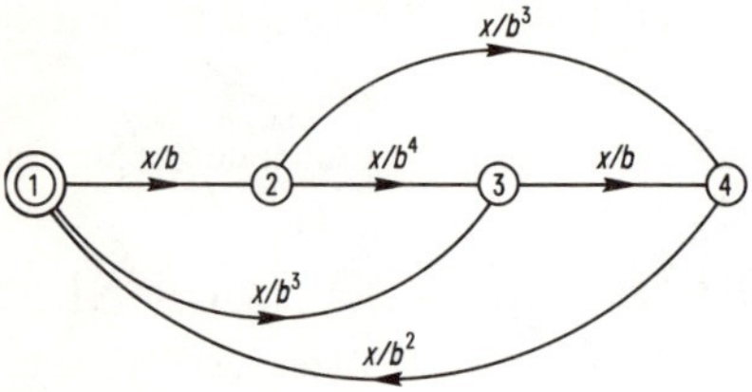


Fig. IV.13

**Example 4.4** Consider the ambiguous transducer of Example 1.3 (Fig. IV.13). Take again the alphabet $Z=\{(x/b),\ (x/b^2),\ (x/b^3),\ (x/b^4)\}$ and the morphism $\alpha$ mapping all letters onto $x$. Then, after a renaming, we are in the situation of Example 3.2. Thus the language

$$B=[(x/b)(x/b^3)(x/b^2)]^*(1\cup(x/b)(x/b^4)(x/b)(x/b^2)\cup[(x/b)(x/b^4)(x/b)(x/b^2)]^2)$$

is a suitable cross-section. This gives the unambiguous transducer in Fig. IV.14. The simpler transducer in Fig. IV.15 cannot be obtained in that way.

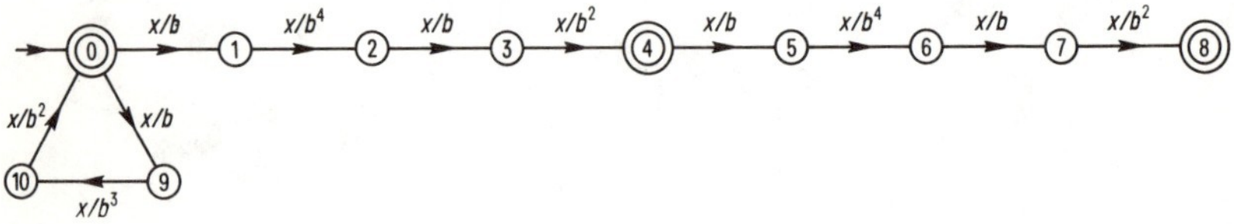


Fig. IV.14

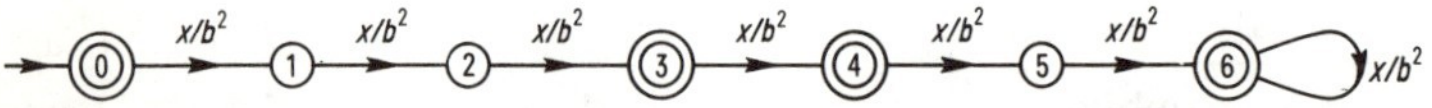


Fig. IV.15

Let $\boldsymbol{T}=\langle X, Y, Q, q_-, Q_+, E\rangle$ be a transducer satisfying (4.1) and (4.2). Define a matrix representation $\boldsymbol{M}=\langle \mu, Q, q_-, Q_+\rangle$ by

$$\mu x_{pq} = \begin{cases} h & \text{if} \quad (p, x, h, q) \in E; \\ 0 & \text{otherwise.} \end{cases}$$

**Proposition 4.4** *Assume $\boldsymbol{T}$ trim. Then $\boldsymbol{T}$ is unambiguous iff $\boldsymbol{M}$ satisfies the two conditions:*

(i) *For $p, q \in Q$, $f, f' \in X^*$, there is at most one $r \in Q$ such that $\mu f_{pr} \neq 0$ and $\mu f'_{rq} \neq 0$.*
(ii) $\operatorname{dom}(\mu_{q_-q_1}) \cap \operatorname{dom}(\mu_{q_-q_2}) = \emptyset$ for $q_1, q_2 \in Q_+$, $q_1 \neq q_2$.

Proof. If $\boldsymbol{T}$ is unambiguous, then (ii) is clearly satisfied. Next, since $\boldsymbol{T}$ is trim, for any $p, q \in Q$, there are $h, h' \in X^*$ such that $\mu h_{q_-p} \neq 0$, $\mu h_{qq_+} \neq 0$ for some $q_+ \in Q_+$. Thus if (i) fails for some $p, q \in Q$, $f, f' \in X^*$, then $hff'h'$ is the input label of at least two successful paths.
Conversely, consider two paths $e_1$ and $e_2$ from $q_-$ to some final states $q_1$ and $q_2$, and assume they have the same input label $f$. Then $q_1 = q_2$ by (ii), and $e_1 = e_2$ by (i). ■

If $\boldsymbol{M}$ is normalized, then (ii) is satisfied by definition. Thus the unambiguity of $\boldsymbol{T}$ is equivalent to condition (i).

**Definition** A morphism $\mu$ satisfying condition (i) of Proposition 4.4 is called unambiguous.

**Example 4.5** If the matrices $\mu f$, $(f \in X^*)$ are row monomial, then $\mu$ is unambiguous. Similarly, if the matrices $\mu f$, $(f \in X^*)$ are column monomial, then $\mu$ is unambiguous.

An unambiguous morphism $\mu : X^* \to \mathrm{Rat}(Y^*)^{Q \times Q}$ is also called a (0, 1)-morphism (Schützenberger [1976], Nivat [1968]) for the following reason. Associate to each matrix $\mu f$, $(f \in X^*)$ a $Q \times Q$-matrix $\theta\mu f$ with integral entries by

$$\theta\mu f_{pq} = \begin{cases} 1 & \text{if} \quad \mu f_{pq} \neq 0; \\ 0 & \text{otherwise.} \end{cases}$$

Then $\theta$ is a morphism iff $\mu$ is unambiguous. Thus a product of two such matrices, computed in $\mathbb{N}^{Q \times Q}$, is still a matrix with entries 0 or 1.

Row monomial and column monomial matrix morphisms are special cases of a more general construction.

**Definition** Let $\mu : X^* \to \mathfrak{P}(Y^*)^{Q \times Q}$ be a morphism. Then $\mu$ is called semimonomial if the set $Q$ is of the form $Q = V \times P$, and if for any $x \in X$, the following hold:

(i) For any $v \in V$, there is at most one $v' \in V$ such that the submatrix

$$(\mu x)_{v\times P, v'\times P} = (\mu x_{(v,p),(v',p')})_{p,p'\in P}$$

is nonzero.
(ii) any submatrix $(\mu x)_{v\times P, v'\times P}$ is column monomial.

Thus the matrix $\mu x$, considered as $V\times V$ matrix whose entries are $P\times P$-matrices, is a row monomial matrix, and each $P\times P$-block is column monomial. Clearly, the product of two semimonomial matrices with same index set $V\times P$ is also semimonomial.

**Example 4.6** For $V=\{1,2,3\}$ and $P=\{1,2,3\}$, the following matrix is semimonomial

$$\left[\begin{array}{c|ccc|ccc} \mathbf{0} & \begin{matrix}0 & b & 1\\ 0 & 0 & 0\\ 0 & 0 & 0\end{matrix} & & & \mathbf{0} \\ \hline \mathbf{0} & \mathbf{0} & & & \begin{matrix}0 & 0 & 0\\ 0 & 0 & 1\\ 0 & 0 & 0\end{matrix} \\ \hline \mathbf{0} & \mathbf{0} & & & \begin{matrix}0 & 0 & 0\\ 0 & 0 & 0\\ 0 & 0 & b\end{matrix} \end{array}\right]$$

Any semimonomial morphism $\mu$ is unambiguous. Consider indeed two words $f$, $f'$, and assume

$$\mu f_{(v,p)(v',p')} \neq 0 \quad \text{and} \quad \mu f'_{(v',p')(v'',p'')} \neq 0.$$

Then $v'$ is uniquely determined by $v$ in view of condition (i), and $p'$ is uniquely determined by $p''$ in view of condition (ii).

The following theorem asserts the existence of a semimonomial representation for any rational function.

**Theorem 4.5** (Schützenberger [1976]) *Let $\tau: X^* \to Y^*$ be a rational function. Then there exists a matrix representation $\mathbf{M} = \langle \mu, Q, q_-, Q_+\rangle$ realizing $\tau$ such that $\mu$ is semimonomial and*

$$\tau = \sum_{q_+\in Q_+} \mu_{q_-q_+}.$$

(Recall that we use the symbol $\sum$ when the domains $\mathrm{dom}(\mu_{q_-q_+})$ are disjoint.)

Proof. Let $\boldsymbol{M}=\langle\mu, Q, q_-, Q_+\rangle$ be a normalized matrix representation realizing $\alpha$. The proof is in two steps. First, the usual powerset algorithm for the determinization of an automaton is employed to obtain the set $V$. Then, "unnecessary" entries in the matrices $\mu x$, $(x\in X)$ are deleted to get the row monomial part.

Let $V$ be the family of subsets of $Q$ defined by

$$v\in V \quad\Leftrightarrow\quad \text{there exists}\quad f\in X^* \quad \text{such that}\quad v=\{q\in Q \mid \mu f_{q_-q}\neq 0\}.$$

Thus, $v\in V$ iff $v$ is the set of states accessible in $\boldsymbol{M}$ by paths starting in $q_-$ and with input label $f$. A "next state" function $V\times X\to V$ is defined by

$$v\cdot x=\{q'\in Q \mid \exists q\in v,\ \mu x_{q,q'}\neq 0\}.$$

This function is extended to $V\times X^*$ by setting, as usual, $v\cdot 1=v$, $v\cdot fx=(v\cdot f)\cdot x$, $(f\in X^*, x\in X)$. Then clearly

$$v\cdot fg=(v\cdot f)\cdot g \qquad f, g\in X^*.$$

For each $v\in V$, $x\in X$, define a $Q\times Q$-matrix $\bar{\mu}_v x$ in the following way. Set $v'=v\cdot x$, and for each $q'\in v'$, let $p(q')$ be an arbitrary element of $v$ such that $\mu x_{p(q'),q'}\neq 0$. Then

$$(\bar{\mu}_v x)_{qq'}=0 \qquad q\in Q, \qquad q'\notin v'; \tag{4.3}$$

$$(\bar{\mu}_v x)_{qq'}=\begin{cases}\mu x_{p(q'),q'} & \text{if}\quad q=p(q');\\ 0 & \text{if}\quad q\neq p(q').\end{cases} \qquad q'\in v'. \tag{4.4}$$

By definition, $\bar{\mu}_v x$ is column monomial. It is obtained from $\mu x$ by deleting all but one nonempty element in each column $q'\in v'$, and by setting equal to zero the other columns.

Next, let $S=V\times Q$, and define the morphism

$$\lambda: X^*\to \mathrm{Rat}(Y^*)^{S\times S}$$

by blocks for $v, v'\in V$, $x\in X$:

$$(\lambda x)_{v\times Q, v'\times Q}=\begin{cases}0 & \text{if}\quad v'\neq v\cdot x;\\ \bar{\mu}_v x & \text{if}\quad v'=v\cdot x.\end{cases}$$

Then the matrices $\lambda x$ are semimonomial. Further, if $f=x_1x_2\cdots x_r$, $(r\geq 1$, $x_i\in X)$, then clearly

$$(\lambda f)_{v\times Q, v'\times Q}=\begin{cases}0 & \text{if}\quad v'\neq v\cdot f;\\ \bar{\mu}_v x_1\bar{\mu}_{v_2}x_2\cdots\bar{\mu}_{v_r}x_r & \text{if}\quad v'=v\cdot f,\end{cases} \tag{4.5}$$

where $v_2=v\cdot x_1,\ldots, v_r=v_{r-1}\cdot x_{r-1}$ and $v'=v_r\cdot x_r$.

Next, we prove that for $v_-=\{q_-\}$, $f\in X^+$, $v=v_-\cdot f$,

$$\lambda f_{(v_-,q_-),(v,q)}=\mu f_{q_-q} \qquad q\in Q, \tag{4.6}$$

For $|f|=1$, (4.6) results from (4.4). Arguing by induction, let $f=gx$ with $g \in X^+$, $x \in X$, and let $v' = v_- \cdot g$. Then $v = v' \cdot x$ and

$$(\lambda gx)_{(v_-,q_-),(v,q)} = \bigcup_{r \in v'} \mu g_{q_- r} (\lambda x)_{(v',r),(v,q)},$$

since $(\lambda g)_{(v_-,q_-),(v',r)} = \mu g_{q_- r} = 0$ for $r \notin v'$. Next if $q \notin v$, then $(\bar{\mu}_{v'} x)_{rq} = \mu x_{rq} = 0$ for $r \in v'$ and $\mu f_{q_- q} = 0$. Thus (4.6) holds if $q \notin v$. If $q \in v$, then there is a unique $p \in v'$ such that $(\bar{\mu}_{v'} x)_{pq} = \mu x_{pq} \neq 0$. Thus

$$(\lambda f)_{(v_-,q_-),(v,q)} = \mu g_{q_- p} \mu x_{pq} = \mu f_{q_- q}.$$

This proves (4.6). Finally, set $V_+ = \{v \in V \mid q_+ \in V\}$ and let

$$S_+ = \begin{cases} V_+ \times \{q_+\} & \text{if } \; Q_+ = \{q_+\}; \\ (V_+ \times \{q_+\}) \cup (v_-, q_-) & \text{if } \; Q_+ = \{q_-, q_+\}. \end{cases}$$

Then $(\lambda 1)_{(v_-,q_-),S_+} = 0$ or 1, according to $(v_-, q_-) \notin S_+$ or $(v_-, q_-) \in S_+$, and for $f \in X^+$:

$$(\lambda f)_{(v_-,q_-),S_+} = \bigcup_{v \in V_+} \lambda f_{(v_-,q_-),(v,q_+)} = \mu f_{q_- q_+}. \tag{4.7}$$

Indeed, (4.7) holds if $v_- \cdot f \in V_+$ (by (4.6)). *If* $v_- \cdot f \notin V_+$, i.e. if $q_+ \notin v_- \cdot f$, then $\mu f_{q_- q_+} = 0$ and $(\lambda f)_{(v_-,q_-),S_+} = 0$ by (4.5). Since there is at most one $v \in V_+$ such that $v = v_- \cdot f$ for any $f \in X^+$, the functions $\lambda_{(v_-,q_-),(v,q_+)}$ have disjoint domains. This achieves the proof. ∎

**Example 4.7** Consider the rational function $\alpha : x^* \to b^*$ of Example 4.3, with matrix

$$\mu x = \begin{bmatrix} 0 & b & 1 \\ 0 & 0 & 1 \\ 0 & 0 & b \end{bmatrix}$$

We first compute $V: v_1 = v_- = \{1\}$, $v_2 = v_1 \cdot x = \{2, 3\}$, $v_3 = v_2 \cdot x = \{3\} = v_3 \cdot x$. The matrices $\bar{\mu}_i$ (we write $\bar{\mu}_i$ instead of $\bar{\mu}_{v_i}$) are:

$$\bar{\mu}_1 x = \underbrace{\begin{bmatrix} 0 & b & 1 \\ 0 & 0 & 0 \\ 0 & 0 & 0 \end{bmatrix}}_{v_2} \Big\} v_1 \qquad \bar{\mu}_2 x = \underbrace{\begin{bmatrix} 0 & 0 & 0 \\ 0 & 0 & 1 \\ 0 & 0 & 0 \end{bmatrix}}_{v_3} \Big\} v_2 \quad \text{or} \quad \bar{\mu}_2 x = \begin{bmatrix} 0 & 0 & 0 \\ 0 & 0 & 0 \\ 0 & 0 & b \end{bmatrix}$$

$$\bar{\mu}_3 x = \underbrace{\begin{bmatrix} 0 & 0 & 0 \\ 0 & 0 & 0 \\ 0 & 0 & b \end{bmatrix}}_{v_3} \Big\} v_3$$

Thus, there are two possible choices for $\bar{\mu}_2$; each choice yields another matrix $\lambda x$:

$$\lambda x = \begin{bmatrix} 0 & \bar{\mu}_1 x & 0 \\ 0 & 0 & \bar{\mu}_2 x \\ 0 & 0 & \bar{\mu}_3 x \end{bmatrix}$$

With the first matrix $\bar{\mu}_2 x$, the matrix $\lambda x$ is the matrix of Example 4.6. Next, $V_+ = \{v_2, v_3\}$. Thus $S_+ = \{(v_2, 3); (v_3, 3)\}$. In the usual notation, these are the columns 6 and 9. For both morphisms $\lambda$, the row with index $(v_-, 1)$ of $\lambda x^n$ is:

$$(0, 0, 0; 0, b, 1; 0, 0, 0) \qquad n = 1;$$
$$(0, 0, 0; 0, 0, 0; 0, 0, b^{n-1}) \qquad n \geq 2.$$

Note that $\lambda$ is not trim. The trim transducer associated to $\lambda$ for the first choice of $\bar{\mu}_2 x$ is given in Fig. IV.16. Thus we get the same transducer as in Example 4.3.

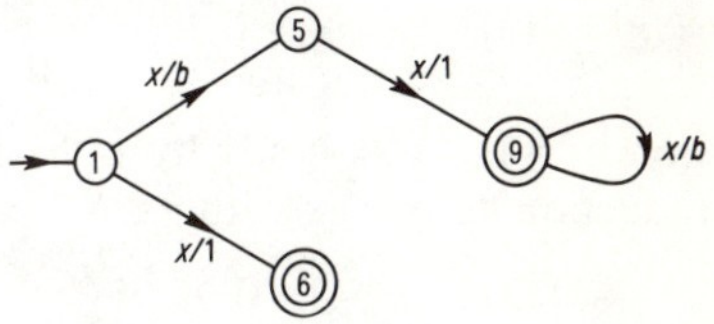


Fig. IV.16

Semimonomial morphisms are particular unambiguous morphisms. Clearly, an unambiguous morphism is not necessarily semimonomial. There is nevertheless an interesting relation between unambiguous and semimonomial morphisms: any unambiguous morphism can be obtained from some semimonomial morphism by choosing in that morphism some fixed rows and by collapsing columns. This yields some procedure for constructing unambiguous morphisms.

**Definition** Let $\mu : X^* \to \mathrm{Rat}(Y^*)^{Q \times Q}$ and $\lambda : X^* \to \mathrm{Rat}(Y^*)^{S \times S}$ be two morphisms. Then $\mu$ is summed up from $\lambda$ if there exist two functions $l : Q \to S$ and $c : Q \to 2^S$ such that

$$\mu f_{pq} = \lambda f_{l(p),c(q)} \quad \left( = \bigcup_{r \in c(q)} \lambda f_{l(p),r} \right) \qquad (p, q \in Q, f \in X^+).$$

Thus $\mu$ is obtained from $\lambda$ by conserving just one row of $\lambda$ for each $p \in Q$, and by summing up elements of that row according to some rule which is independent from $p$.

**Proposition 4.6** *Let $M = \langle \mu, Q, q_-, Q_+ \rangle$ be a trim matrix representation from $X^*$ into $Y^*$. If $\mu$ is unambiguous, then $\mu$ is summed up from some semimonomial morphism.*

Proof. Assume $\mu$ is unambiguous. We shall verify that $\mu$ can be extracted from the semimonomial morphism $\lambda$ constructed in the proof of Theorem 4.5. We keep the notations of that proof. First, we claim that for the matrix $\bar{\mu}_v x$ defined by (4.3) and (4.4), we have

$$(\bar{\mu}_v x)_{qq'} = \begin{cases} \mu x_{qq'} & \text{if } q \in v, \quad q' \in Q; \\ 0 & \text{if } q \notin v, \quad q' \in Q. \end{cases} \tag{4.8}$$

Thus $\bar{\mu}_v x$ and $\mu x$ have same rows with index $q \in v$. Indeed, $(\bar{\mu}_v x)_{qq'} = 0$ for $q \notin v$ by (4.4), and $(\bar{\mu}_v x)_{qq'} = \mu x_{qq'} = 0$ for $q \in v$, $q' \in Q \backslash v'$ by definition of $v' = v \cdot x$. Next, for any $q' \in v'$, there exists a unique $p \in v$ such that $\mu x_{pq'} \neq 0$, since for any $f$ such that $v_- \cdot f = v$, one has $\mu f_{q\_p} \neq 0$ and $\mu x_{pq'} \neq 0$, and thus $p$ is unique by Proposition 4.4(i). Therefore $(\bar{\mu}_v x)_{pq'} = \mu x_{pq'}$ for that $p$, and $(\bar{\mu}_v x)_{qq'} = \mu x_{qq'} = 0$ for all $q \in v \backslash p$. Thus (4.8) is proved.

Next, we prove that for $f \in X^+$, $v \in V$, $v' = v \cdot f$,

$$(\lambda f)_{(v,p),(v',q)} = \begin{cases} \mu f_{pq} & \text{if } p \in v, \quad q \in Q; \\ 0 & \text{if } p \notin v, \quad q \in Q. \end{cases} \tag{4.9}$$

For $|f| = 1$, (4.9) holds by (4.8) in view of the definition of $\lambda$. Arguing by induction, let $f = gx$ with $g \in X^+$, $x \in X$, $v'' = v \cdot g$, thus $v' = v'' \cdot x$. Then clearly

$$(\lambda f)_{(v,p)(v',q)} = 0 \qquad \text{if } p \notin v, \quad q \in Q.$$

Thus assume $p \in v$. Then

$$(\lambda f)_{(v,p),(v',q)} = \bigcup_{r \in Q} (\lambda g)_{(v,p),(v'',r)} (\lambda x)_{(v'',r),(v',q)},$$

and since $\lambda x_{(v'',r),(v',q)} = 0$ if $r \notin v''$, it follows from (4.8) that

$$(\lambda f)_{(v,p),(v',q)} = \bigcup_{r \in v''} \mu g_{p,r} \mu x_{r,q} = \mu f_{p,q}.$$

Note that in view of (4.5)

$$\lambda f_{(v,p),(v',q)} = 0 \quad \text{if} \quad v' \neq v \cdot f \qquad (p, q \in Q).$$

Thus it follows from (4.9) that

$$\mu f_{p,q} = \bigcup_{v' \in V} \lambda f_{(v,p)(v',q)} \qquad (p \in v). \tag{4.10}$$

Now define a function $l: Q \to S$ as follows. For any $p \in Q$, choose a $v \in V$ such that $p \in v$ and set $l(p) = (v, p)$. Such a $v$ exists since $\mu$ is trim. Next, define $c$ by

$$c(q) = \{(v', q) : v' \in V\}.$$

Then by (4.10)

$$\mu f_{p,q} = \lambda f_{l(p),c(q)}. \qquad ∎$$

For a more exhaustive treatment of related questions, especially in connection with codes, see Boë [1976], Boë, Boyat et al. [1978] Césari [1974], Perrin and Schützenberger [1977].

## Exercises

**4.1** Compute the trim transducer associated to the second of the two semimonomial morphisms $\lambda$ of Example 4.7.

**4.2** Use Exercise 3.1 to give a second unambiguous transducer for the transduction of Example 4.3, and compare with the transducer of Exercise 4.1.

**4.3** A partial function $\beta: X^* \to Z^*$ is length preserving if $\beta f \neq 0$ implies $|f| = |\beta f|$. Show that any rational function $\alpha: X^* \to Y^*$ can be written in the form $\alpha = \gamma \circ \beta$, where $\beta: X^* \to Z^*$ is a length preserving rational function and $\gamma: Z^* \to Y^*$ is a morphism.

**4.4** Let $M$ be a monoid. The family of unambiguous rational subsets of $M$ is the least family of subsets of $M$ containing $\emptyset$ and the singletons $\{m\}$, $(m \in M)$ and closed under the following operations: unambiguous union, unambiguous product, unambiguous star. (A union $A \cup B$ is unambiguous if $A \cap B = \emptyset$. A product $AB$ is unambiguous if $a$, $a' \in A$, $b$, $b' \in B$, $ab = a'b'$ imply $a = a'$ and $b = b'$. A star $A^*$ is unambiguous if $A^*$ is a free submonoid freely generated by $A$.)

Show that if $\alpha: X^* \to Y^*$ is a rational function, then the graph of $\alpha$ is an unambiguous rational subset of $X^* \times Y^*$.

**4.5** (Choffrut) Show that for any rational function $\alpha: X^* \to Y^*$, there is a rational subset $C$ of dom($\alpha$) such that $\alpha$ maps bijectively $C$ onto $\alpha(X^*) = \text{im}(\alpha)$. (This is an extension of the Cross-Section Theorem to rational functions.)

**4.6** (Choffrut) Show that for any rational function $\alpha: X^* \to Y^*$ there exists a rational function $\beta: Y^* \to X^*$ such that $\alpha \circ \beta: \alpha(X^*) \to \alpha(X^*)$ is the identity function (Hint. Use Exercise 4.5.)

**4.7** (Choffrut) Use the previous exercise to show that if $\alpha: X^* \to Y^*$ and $\beta: Y^* \to Z^*$ are partial functions, and if $\alpha$ and $\beta \circ \alpha$ and $\beta$ are rational, then the restriction $\beta|_{\alpha(X^*)}$ is rational. Show that if $\beta \circ \alpha$ and $\beta$ are rational, then $\alpha$ needs not to be rational.

## IV.5 Bimachines

Bimachines are, in some sense, simultaneously left sequential and right sequential transducers. We show that a partial function is rational iff it is realized by a bimachine, and use this fact to prove that any rational function can be obtained as the composition of a left sequential function followed by a right sequential function.

**Definition** A bimachine $\boldsymbol{B} = \langle Q, q_-, P, p_-, \gamma \rangle$ over $X$ and $Y$ is composed of two finite sets of states $Q$, $P$, two initial states $q_- \in Q$, $p_- \in P$, of two partial next state functions $Q \times X \to Q$ and $X \times P \to P$ denoted by dots, and a partial output function $\gamma: Q \times X \times P \to Y^*$.

The next state functions are extended to $Q \times X^*$ and $X^* \times P$ in the usual way by setting:

$$q \cdot 1 = q, \qquad 1 \cdot p = p$$
$$q \cdot (fx) = (q \cdot f) \cdot x, \qquad (xf) \cdot p = x \cdot (f \cdot p)$$

for $q \in Q$, $p \in P$, $f \in X^*$, $x \in X$. Next the output function $\gamma$ is extended to $Q \times X^* \times P$ by

$$\gamma(q, 1, p) = 1;$$
$$\gamma(q, fx, p) = \gamma(q, f, x \cdot p)\gamma(q \cdot f, x, p)$$

for $f \in X^*$, $x \in X$, $q \in Q$, $p \in P$. Then it is easily verified that

$$\gamma(q, fg, p) = \gamma(q, f, g \cdot p)\gamma(q \cdot f, g, p) \qquad (f, g \in X^*),$$

and if $f = x_1 x_2 \cdots x_n$, $(x_i \in X)$, then

$$\gamma(q, f, p) = \gamma(q, x_1, x_2 \cdots x_n \cdot p)\gamma(q \cdot x_1, x_2, x_3 \cdots x_n \cdot p) \cdots \gamma(q \cdot x_1 \cdots x_{n-1}, x_n, p).$$

The partial function $X^* \to Y^*$ realized by $\boldsymbol{B}$ is defined by

$$|\boldsymbol{B}|(f) = \gamma(q_-, f, p_-).$$

If $P = \{p_-\}$, then $\boldsymbol{B}$ is, up to considerations concerning the domains, a left sequential transducer. Similarly, if $Q = \{q_-\}$, then $\boldsymbol{B}$ is a right sequential transducer.

Bimachines were introduced by Schützenberger [1961b]. See also Nivat [1968].

**Example 5.1** Let $\alpha: x^* \to \{a, b\}^*$ be given by

$$\alpha(x^n) = \begin{cases} a^n & \text{if } n \text{ even;} \\ b^n & \text{if } n \text{ odd.} \end{cases}$$

Consider $P = \{p_-, p_1\}$, $Q = \{q_-, q_1\}$ and define the next state functions by $x \cdot p_- = p_1$, $x \cdot p_1 = p_-$, and $q_- \cdot x = q_1$, $q_1 \cdot x = q_-$. Further, let $\gamma$ be given by the table

| | $p_-$ | $p_1$ |
|---|---|---|
| $q_-$ | $b$ | $a$ |
| $q_1$ | $a$ | $b$ |

Then a simple calculation shows that $\alpha(x^n) = \gamma(q_-, x^n, p_-)$ for $n \geqslant 0$.

Note that in the above definition, no assumption was made about the domains of the next state functions and $\gamma$. Call a bimachine state complete if both next state functions $Q \times X \to Q$ and $X \times P \to P$ are total functions.

**Theorem 5.1** (Eilenberg [1974]) *Let $\alpha : X^* \to Y^*$ be a partial function with $\alpha(1) = 1$. Then $\alpha$ is rational iff it is realized by some bimachine over $X$ and $Y$.*

We shall see that a rational function can always be realized by a state complete bimachine.

Proof. Let $\boldsymbol{B} = \langle Q, q_-, P, p_-, \gamma \rangle$ be a bimachine over $X$ and $Y$, define $S = Q \times P$, and consider the transducer $\boldsymbol{T}$ with set of states $S$, and set of transitions $E \subset S \times X \times Y^* \times S$ given by:

$$((q, p), x, h, (q', p')) \in E$$

iff $q \cdot x = q'$, $p = x \cdot p'$, and $h = \gamma(q, x, p') \neq 0$. Consider any path

$$((q_1, p_1), x_1, h_1, (q'_1, p'_1)) \cdots ((q_n, p_n), x_n, h_n, (q'_n, p'_n))$$

in $\boldsymbol{T}$, with $f = x_1 x_2 \cdots x_n$, $g = h_1 h_2 \cdots h_n$. Then clearly $q_1 \cdot f = q'_n$, $p_1 = f \cdot p'_n$, and $g = \gamma(q_1, f, p'_n)$. For any $q \in Q$, $p \in P$, define the rational transduction $\tau_{q,p} : X^* \to Y^*$ by

$$\tau_{q,p}(f) = g$$

iff there is a path from $(q_-, p)$ to $(q, p_-)$ with input label $f$ and output label $g$, and set $\tau_{q,p}(f) = 0$ otherwise. Then $\tau_{q,p}(f) = g \neq 0$ iff $g = \gamma(q_-, f, p_-)$, and

$$\alpha = \sum_{(q,p) \in S} \tau_{q,p}.$$

Thus $\alpha$ is a rational function.

Conversely, let $\alpha$ be realized by an unambiguous normalized matrix representation $\boldsymbol{M} = \langle \mu, Q, q_-, \{q_-, q_+\} \rangle$. Define two families $V$, $W$ of subsets of $Q$ as follows:

$$v \in V \quad \Leftrightarrow \quad \exists f \in X^* : v = \{q \in Q \mid \mu f_{q_- q} \neq 0\};$$
$$w \in W \quad \Leftrightarrow \quad \exists f \in X^* : w = \{q \in Q \mid \mu f_{q q_+} \neq 0\}.$$

Then define functions $V \times X \to V$, $X \times W \to W$ as follows:

$$v \cdot x = \{q' \in Q \mid \exists q \in v : \mu x_{qq'} \neq 0\} \qquad v \in V;$$
$$x \cdot w = \{q' \in Q \mid \exists q \in w : \mu x_{q'q} \neq 0\} \qquad w \in W.$$

Extend them to words in the usual way by setting:

$$v \cdot 1 = v, \qquad v \cdot (fx) = (v \cdot f) \cdot x;$$
$$1 \cdot w = w, \qquad (xf) \cdot w = x \cdot (f \cdot w)$$

for $f \in X^*$, $x \in X$. Then clearly for $f \in X^*$

$$\begin{aligned} v \cdot f &= \{q' \in Q \mid \exists q \in v : \mu f_{qq'} \neq 0\} \qquad v \in V; \\ f \cdot w &= \{q' \in Q \mid \exists q \in w : \mu f_{q'q} \neq 0\} \qquad w \in W. \end{aligned} \tag{5.1}$$

Next we prove

$$\mathrm{Card}(v \cap w) \leq 1 \qquad \text{for } v \in V, \quad w \in W. \tag{5.2}$$

Assume indeed $r, r' \in v \cap w$. By definition, there exists a word $f$ such that $\mu f_{q_-r} \neq 0$, $\mu f_{q_-,r'} \neq 0$, and similarly there is a $g \in X^*$ such that $\mu g_{rq_+} \neq 0$, $\mu g_{r'q_+} \neq 0$. Then $r = r'$ by Proposition 4.4.
Define a partial function

$$\gamma : V \times X^* \times W \to Y^*$$

by $\quad \gamma(v, 1, w) = 1,$

and for $f \in X^+$,

$$\gamma(v, f, w) = \begin{cases} 0 & \text{if } \; v \cap f \cdot w = \emptyset \; \text{ or } \; v \cdot f \cap w = \emptyset; \\ \mu f_{pq} & \text{if } \; v \cap f \cdot w = p \; \text{ and } \; v \cdot f \cap w = q. \end{cases} \tag{5.3}$$

We claim

$$\gamma(v, hh', w) = \gamma(v, h, h' \cdot w)\gamma(v \cdot h, h', w) \qquad h, h' \in X^*. \tag{5.4}$$

Clearly, (5.4) holds if $h = 1$ or $h' = 1$. Thus we may assume $h, h' \in X^+$. Next, if $v \cap hh' \cdot w = \emptyset$ or $v \cdot hh' \cap w = \emptyset$, then both sides of (5.4) are empty. If $p = v \cap hh' \cdot w$ and $q = v \cdot hh' \cap w$, then by definition $\gamma(v, hh', w) = \mu hh'_{pq} \neq 0$. Since $\mu$ is unambiguous, there is exactly one $r \in Q$ such that

$$\mu hh'_{pq} = \mu h_{pr} \mu h'_{rq}. \tag{5.5}$$

Thus by (5.1) $r \in v \cdot h$ and $r \in h' \cdot w$. Consequently $r = v \cdot h \cap h' \cdot w$ by (5.2), and therefore $\gamma(v, h, h' \cdot w) = \mu h_{pr}$ and $\gamma(v \cdot h, h', w) = \mu h_{rq}$. Thus (5.4) follows from (5.5).
Define $v_- = \{q_-\}$, $w_+ = \{q_+\}$. Then in view of (5.4),

$$\boldsymbol{B} = \langle V, v_-, W, w_+, \gamma \rangle$$

is a bimachine over $X$ and $Y$, and by construction $\boldsymbol{B}$ is state complete. Next let $f \in X^+$. Then by (5.1)

$$q_+ \in v_- \cdot f \;\Leftrightarrow\; \mu f_{q_-q_+} \neq 0 \;\Leftrightarrow\; q_- \in f \cdot w_+ \;\Leftrightarrow\; f \in \mathrm{dom}(\alpha).$$

Thus (5.3) implies

$$\gamma(v_-, f, w_+) = \mu f_{q_-q_+} \qquad (f \in X^+).$$

Since $\quad \gamma(v_-, 1, w_+) = 1,$

it follows that $\alpha = |\boldsymbol{B}|$. ∎

We conclude this section by the following nice "decomposition theorem".

**Theorem 5.2** (Elgot and Mezei [1965]) *Let $\alpha : X^* \to Y^*$ be a partial function with $\alpha(1) = 1$. Then $\alpha$ is rational iff there are a left sequential function*

*$\lambda: X^* \to Z^*$, and a right sequential function $\rho: Z^* \to Y^*$ such that $\alpha = \rho \circ \lambda$. Moreover, $\lambda$ can be chosen to be total and length preserving (i.e. $|\lambda(f)| = |f|$ for all $f \in X^*$).*

Thus in order to compute $\alpha(f)$ for some $f \in X^*$, one first reads $f$ sequentially from left to right and transforms it into a word $g$ by some left sequential transducer; then the resulting word $g$ is read from right to left and transformed into $\alpha(f)$ by a right sequential transducer.

Proof. If $\alpha = \rho \circ \lambda$, then $\alpha$ is a partial function and $\alpha$ is rational since the composition of two rational transductions is a rational transduction.

Conversely, consider a bimachine $\boldsymbol{B} = \langle Q, q_-, P, p_-, \gamma \rangle$ over $X$ and $Y$ realizing $\alpha$. We may assume that $\boldsymbol{B}$ is state complete, i.e. the next state functions $Q \times X \to Q$ and $X \times P \to P$ are total. Set $Z = Q \times X$, and define a left sequential transducer

$$\boldsymbol{L} = \langle X, Z, Q, q_- \rangle$$

as follows. The next state function of $\boldsymbol{L}$ is the next state function $Q \times X \to Q$ of $\boldsymbol{B}$, and for $q \in Q$, $x \in X$,

$$q * x = (q, x).$$

Define $\lambda = |\boldsymbol{L}|$. Then $\lambda$ is length preserving.

Next define a right sequential transducer

$$\boldsymbol{R} = \langle Z, Y, P, p_- \rangle$$

by

$$(q, x) * p = \gamma(q, x, p)$$

$$(q, x) \cdot p = \begin{cases} 0 & \text{if} \quad \gamma(q, x, p) = 0; \\ x \cdot p & \text{otherwise,} \end{cases}$$

where $x \cdot p$ is the next state of $p$ in $\boldsymbol{B}$. Thus the next state function and the output function of $\boldsymbol{R}$ have same domain. Set $\rho = |\boldsymbol{R}|$.

Let $f = x_1 x_2 \cdots x_n$, $(n \geqslant 1, x_i \in X)$. Then

$$\begin{aligned} \lambda(f) &= (q_- * x_1)(q_- \cdot x_1 * x_2) \cdots (q_- \cdot x_1 x_2 \cdots x_{n-1} * x_n) \\ &= (q_-, x_1)(q_1, x_2) \cdots (q_{n-1}, x_n), \end{aligned}$$

where $\quad q_i = q_- \cdot x_1 x_2 \cdots x_i \quad \text{for} \quad i = 1, \ldots, n-1.$

Consequently,

$$\begin{aligned} \rho(\lambda(f)) &= \lambda(f) * p_- \\ &= ((q_-, x_1) * p_{n-1})((q_1, x_2) * p_{n-2}) \cdots ((q_{n-2}, x_{n-1}) * p_1)((q_{n-1}, x_n) * p_-) \end{aligned}$$

where $p_i = (q_{n-i}, x_{n-i+1}) \cdots (q_{n-1}, x_n) \cdot p_-$ for $i = 1, \ldots, n-1$. Thus

$$\rho(\lambda(f)) = \gamma(q_-, x_1, p_{n-1}) \cdots \gamma(q_{n-1}, x_n, p_-) = \alpha(f).$$

(This computation holds also if $\alpha(f)=0$ with the usual convention that $x \cdot 0 = 0$.) Thus $\alpha = \rho \circ \lambda$, and the theorem is proved. ■

There exists another characterization of rational functions by a property similar to bimachines. See Schützenberger [1977b].

### Exercises

**5.1** Prove that a partial function $\alpha: X^* \to Y^*$ is rational iff $\alpha = \lambda \circ \rho$, where $\rho: X^* \to Z^*$ is a right sequential function and $\lambda: Z^* \to Y^*$ is left sequential.

**5.2** Let $x$, $y$ be letters. A partial function $\alpha: x^* \to y^*$ can be viewed as a partial function $\alpha: \mathbb{N} \to \mathbb{N}$ by identifying a word with its length. Show that $\alpha: \mathbb{N} \to \mathbb{N}$ is rational iff $\alpha = \alpha_1 + \cdots + \alpha_k$, where each $\alpha_i$ is a partial function with domain $r_i\mathbb{N} + s_i$, $(r_i, s_i \in \mathbb{N})$ given by $\alpha_i(r_i n + s_i) = r'_i n + s'_i$, $(n \in \mathbb{N})$ for some $r'_i$, $s'_i \in \mathbb{N}$.

## IV.6 A Decidable Property

In this section, we continue the investigation of sequential and subsequential functions started in Section 2.

**Theorem 6.1** (Choffrut [1977]) *Given a matrix representation $\boldsymbol{M}$ from $X^*$ into $Y^*$, it is decidable whether $|\boldsymbol{M}|$ is subsequential, and whether $|\boldsymbol{M}|$ is sequential.*

According to Proposition III.7.4 (and Exercise III.8.3), $\boldsymbol{M}$ can be supposed to be trim. In view of Theorem 1.2, it is decidable whether $|\boldsymbol{M}|$ is a rational function. Further, the results of Section 4 show that then an unambiguous representation realizing $|\boldsymbol{M}|$ can effectively be constructed. Thus we may assume that the representation $\boldsymbol{M}$ in Theorem 6.1 is unambiguous and normalized.

We use the notations and definitions of Section 2. We consider $Y^*$ as a submonoid of the free group $Y^{(*)}$, according to the discussion of Section II.3. Let $\boldsymbol{M} = \langle \mu, Q, q_-, Q_+ \rangle$ be an unambiguous normalized matrix representation from $X^*$ into $Y^*$, and set $\alpha = |\boldsymbol{M}|$. Then in particular $Q_+ = \{q_-, q_+\}$ or $Q_+ = \{q_+\}$ according to $1 \in \mathrm{dom}(\alpha)$ or $1 \notin \mathrm{dom}(\alpha)$. First we define a property on $\boldsymbol{M}$ which will appear to express that $\alpha$ has bounded variation.

**Definition** Two states $q_1, q_2 \in Q$ are twinned iff for all $f, u \in X^*$ the following condition holds:

$$\left.\begin{array}{ll} 0 \neq a_1 = \mu f_{q_- q_1}, & 0 \neq b_1 = \mu u_{q_1 q_1} \\ 0 \neq a_2 = \mu f_{q_- q_2}, & 0 \neq b_2 = \mu u_{q_2 q_2} \end{array}\right\} \Rightarrow a_1 b_1 a_1^{-1} = a_2 b_2 a_2^{-1} \tag{6.1}$$

$\boldsymbol{M}$ has the twinning property if any two states are twinned.

A pair $f$, $u \in X^*$ which satisfies the hypotheses of (6.1) is called **admissible** for $q_1$, $q_2$. The conclusion of (6.1) can be formulated as follows without use of inverses.

**Proposition 6.2** *Let $a_1, a_2, b_1, b_2 \in Y^*$. Then*

$$a_1 b_1 a_1^{-1} = a_2 b_2 a_2^{-1} \tag{6.2}$$

*iff one of the following conditions is verified:*

(i) $b_1 = b_2 = 1$;

(ii) $b_1 \neq 1 \neq b_2$, *and there exists* $e \in Y^*$ *such that either*

(ii.1) $a_2 = a_1 e$ *and* $e b_2 = b_1 e$; *or*

(ii.2) $a_1 = a_2 e$ *and* $e b_1 = b_2 e$.

Proof. Assume (i) holds. Then (6.2) is obvious. Next, suppose for instance (ii.1). Then $a_2 b_2 a_2^{-1} = a_1 e b_2 e^{-1} a_1^{-1} = a_1 b_1 e e^{-1} a_1^{-1} = a_1 b_1 a_1^{-1}$.
Conversely, suppose that (6.2) holds. Then $b_1 = 1$ iff $b_2 = 1$. Thus assume $b_1 \neq 1$, $b_2 \neq 1$, and let $a$ be the longest left factor common to $a_1$ and $a_2$. Set $a_1 = ac_1$, $a_2 = ac_2$. Then (6.2) becomes: $c_1 b_1 c_1^{-1} = c_2 b_2 c_2^{-1}$. If $c_1 = 1$, then (ii.1) holds with $e = c_2$; if $c_2 = 1$, then (ii.2) holds with $e = c_1$. If both $c_1, c_2 \neq 1$, then they differ by their initial letter by definition of $a$. Thus the equation $c_1 b_1 c_1^{-1} = c_2 b_2 c_2^{-1}$ implies $b_1 = b_2 = 1$, contrary to the assumption. ■

**Example 6.1** Consider the following unambiguous normalized transducer (Fig. IV.17). The function $\alpha : x^* \to \{y, z, t\}^*$ realized by this transducer is given by

$$\alpha(1) = 1; \quad \alpha(x^{2n}) = t(zy)^n, \quad n \geq 1; \quad \alpha(x^{2n+1}) = t(zy)^{n+1}, \ n \geq 0.$$

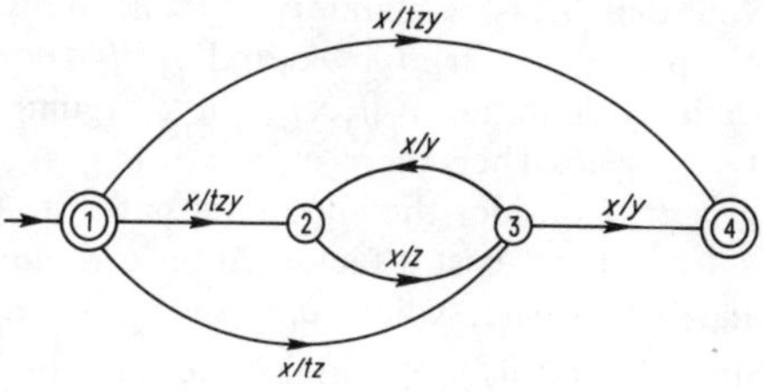


Fig. IV.17

In order to verify that the matrix representation $\boldsymbol{M}$ associated to the transducer has the twinning property, it suffices to show that the states 2 and 3 are twinned. For this, let $f = x^{2n+1}$, $u = x^{2m}$ be an admissible pair for 2, 3. Then $a_1 = \mu f_{1,2} = t(zy)^n$, $b_1 = \mu u_{22} = (zy)^m$, and $a_2 = \mu f_{1,3} = tz(yz)^n = t(zy)^n z = a_1 e$ with $e = z$, and $b_2 = \mu u_{33} = (yz)^m$, whence $e b_2 = b_1 e$. Thus $a_1 b_1 a_1^{-1} = a_2 b_2 a_2^{-1}$ by Proposition 6.2, and 2, 3 are twinned.

We note the following corollary.

**Corollary 6.3** *Let $a_1, a_2, b_1, b_2 \in Y^*$. If $a_1 b_1^k a_1^{-1} = a_2 b_2^k a_2^{-1}$ for some $k > 0$, then $a_1 b_1 a_1^{-1} = a_2 b_2 a_2^{-1}$.*

Proof. We may assume $b_1, b_2 \neq 1$ and for instance $|a_2| \geqslant |a_1|$. Then there exists, in view of Proposition 6.2, a word $e \in Y^*$ such that $a_2 = a_1 e$, and $e b_2^k = b_1^k e$. We prove that this implies $e b_2 = b_1 e$ by induction on $|e|$, the case $|e| = 0$ being immediate. If $|e| \leqslant |b_1|$, then $b_1 = ef$ for some word $f$, hence $e b_2^k = (ef)^k e = e(fe)^k$. Therefore $b_2 = fe$ and $e b_2 = efe = b_1 e$. If $|e| > |b_1|$, then $e = b_1 e'$ for some $e'$. Next $e b_2^k = b_1 e' b_2^k = b_1^k b_1 e'$, thus $e' b_2^k = b_1^k e'$ and $e' b_2 = b_1 e'$ by induction. Thus $e b_2 = b_1 e$. ■

We note also that if $a_1 b_1 a_1^{-1} = a_2 b_2 a_2^{-1}$, then for all $c_1, c_2$

$$\|a_1 b_1 c_1, a_2 b_2 c_2\| \leqslant \|a_1 c_1, a_2 c_2\|. \tag{6.3}$$

Indeed, (6.3) is obvious if $b_1 = b_2 = 1$. Otherwise, we may assume by Proposition 6.2 that for instance $a_2 = a_1 e$, and $e b_2 = b_1 e$ for some word $e$. Then $a_2 b_2 = a_1 b_1 e$, and consequently $|a_1 b_1 c_1 \wedge a_2 b_2 c_2| \geqslant |a_1 b_1|$. This proves (6.3).

**Proposition 6.4** *The two following conditions are equivalent:*

*(i) $\boldsymbol{M}$ has the twinning property;*
*(ii) $\alpha$ has bounded variation.*

Proof. Assume that $\boldsymbol{M}$ has the twinning property. Let $n$ be the number of states of $Q$. Consider an integer $k \geqslant 0$, and define

$$K = \max\{\|\alpha(f_1), \alpha(f_2)\| : f_1, f_2 \in \mathrm{dom}(\alpha), \|f_1, f_2\| \leqslant k, |f_1 \wedge f_2| \leqslant n^2\}.$$

Note that $\|f_1, f_2\| \leqslant k$ and $|f_1 \wedge f_2| \leqslant n^2$ imply $|f_1| + |f_2| \leqslant k + 2n^2$. Thus $K$ is finite. We prove that $\|f_1, f_2\| \leqslant k$ and $f_1, f_2 \in \mathrm{dom}(\alpha)$ imply: $\|\alpha(f_1), \alpha(f_2)\| \leqslant K$. This holds by definition if $|f_1 \wedge f_2| \leqslant n^2$. Arguing by induction on $|f_1 \wedge f_2|$, we assume $|f_1 \wedge f_2| > n^2$. Then there exist words $g, h_1, h_2$ with $g = f_1 \wedge f_2$, $f_i = g h_i$, $i = 1, 2$, $|g| > n^2$. Consider the successful paths in $\boldsymbol{M}$ with input labels $f_1$ and $f_2$. Since $|g| > n^2$, there exist a factorization $g = wuv$, $|u| > 0$, and two states $q_1, q_2$, such that $\alpha(f_i) = a_i b_i c_i$, where $a_i = \mu w_{q_- q_i}$, $b_i = \mu u_{q_i q_i}$, $c_i = \mu(v h_i)_{q_i q_+}$, $(i = 1, 2)$. Since $q_1$ and $q_2$ are twinned, we have by (6.3),

$$\|\alpha(f_1), \alpha(f_2)\| \leqslant \|a_1 c_1, a_2 c_2\| = \|\alpha(f_1'), \alpha(f_2')\|$$

where $f_1' = w v h_1$, $f_2' = w v h_2 \in \mathrm{dom}(\alpha)$. Further $f_1' \wedge f_2' = wv$ is strictly shorter than $g$. Consequently $\|\alpha(f_1), \alpha(f_2)\| \leqslant K$ and $\alpha$ has bounded variation.

Conversely, let $q_1, q_2$ be two states in $\boldsymbol{M}$, and consider a pair $f, u$ of words which is admissible for $q_1, q_2$, i.e. satisfying the hypotheses of (6.1). Since $\boldsymbol{M}$ is trim, there are words $h_1, h_2 \in X^*$ such that $c_i = (\mu h_i)_{q_i q_+} \neq 0$ for $i = 1, 2$. Consequently $f u^m h_i \in \mathrm{dom}(\alpha)$ for $m \geqslant 0$, $i = 1, 2$. Next $\|f u^m h_1, f u^m h_2\| = \|h_1, h_2\|$, and since $\alpha$ has bounded variation, there is an integer $K$ such that

$$\|a_1b_1^mc_1, a_2b_2^mc_2\| \leqslant K \qquad m \geqslant 0.$$

Consequently, there exist words $d_1$, $d_2$, with $|d_1|+|d_2| \leqslant K$, such that $d_i$ is a right factor of $a_ib_i^mc_i$, $(i=1,2)$ for an infinity of exponents $m$. In particular, there are integers $p \geqslant 0$, $k > 0$ such that:

$$a_1b_1^pc_1d_1^{-1} = a_2b_2^pc_2d_2^{-1}; \tag{6.4}$$

$$a_1b_1^{k+p}c_1d_1^{-1} = a_2b_2^{k+p}c_2d_2^{-1}. \tag{6.5}$$

(6.5) can be written as:

$$a_1b_1^ka_1^{-1}a_1b_1^pc_1d_1^{-1} = a_2b_2^ka_2^{-1}a_2b_2^pc_2d_2^{-1}.$$

In view of (6.4), this implies:

$$a_1b_1^ka_1^{-1} = a_2b_2^ka_2^{-1},$$

and by Corollary 6.3, $a_1b_1a_1^{-1} = a_2b_2a_2^{-1}$. Thus $q_1$ and $q_2$ twinned. This completes the proof. ∎

The following proposition yields the main argument for the proof of Theorem 6.1.

**Proposition 6.5** Let $n = \text{Card}(Q)$. *Then **M** has the twinning property iff for all* $q_1, q_2 \in Q$, (6.1) *holds for all pairs* $f, u \in X^*$ *with* $|fu| \leqslant 2n^2$.

Proof. We argue by induction on $|fu|$, i.e. we assume that (6.1) holds for all $q_1$, $q_2 \in Q$ and for all pairs $f'$, $u'$ of words admissible for $q_1$, $q_2$ such that $|f'u'| < |fu|$. Consider $q_1$, $q_2 \in Q$, and consider a pair $f$, $u$ of words such that the hypotheses of (6.1) hold. Clearly we may assume $|fu| \geqslant 2n^2$ and $|u| > 0$. Next either $|f| > n^2$ or $|u| > n^2$. If $|f| \geqslant n^2+1$ then there exist a factorization $f = f_1vf_2$, with $v \neq 1$, and $r_1$, $r_2 \in Q$, $c_1$, $d_1$, $e_1$, $c_2$, $d_2$, $e_2 \in Y^*$ such that $a_1 = c_1d_1e_1$, $a_2 = c_2d_2e_2$ and

$$c_1 = \mu(f_1)_{q_-r_1}, \quad d_1 = \mu v_{r_1r_1}, \quad e_1 = \mu(f_2)_{r_1q_1},$$

$$c_2 = \mu(f_1)_{q_-r_2}, \quad d_2 = \mu v_{r_2r_2}, \quad e_2 = \mu(f_2)_{r_2q_2}.$$

Thus $f_1$, $v$ is an admissible pair for $r_1$, $r_2$ and $f_1f_2$, $u$ is an admissible pair for $q_1$, $q_2$. Consequently by induction

$$c_1d_1c_1^{-1} = c_2d_2c_2^{-1} \qquad \text{and} \qquad c_1e_1b_1e_1^{-1}c_1^{-1} = c_2e_2b_2e_2^{-1}c_2^{-1}.$$

Hence $a_1b_1a_1^{-1} = (c_1d_1c_1^{-1})(c_1e_1b_1e_1^{-1}c_1^{-1})(c_1d_1^{-1}c_1^{-1})$

$$= c_2d_2e_2b_2e_2^{-1}d_2^{-1}c_2^{-1} = a_2b_2a_2^{-1}.$$

Next, assume $|u| \geqslant n^2+1$. Then similarly, there exist a factorization $u = u_1vu_2$, with $v \neq 1$, and $r_1$, $r_2 \in Q$, $c_1$, $d_1$, $e_1$, $c_2$, $d_2$, $e_2 \in Y^*$ such that $b_1 = c_1d_1e_1$,

$b_2=c_2d_2e_2$, and

$$c_1=\mu(u_1)_{q_1r_1},\qquad d_1=\mu v_{r_1r_1},\qquad e_1=\mu(u_2)_{r_1q_1},$$
$$c_2=\mu(u_1)_{q_2r_2},\qquad d_2=\mu v_{r_2r_2},\qquad e_2=\mu(u_2)_{r_2q_2}.$$

If $u_2=1$, then $r_1=q_1$, $r_2=q_2$, $e_1=e_2=1$. Thus $(f, u_1)$ and $(f, v)$ are admissible pairs for $q_1$, $q_2$, and by induction

$$a_1c_1a_1^{-1}=a_2c_2a_2^{-1}\qquad\text{and}\qquad a_1d_1a_1^{-1}=a_2d_2a_2^{-1}.$$

Then $a_1b_1a_1^{-1}=a_1c_1d_1a_1^{-1}=(a_1c_1a_1^{-1})(a_1d_1a_1^{-1})=(a_2c_2a_2^{-1})(a_2d_2a_2^{-1})$
$=a_2b_2a_2^{-1}$.

Finally if $u_2\neq 1$, then $f$, $u_1u_2$ is an admissible pair for $q_1$, $q_2$, and $fu_1$, $v$ is an admissible pair for $r_1$, $r_2$. By induction

$$a_1c_1e_1a_1^{-1}=a_2c_2e_2a_2^{-1}\qquad\text{and}\qquad a_1c_1d_1c_1^{-1}a_1^{-1}=a_2c_2d_2c_2^{-1}a_2^{-1}.$$

It follows that

$$\begin{aligned}a_1b_1a_1^{-1}&=a_1c_1d_1e_1a_1^{-1}=(a_1c_1d_1c_1^{-1}a_1^{-1})(a_1c_1e_1a_1^{-1})\\&=(a_2c_2d_2c_2^{-1}a_2^{-1})(a_2c_2e_2a_2^{-1})=a_2c_2d_2e_2a_2^{-1}=a_2b_2a_2^{-1}.\end{aligned}$$ ■

**Proposition 6.6** *If **M** has n states and has the twinning property, then $\alpha$ preserves left factors iff $\alpha(1)=1$ and for any $f\in X^*$ with $|f|\leq n^2$, and for any $x\in X$, $\alpha(fx)\neq 0$ implies $\alpha(fx)\in\alpha(f)Y^*$.*

Proof. The conditions are obviously necessary. Conversely, let $f\in X^*$, $x\in X$ such that $fx\in\mathrm{dom}(\alpha)$. Arguing by induction, we may assume that $|f|>n^2$. There exists a factorization $f=f_1vf_2$ with $v\neq 1$, and $q_1, q_2\in Q$, $a_1, b_1, c_1, a_2, b_2, c_2\in Y^*$ such that

$$\alpha(f)=a_1b_1c_1,\qquad \alpha(fx)=a_2b_2c_2,$$
$$a_1=(\mu f_1)_{q_-q_1},\qquad b_1=\mu v_{q_1q_1},\qquad c_1=\mu(f_2)_{q_1q_+},$$
$$a_2=(\mu f_1)_{q_-q_2},\qquad b_2=\mu v_{q_2q_2},\qquad c_2=\mu(f_2x)_{q_2q_+}.$$

It follows that $\alpha(f_1f_2)=a_1c_1$, $\alpha(f_1f_2x)=a_2c_2$. Since **M** has the twinning property, and since $f_1$, $v$ is an admissible pair for $q_1$, $q_2$,

$$a_1b_1a_1^{-1}=a_2b_2a_2^{-1}. \tag{6.6}$$

Next since $|f_1f_2|<|f|$, there is a word $u\in Y^*$ such that

$$a_2c_2=\alpha(f_1f_2x)=\alpha(f_1f_2)u=a_1c_1u. \tag{6.7}$$

Combining (6.6) and (6.7), we obtain

$$\alpha(fx)=a_2b_2c_2=a_2b_2a_2^{-1}a_2c_2=a_1b_1a_1^{-1}a_1c_1u=a_1b_1c_1u=\alpha(f)u.$$ ■

Proof of Theorem 6.1 Since $\alpha$ is realized by $\boldsymbol{M}$, $\alpha$ is rational. Consequently $\alpha^{-1}: Y^* \to X^*$ is a rational transduction and by Corollary III.4.2, $\alpha^{-1}$ preserves rational languages. Thus in view of Theorem 2.7, $\alpha$ is subsequential iff $\alpha$ has bounded variation, and by Proposition 6.4 this holds iff $\boldsymbol{M}$ has the twinning property which is decidable by Proposition 6.5. Thus it is decidable whether $\alpha$ is subsequential. Further $\alpha$ is sequential iff $\boldsymbol{M}$ has the twinning property and $\alpha$ preserves left factors. By Proposition 6.6, this is decidable. Thus the proof is complete. ■

# V Families of Languages

This chapter introduces basic notions on families of languages. Rational cones are defined in Section 2, and closure properties of cones are derived from the examples of Chapter III. The second important type of families of languages considered in this book, namely full AFLs, is defined in Section 4. The next section contains basic facts about substitutions in relation with cones and full AFLs. In Section 6, we study the family of linear languages which is shown to be a principal cone, but not a full AFL. In the last section, we prove the incomparability of some special languages.

## V.1 Definition

A family of languages should contain all languages having the same, or similar "structure". This structure does not depend of the alphabet chosen to represent the language. More precisely, if $X$ is an alphabet and $A \subset X^*$, and if $A'$ is a copy of $A$ on an alphabet $X'$, then $A$ and $A'$ have the same structure. We thus require a family of languages to contain "all" copies of its elements, with the necessary set-theoretic precautions to avoid any paradox. We therefore introduce a fixed infinite alphabet $\Sigma$, and restrict alphabets, languages and the above requirements to subsets of $\Sigma$ and $\Sigma^*$.
Let $\Sigma$ be a fixed, infinite alphabet. In the sequel, an *alphabet* is a finite, non-empty subset $X$ of $\Sigma$. A *formal language* $L$ is a set $L \subset X^*$, where $X$ is an alphabet. For any language $L$, we denote by $X_L$ the least alphabet such that $L \subset X_L^*$. If $L = \emptyset$ or $L = \{1\}$, then $X_L$ is an arbitrary alphabet. Whenever, in the sequel, new letters are used or constructed, they will be assumed to be in $\Sigma$.

**Definition** A *family of languages* $\mathscr{L}$ is a nonempty set of languages, distinct from $\{\emptyset\}$, and satisfying the following property:

if $L \in \mathscr{L}$, and if $L'$ is a copy of $L$, then $L' \in \mathscr{L}$.

Recall (see Section I.4) that $L'$ is a copy of $L$ if $L' = \alpha(L)$, where $\alpha$ is an isomorphism from $X_L^*$ onto $X_{L'}^*$.

**Example 1.1** We introduce some fundamental families of languages. Rat is the family of all regular or rational languages; Alg is the family of context-free or algebraic languages; Lin is the family of linear languages (see Section 6). There are several simple families of languages which will be used to represent some closure properties by means of substitution. The family Elm consists of all finite subsets of $\Sigma$, that is of all alphabets and of the empty set. The family Fin is the family of all finite languages (including $\emptyset$ and $\{1\}$).

**Example 1.2** The set of all languages is a family of languages. We denote it by $\hat{\Sigma}$.

Let $\mathscr{L}$ be a family of languages. Then $\mathscr{L}$ is finite if there exist finitely many languages $L_1, \ldots, L_n$ such that any language $L$ of $\mathscr{L}$ is a copy of one of the languages $L_1, \ldots, L_n$. In this case, we write $\mathscr{L} = \{L_1, \ldots, L_n\}$. If all languages of $\mathscr{L}$ are copies of a single language $L$, that is $\mathscr{L} = \{L\}$, then $\mathscr{L}$ is said to have just one element.

The union of an arbitrary number of families of languages is a family of languages; the same is true for intersection, provided the intersection is nonempty and distinct from $\{\emptyset\}$. Families of languages are ordered by inclusion. Two families $\mathscr{L}$ and $\mathscr{M}$ are incomparable if neither $\mathscr{L} \subset \mathscr{M}$ nor $\mathscr{M} \subset \mathscr{L}$.

**Exercise**

**1.1** Describe the minimal families of languages with respect to the inclusion order.

## V.2 Rational Equivalence, Rational Cones

Rational domination and rational equivalence are the basic tools for comparison of formal languages. These relations are transformed into inclusion and equality of families of languages of a special type, called rational cones. Two languages are rationally equivalent if each one can be transformed into the other by a rational transduction. Since these transformations involve only finite memory, the "essential" features of two rationally equivalent languages are the same in a sense which will be precised in Chapter VIII. Rational cones are families of languages closed under morphism, inverse morphism and intersection with regular sets. Further closure properties follow from the properties of rational transductions.

**Definition** Let $L$ and $L'$ be languages. Then $L'$ is a rational image of $L$, or $L$ dominates rationally $L'$ iff there exists a rational transduction $\tau: X_L^* \to X_{L'}^*$ such that $L' = \tau(L)$. We then write $L \geq L'$ or $L' \leq L$.

The composition of two rational transductions is still a rational transduction; thus $\geq$ is a preorder. If $L \geq L'$ and $L' \geq L$, then $L$ and $L'$ are rationally equivalent. We denote this fact by $L \approx L'$, and the contrary by $L \not\approx L'$. We write $L > L'$ when $L \geq L'$ and $L \not\approx L'$ and say that $L$ strictly dominates $L'$, and we write $L \not\geq L'$ for: not $(L \geq L')$. Finally, $L$ and $L'$ are (rationally) incomparable, if neither $L \geq L'$ nor $L' \geq L$.

**Example 2.1** The languages $S_=\{a^nb^n|n\geq 0\}$ and $A=\{(x^+y)^nz^n|n\geq 0\}$ are rationally equivalent: $S_=\approx A$. Indeed, $A$ is obtained from $S_=$ by the rational substitution which maps $a$ into $x^+y$ and $b$ into $z$. The converse transformation is performed by the following transducer with initial and final state 1 (Fig. V.1).

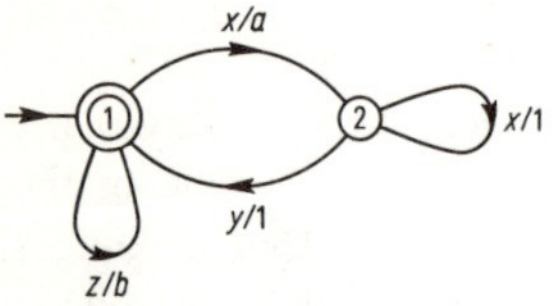


Fig. V.1

**Example 2.2** Let $S_\leq=\{a^nb^m \mid 0\leq n\leq m\}$. Then $S_=>S_\leq$. Indeed, $S_\leq$ is the image of $S_=$ under the substitution which substitutes $a$ to $a$ and $b^+$ to $b$. The nonequivalence of the two languages will be shown in Section 7.

**Example 2.3** Let $S_\geq=\{a^nb^m \mid n\geq m\geq 0\}$. The languages $S_\geq$ and $S_\leq$ are rationally incomparable. This will be proved in Section 7, too.

**Definition** A family of languages $\mathscr{L}$ is a rational cone, or shorter a cone, if it satisfies the following condition:

$$L\in\mathscr{L}, L\geq L' \text{ imply } L'\in\mathscr{L}.$$

Thus, a rational cone is a family of languages closed under rational transduction. A rational cone is called a "full trio" by Ginsburg [1975], since it is closed under three operations: morphism, inverse morphism, and intersection with regular sets. Conversely, a family of languages closed under these three operations is, by Nivat's theorem, also closed under rational transduction. Note that in order to prove that a given family of languages is a cone, it suffices to show that it is closed under projection, inverse projection, and intersection with rational languages. This frequently simplifies the verification.

**Example 2.4** The families $\hat{\Sigma}$, Rat, Alg, Lin are cones; neither Elm nor Fin is a cone.

The cone Rat is the least cone, since the following property holds

**Proposition 2.1** *Any rational cone contains* Rat.

Proof. Let $\mathscr{L}$ be a cone, and $L\in\mathscr{L}$, $L\neq\emptyset$. Let $K\in$ Rat. The transduction $\tau_K: X_L^*\to X_K^*$ defined for $w\in X_L^*$ by $\tau_K(w)=K$ is rational (see Example III.5.2). Since $\tau_K(L)=K$ and $\mathscr{L}$ is a cone, it follows that $K\in\mathscr{L}$. ■

By this proposition, the intersection of a family of rational cones is nonempty; it is easily seen to be again a rational cone.

**Definition** Let $\mathscr{L}$ be a family of languages. The cone generated by $\mathscr{L}$ is the intersection of all cones containing $\mathscr{L}$, that is the least cone containing $\mathscr{L}$. It is denoted by $\mathscr{L}\Gamma$. If $\mathscr{L}=\{L\}$, we write $L\Gamma$ instead of $\{L\}\Gamma$.

**Proposition 2.2** *Let $\mathscr{L}$ be a family of languages; then*

$$\mathscr{L}\Gamma = \{L' \mid L' \leqslant L \quad \text{for some} \quad L \in \mathscr{L}\}.$$

Proof. Set $\mathscr{M} = \{L' | L' \leqslant L$ for some $L \in \mathscr{L}\}$. Then $M \subset \mathscr{L}\Gamma$ since $\mathscr{L}\Gamma$ is a cone. In order to show the converse inclusion, it suffices to show that $\mathscr{M}$ is a cone, since $\mathscr{L} \subset \mathscr{M}$. Let $M \in \mathscr{M}$, and $M'$ such that $M \geqslant M'$. There exists $L \in \mathscr{L}$ such that $L \geqslant M$, and since $\geqslant$ is transitive, this implies $L \geqslant M'$ and thus $M' \in \mathscr{M}$. Thus $\mathscr{M}$ is a cone. ∎

**Proposition 2.3** *Let $\mathscr{L}$ and $\mathscr{L}'$ be two families of languages; then*

$$\mathscr{L} \subset \mathscr{L}\Gamma; \qquad \mathscr{L} \subset \mathscr{L}' \Rightarrow \mathscr{L}\Gamma \subset \mathscr{L}'\Gamma; \qquad \mathscr{L}\Gamma\Gamma = \mathscr{L}\Gamma.$$ ∎

We now summarize some closure properties of rational cones.

**Proposition 2.4** *Let $\mathscr{L}$ be a cone; if $L \in \mathscr{L}$ and $K \in \mathrm{Rat}$, then*

$$LK, \quad KL, \quad LK^{-1}, \quad K^{-1}L, \quad L \cup K, \quad L \backslash K \in \mathscr{L}.$$

Proof. Let $X$ be an alphabet such that $L, K \subset X^*$. The transduction $w \mapsto wK$ from $X^*$ into $X^*$ is rational by Example III.5.6; thus $L \geqslant LK$ and $LK \in \mathscr{L}$. Similarly, $KL \in \mathscr{L}$. The transductions $w \mapsto K^{-1}w$ and $w \mapsto w \cup K$ are also rational (Examples III.5.7 and 5). Finally $L \backslash K = L \cap (X^* \backslash K) \in \mathscr{L}$. ∎

We single out a special case of proposition 2.4 which will be of constant use in the sequel.

**Corollary 2.5** *Let $\mathscr{L}$ be a cone, let $L$ be a language and # a letter not in $X_L$; then*

$$L \in \mathscr{L} \quad \Leftrightarrow \quad \#L \in \mathscr{L} \quad \Leftrightarrow \quad L\# \in \mathscr{L}.$$ ∎

**Definition** A rational cone $\mathscr{L}$ is principal if there exists a language $L_0 \in \mathscr{L}$ such that any language of $\mathscr{L}$ is a rational image of $L_0$, that is if $\mathscr{L} = L_0\Gamma$. The language $L_0$ is a cone generator of $\mathscr{L}$.

Two generators of a cone $\mathscr{L}$ are rationally equivalent. More precisely, $L \geqslant L'$ holds iff $L\Gamma \supset L'\Gamma$, and the languages $L$ and $L'$ are incomparable iff the cones $L\Gamma$ and $L'\Gamma$ are incomparable families of languages.

In general, a cone is not closed under union (Exercise 2.3). For principal cones we have

**Proposition 2.6** *A principal cone is closed under union.*

Proof. Let $L_0$ be a generator of a principal cone $\mathscr{L}$, and let $L, M \in \mathscr{L}$. Let $X$ be an alphabet such that $L, M \subset X^*$, and set $Y = X_{L_0}$. There exist two rational transductions $\tau, \tau' : Y^* \to X^*$ such that $\tau(L_0) = L$, $\tau'(L_0) = M$. Then the transduction $\tau'' = \tau \cup \tau'$ is also rational, and $\tau''(L_0) = L \cup M$; thus $L \cup M \in L_0\Gamma = \mathscr{L}$. ∎

**Exercises**

**2.1** Show that Rat is a principal cone and that any nonempty rational language is a generator of the cone Rat.

**2.2** Show that the union of a family of cones is a cone.

**2.3** Let $\mathscr{L}$ and $\mathscr{M}$ be two cones, and define

$$\mathscr{L} \vee \mathscr{M} = \{L \cup M \mid L \in \mathscr{L}, M \in \mathscr{M}\}.$$

a) Show that $\mathscr{L} \vee \mathscr{M}$ is a cone, and that $\mathscr{L} \cup \mathscr{M} \subset \mathscr{L} \vee \mathscr{M}$.
b) Show that $\mathscr{L} \cup \mathscr{M} \not\subset \mathscr{L} \vee \mathscr{M}$ if $\mathscr{L}$ and $\mathscr{M}$ are incomparable (Hint. Show that there are languages $L \in \mathscr{L} \backslash \mathscr{M}$ and $M \in \mathscr{M} \backslash \mathscr{L}$ such that $M \subset X_M^+$, $L \subset X_L^+$, $X_M \cap X_L = \emptyset$, and take the language $A = L \cup M$.) Conclude that if $\mathscr{L}$ and $\mathscr{M}$ are incomparable, then $\mathscr{L} \cup \mathscr{M}$ is not closed under union.
c) Assume that $\mathscr{L}$ and $\mathscr{M}$ are principal cones; show that $\mathscr{L} \cup \mathscr{M}$ is principal iff $\mathscr{L}$ and $\mathscr{M}$ are comparable.

**2.4** Let $X = \{a, b\}$, and $\Theta = \{=, \neq, \geqslant, \leqslant, >, <\}$. The language of Goldstine (from Goldstine [1972]) $G_\theta$ is defined, for $\theta \in \Theta$ by:

$$G_\theta = \{a^{n_1} b a^{n_2} b \cdots a^{n_p} b \mid p \geqslant 1, n_i \geqslant 0, \exists j, 1 \leqslant j \leqslant p : n_j \theta j\}.$$

Let $S_=$ be as in Example 2.1.

a) Show that $S_= \geqslant G_\theta$ for $\theta \in \Theta$. (This shows that $G_\theta$ is context-free.)
b) Show that $S_= \approx G_=$, $G_< \approx G_\leqslant$, $G_\geqslant \approx G_>$.

## V.3 Rationally Closed Families

The second important class of families of languages is the class of rationally closed families. Rational closure means closure under the three "rational" operations, namely union, product and the star (or the plus) operation. In connection with closure under rational transductions, this leads to the notion of full AFL defined in the next section.

**Definition** A family of languages $\mathscr{L}$ is rationally closed if the following conditions are satisfied:

(i) $\emptyset \in \mathscr{L}, \{1\} \in \mathscr{L}$;

(ii) $L, M \in \mathscr{L} \Rightarrow L \cup M, L \cdot M \in \mathscr{L}$;

(iii) $L \in \mathscr{L} \Rightarrow L^+ \in \mathscr{L}$.

Conditions (i), (ii), (iii) imply that $L^* = \{1\} \cup L^+ \in \mathscr{L}$ for $L \in \mathscr{L}$; conversely, if (ii) holds, then $L, L^* \in \mathscr{L}$ imply $L^+ = LL^* \in \mathscr{L}$.

The above definition slightly differs from the definition of Eilenberg [1974] by the fact that we require the additional condition (i). This difference is not essential but simplifies alternative representations (see Proposition 3.2).

**Example 3.1** The families Rat, Alg, $\hat{\Sigma}$ are rationally closed and Lin is not rationally closed (see Section 6).

The intersection of a family of rationally closed families of languages is again a rationally closed family of languages. The rational closure of a family $\mathscr{L}$ is the intersection of all rationally closed families containing $\mathscr{L}$, and thus the least rationally closed family of languages containing $\mathscr{L}$. The rational closure of $\mathscr{L}$ is denoted by $\mathscr{L}\,\underline{\mathrm{Rat}}$. Clearly, we have

**Proposition 3.1** *Let $\mathscr{L}$ and $\mathscr{L}'$ be two families of languages; then*

$$\mathscr{L} \subset \mathscr{L}\,\underline{\mathrm{Rat}}; \qquad \mathscr{L} \subset \mathscr{L}' \Rightarrow \mathscr{L}\,\underline{\mathrm{Rat}} \subset \mathscr{L}'\,\underline{\mathrm{Rat}}; \qquad \mathscr{L}\,\underline{\mathrm{Rat}}\,\underline{\mathrm{Rat}} = \mathscr{L}\,\underline{\mathrm{Rat}}. \qquad \blacksquare$$

Let $\mathscr{L}$ be a family of languages and let $\sigma: X^* \to Y^*$ be a substitution. Then $\sigma$ is a $\mathscr{L}$-substitution if $\sigma(x) \in \mathscr{L}$ for all $x \in X$. The following result links together rational closure and $\mathscr{L}$-substitutions.

**Proposition 3.2** *Let $\mathscr{L}$ be a family of languages; then*

$$\mathscr{L}\,\underline{\mathrm{Rat}} = \{\sigma(K) \mid K \in \mathrm{Rat}, \sigma \text{ a } \mathscr{L}\text{-substitution}\}.$$

Proof. Set

$$\mathscr{M} = \{\sigma(K) \mid K \in \mathrm{Rat}, \sigma \text{ a } \mathscr{L}\text{-substitution}\}.$$

We first show the inclusion $\mathscr{L}\,\underline{\mathrm{Rat}} \subset \mathscr{M}$, by proving that $\mathscr{M}$ contains $\mathscr{L}$ and is rationally closed. Let $L \in \mathscr{L}$, let $x$ be a letter, and define $\sigma: x^* \to X_L^*$ by $\sigma(x) = L$. Then $\sigma$ is a $\mathscr{L}$-substitution, and $x \in \mathrm{Rat}$, thus $L = \sigma(x) \in \mathscr{M}$. This shows that $\mathscr{L} \subset \mathscr{M}$.

Next, consider $L, M \in \mathscr{M}$, and let $Y$ be an alphabet such that $L, M \subset Y^*$. There exist two $\mathscr{L}$-substitutions $\sigma_1: X_1^* \to Y^*$ and $\sigma_2: X_2^* \to Y^*$ such that $L = \sigma_1(K_1)$, $M = \sigma_2(K_2)$ for some $K_1, K_2 \in \mathrm{Rat}$. After a copy if necessary, we may assume $X_1 \cap X_2 = \emptyset$. Set $X = X_1 \cup X_2$ and define a substitution $\sigma: X^* \to Y^*$ by

$$\sigma(x) = \begin{cases} \sigma_1(x_1) & \text{if } x_1 \in X_1; \\ \sigma_2(x_2) & \text{if } x_2 \in X_2. \end{cases}$$

Then $\sigma$ is a $\mathscr{L}$-substitution, and $L = \sigma(K_1)$, $M = \sigma(K_2)$. This implies

$$L \cup M = \sigma(K_1 \cup K_2), \qquad LM = \sigma(K_1 K_2), \qquad L^+ = \sigma(K_1^+) \in \mathscr{M}.$$

Since $\emptyset = \sigma(\emptyset)$, $\{1\} = \sigma(1) \in \mathscr{M}$, the family $\mathscr{M}$ is rationally closed. In order to prove the converse inclusion $\mathscr{M} \subset \mathscr{L}\,\underline{\mathrm{Rat}}$, consider the set $\mathscr{N}$ of all languages $K$ such that $\sigma(K) \in \mathscr{L}\,\underline{\mathrm{Rat}}$ for any $\mathscr{L}$-substitution $\sigma$. By definition $\mathscr{L}\,\underline{\mathrm{Rat}} \supset \{\sigma(K) \mid K \in \mathscr{N}, \sigma \text{ a } \mathscr{L}\text{-substitution}\}$ and it suffices to show that $\mathrm{Rat} \subset \mathscr{N}$.

Clearly $\mathscr{N}$ is closed under copy and thus $\mathscr{N}$ is a family of languages. Next $\emptyset$, $\{1\} \in \mathscr{N}$ and $\{x\} \in \mathscr{N}$ for all $x \in \Sigma$. Assume now $K, K' \in \mathscr{N}$. Set $X = X_K \cup X_{K'}$ and let $\sigma: X^* \to Y^*$ be an arbitrary $\mathscr{L}$-substitution. By assumption,

$\sigma(K), \sigma(K') \in \mathscr{L}\ \underline{\text{Rat}}$; thus $K \cup K'$, $K \cdot K'$, $K^+ \in \mathscr{N}$ since $\mathscr{L}\ \underline{\text{Rat}}$ is rationally closed. This shows that $\mathscr{N} \supset \text{Rat}$, and proves the inclusion. ∎

Proposition 3.2 is very useful for proving properties of families of languages constructed by substitutions. We give now an example of this fact.

**Definition** A language $A$ has no infinite rational subset if $K \in \text{Rat}$ and $K \subset A$ imply that $K$ is finite.

Such a language is said to satisfy the IRS-condition by Greibach [1975].

**Example 3.2** The language $\{a^{k+n}b^n c^p d^{p+k} \mid k, p, n \geq 1\}$ has no infinite rational subset; on the contrary, the language $S_< = \{x^n y^m \mid 0 \leq n < m\}$ contains infinite rational languages, for instance $y^+$.

**Proposition 3.3** *Let A be a language having no infinite rational subset, and let $\mathscr{L}$ be a family of languages; if $A \in \mathscr{L}\ \underline{\text{Rat}}$, then A is a finite union of products of languages in $\mathscr{L}$.*

Proof. Since $A \in \mathscr{L}\ \underline{\text{Rat}}$, there exist a rational language $K$, and a $\mathscr{L}$-substitution $\sigma: X_K^* \to X_A^*$ such that $A = \sigma(K)$. We always may assume that $\sigma(x) \neq 1$ for any letter $x \in X_K$. Indeed, let $X_0 = \{x \in X_K : \sigma(x) = 1\}$, set $Y = X_K \setminus X_0$, and let $\pi$ be the projection of $X_K^*$ onto $Y^*$; then $A = \sigma'(\pi(K))$ where $\sigma'$ is the restriction of $\sigma$ to $Y^*$; $\sigma'(x) \neq 1$ for $x \in Y$ and $\pi(K) \in \text{Rat}$.
We have to prove that $K$ is finite. Assume $K$ infinite; then by the iteration lemma, there exist words $a, u, b$ such that $au^*b \subset K$ and $u \neq 1$. Let $f \in \sigma(a)$, $w \in \sigma(u)$, $g \in \sigma(b)$; since $\sigma(u) \neq 1$, we can choose $w \neq 1$. Then $fw^*g \subset A$, and $A$ contains an infinite rational set, contrary to the hypothesis. ∎

## Exercises

**3.1** Let $\mathscr{L}$ be a family of languages. Show that

$$\{\sigma(K) \mid K \in \text{Elm}, \sigma \text{ a } \mathscr{L}\text{-substitution}\}$$

is the least family of languages containing $\mathscr{L} \cup \{\emptyset\}$ and closed under union. Show that

$$\{\sigma(K) \mid K \in \text{Fin}, \sigma \text{ a } \mathscr{L}\text{-substitution}\}$$

is the least family of languages containing $\mathscr{L} \cup \{\emptyset, \{1\}\}$ and closed under union and product.

**3.2** A language $A$ is said to contain no infinite product if for any two languages $L, L'$ such that $LL' \subset A$, either $L$ or $L'$ is finite.

a) Show that a language which contains no infinite product contains no infinite regular set; show that the converse is false.

b) Show that if $A$ contains no infinite product, and if $A \in \mathscr{L}\ \underline{\text{Rat}}$, where $\mathscr{L}$ is a cone, then $A$ is a finite union of languages in $\mathscr{L}$.

## V.4 Full AFLs

Families of languages which are both rational cones and rationally closed are called full AFLs. In this section, we prove some general properties of full AFLs. The main "decomposition" theorem stating that any full AFL is equal to the rational closure of a rational cone will be proved in Section VI.4.

**Definition** A family of languages $\mathscr{L}$ is a full AFL if $\mathscr{L}$ is a cone and is rationally closed, that is:

$$\mathscr{L} = \mathscr{L}\Gamma \quad \text{and} \quad \mathscr{L} = \mathscr{L}\,\underline{\text{Rat}}.$$

A family $\mathscr{L}$ is a full AFL iff it is closed under the following six operations: morphism, inverse morphism, intersection with regular sets, union, product, and star or plus operation.

The term AFL is an abbreviation of abstract family of languages. The adjective "full" is used to distinguish these families from AFL which are required to be closed only under continuous ($\varepsilon$-free) morphisms instead of arbitrary morphisms. This distinction is necessary for instance if context-sensitive languages have to be included into the general theory, and leads also to further refinements of the theory (see Ginsburg [1975]). In this book, we restrict ourselves to full AFL.

Sometimes, it is convenient to consider a notion which is intermediate between cones and full AFLs:

**Definition** A family of languages $\mathscr{L}$ is a full semi-AFL if it is a cone and is closed under union.

**Example 4.1** The families Rat, Alg, $\hat{\Sigma}$ are full AFLs.

**Example 4.2** Any principal rational cone is a full semi-AFL.

The intersection of a family of full AFLs is again a full AFL. The full AFL generated by a family of languages $\mathscr{L}$ is the intersection of all full AFLs containing $\mathscr{L}$, that is the least full AFL containing $\mathscr{L}$. It is denoted by $\mathscr{L}\hat{\Gamma}$.

**Proposition 4.1** *Let $\mathscr{L}$ and $\mathscr{L}'$ be two families of languages; then*

$$\mathscr{L} \subset \mathscr{L}\hat{\Gamma}; \qquad \mathscr{L} \subset \mathscr{L}' \Rightarrow \mathscr{L}\hat{\Gamma} \subset \mathscr{L}'\hat{\Gamma}; \qquad \mathscr{L}\hat{\Gamma}\hat{\Gamma} = \mathscr{L}\hat{\Gamma}. \qquad \blacksquare$$

The six operations defining full AFLs are not independent.

**Proposition 4.2** *Let $\mathscr{L}$ be a rational cone closed under union and plus-operation; then $\mathscr{L}$ is a full* AFL.

Thus a family of languages is closed under product provided it is closed under the other five of the six AFL operations. For other relations between the six axioms for full AFLs see Ginsburg [1975].

Proof. Let $L_1, L_2$ be languages in $\mathscr{L}$. We have to show that $L_1L_2 \in \mathscr{L}$. Set $X = X_{L_1} \cup X_{L_2}$ and let $\#, \flat$ be two new letters. Then $L_1\#, L_2\flat \in \mathscr{L}$ by Corollary 2.5, and $L_1\# \cup L_2\flat$, $(L_1\# \cup L_2\flat)^+ \in \mathscr{L}$ by the assumptions on $\mathscr{L}$. Since $\mathscr{L}$ is a cone, it follows

$$L = (L_1\# \cup L_2\flat)^+ \cap X^*\#X^*\flat \in \mathscr{L}.$$

Now $L = L_1\#L_2\flat$. Consider the projection $\pi : (X \cup \# \cup \flat)^* \to X^*$. Then $\pi(L) = L_1L_2 \in \mathscr{L}$. ∎

**Definition** Let $\mathscr{L}$ be a full AFL. Then $\mathscr{L}$ is a principal full AFL if there exists a language $L_0$ such that $\mathscr{L} = L_0\hat{\Gamma}$. The language $L_0$ is a full AFL generator of $\mathscr{L}$.

Full AFL generators and cone generators must be carefully distinguished. Consider indeed a principal full AFL $\mathscr{L}$, and let $L$ be a full AFL generator of $\mathscr{L}$; thus $L\hat{\Gamma} = \mathscr{L}$. Since $\mathscr{L}$ is a full AFL, it is also a cone.
Consider the cone $L\Gamma$ generated by $L$. Then $L\Gamma \subset \mathscr{L}$, and in general the inclusion is strict. On the other hand, we shall prove that any full principal AFL $\mathscr{L}$ is also a principal cone (Proposition 4.4 below). Let $L'$ be a cone generator of $\mathscr{L} : L'\Gamma = \mathscr{L}$. Then $\mathscr{L} = L'\Gamma \subset L'\hat{\Gamma} = \mathscr{L}$. Thus any cone generator of a principal full AFL is also a full AFL generator. It is quite remarkable that the converse is also true in the special case of substitution-closed full AFL, and thus for the family Alg. This will be shown in Section VII.3.
We first prove the following:

**Proposition 4.3** *Let $\mathscr{L}$ be a finite family of languages; then the full* AFL $\mathscr{L}\hat{\Gamma}$ *generated by $\mathscr{L}$ is principal.*

Proof. Let $\mathscr{L} = \{L_1, L_2, \ldots, L_n\}$, set $X = X_{L_1} \cup X_{L_2} \cup \cdots \cup X_{L_n}$, and let $\#_1, \#_2, \ldots, \#_n \notin X$ be $n$ new letters. Define

$$L = L_1\#_1 \cup L_2\#_2 \cup \cdots \cup L_n\#_n;$$

Each language $L_i\#_i$ is in $\mathscr{L}\Gamma$ by Corollary 2.5, thus $L \in \mathscr{L}\hat{\Gamma}$ and $L\hat{\Gamma} \subset \mathscr{L}\hat{\Gamma}$. In order to show the converse inclusion, note first that

$$L_i\#_i = L \cap X^*\#_i;$$

thus $L_i\#_i \in L\Gamma$ and $L_i \in L\Gamma$ by Corollary 2.5 for $i = 1, \ldots, n$. This implies that $\mathscr{L} \subset L\Gamma$, and hence $\mathscr{L}\hat{\Gamma} \subset L\hat{\Gamma}$. ∎

Note that Proposition 4.3 also holds for full semi-AFL, but is false for cones (Exercise 4.1).

**Theorem 4.4** (Ginsburg and Greibach [1970]) *Let $L$ be a language, and let $\#$ be a letter not in $X_L$; then*

$$L\hat{\Gamma} = (L\#)^+\Gamma.$$

This theorem supplies explicitly a cone generator for any principal full AFL. For the proof, we need two preliminary results.

**Proposition 4.5** *Let $L, M$ be two languages, and let $\#$ be a letter not in $X_L \cup X_M$; if $L \geqslant M$, then $(L\#)^+ \geqslant (M\#)^+$.*

Proof. Let $\tau: X_L^* \to X_M^*$ be a rational transduction such that $\tau(L) = M$, and let $R \in \text{Rat}(X_L^* \times X_M^*)$ be the graph of $\tau$. Consider now the relation

$$S \subset (X_L \cup \#)^* \times (X_M \cup \#)^*$$

defined by $S = R \cdot \{(\#, \#)\}$. Then $S$ is rational. Let

$$\bar{\tau}: (X_L \cup \#)^* \to (X_M \cup \#)^*$$

be the rational transduction defined by $S$. Then $\bar{\tau}(w) \neq \emptyset$ only if $w = w'\#$ for some $w' \in X_L^*$, and then

$$\bar{\tau}(w'\#) = \tau(w')\#. \tag{4.1}$$

Consider now the rational tranduction $\bar{\tau}^+$ defined by $S^+$. Then

$$f \in \bar{\tau}^+(w)$$

iff $w = w_1\#w_2\# \cdots w_n\#$ for some $n \geqslant 1$, $w_1, w_2, \ldots, w_n \in X_L^*$, and by (4.1):

$$f \in \tau(w_1)\#\tau(w_2)\# \cdots \tau(w_n)\#.$$

This shows that $\bar{\tau}^+((L\#)^+) = (M\#)^+$ and achieves the proof. ■

**Lemma 4.6** *Let $L$ be a language, and let $\#, \$$ be two letters not in $X_L$; then $(L\#)^+ \geqslant [(L\#)^+\$]^+$.*

Proof. Set $Y = X_L \cup \#$, consider a third letter $\flat$ not in $Y$, and define a morphism

$$\theta: (Y \cup \flat)^* \to Y^*$$

by setting $\theta(y) = y$ $(y \in Y)$, $\theta(\flat) = \#$. Then $\theta$ is strictly alphabetic, and

$$\theta^{-1}[(L\#)^+] = (L\# \cup L\flat)^+,$$

thus $\quad \theta^{-1}[(L\#)^+] \cap [(X_L{}^*\#)^* X_L \flat]^+ = [(L\#)^* L\flat]^+.$

Hence $\quad (L\#)^+ \geqslant [(L\#)^* L\flat]^+. \tag{4.2}$

Consider now the morphism

$$\psi: (Y \cup \flat)^* \to (Y \cup \$)^*$$

defined by

$$\psi(y) = y (y \in Y), \qquad \psi(\flat) = \#\$$$

Then $\psi[(L\#)^* L\flat]^+ = [(L\#)^* L\#\$]^+ = [(L\#)^+\$]^+$;

together with (4.2), this shows that $(L\#)^+ \geqslant [(L\#)^+\$]^+$.

Proof. of Theorem 4.4 Clearly $(L\#)^+ \in L\hat{\Gamma}$, and thus

$$(L\#)^+\Gamma \subset L\hat{\Gamma}.$$

In order to prove the converse inclusion, we first note that $L \in (L\#)^+\Gamma$ since

$$(L\#)^+ \geqslant (L\#)^+ \cap X_L^*\# = L\# \geqslant L.$$

Thus, it suffices to show that $(L\#)^+\Gamma$ is a full AFL. Now $(L\#)^+\Gamma$ is already a principal cone, and closure under union follows from Proposition 2.6 In view of Proposition 4.2, it suffices to show that $(L\#)^+\Gamma$ is closed under plus operation. Let $M \in (L\#)^+\Gamma$, and let \$ be a letter not in $X_L \cup X_M \cup \#$. By Proposition 4.5, we have

$$[(L\#)^+\$]^+ \geqslant (M\$)^+,$$

and by Lemma 4.6, $(L\#)^+ \geqslant [(L\#)^+\$]^+$; thus

$$(L\#)^+ \geqslant (M\$)^+.$$

Clearly $(M\$)^+ \geqslant M^+$, showing that $M^+ \in (L\#)^+\Gamma$. Thus $(L\#)^+\Gamma$ is a full AFL, and the theorem is proved. ■

## Exercises

**4.1** Show that Proposition 4.3 remains true for full semi-AFLs, but not for cones.

**4.2** Let $\mathscr{L}$ and $\mathscr{M}$ be full AFLs.

a) Show that $\mathscr{L} \cup \mathscr{M}$ is closed under plus operation.
b) Show that $\mathscr{L} \cup \mathscr{M}$ is a full AFL iff $\mathscr{L}$ and $\mathscr{M}$ are comparable.
c) Show that $\mathscr{L} \cup \mathscr{M}$ is closed under product iff $\mathscr{L}$ and $\mathscr{M}$ are comparable.

**4.3** Prove the formulas

$$(\mathscr{L}_1 \cup \mathscr{L}_2)\,\underline{\mathrm{Rat}} = (\mathscr{L}_1\,\underline{\mathrm{Rat}} \cup \mathscr{L}_2\,\underline{\mathrm{Rat}})\,\underline{\mathrm{Rat}};$$
$$(\mathscr{L}_1 \cup \mathscr{L}_2)\hat{\Gamma} = (\mathscr{L}_1\hat{\Gamma} \cup \mathscr{L}_2\hat{\Gamma})\hat{\Gamma}.$$

**4.4** Show that the converse of Proposition 4.5 is false.

**4.5** Let $L_1$, $L_1'$, $L_2$, $L_2'$ be languages and let # be a new letter. Show that if $L_1 \geqslant L_1'$, $L_2 \geqslant L_2'$, then $L_1\#L_2 \geqslant L_1'\#L_2'$, and give an example showing that the conclusion $L_1L_2 \geqslant L_1'L_2'$ is in general false.

**4.6** Give an example showing that $L_1 \geqslant L_2$ does not imply $L_1^+ \geqslant L_2^+$.

## V.5 Substitution

Substitutions are a fundamental tool for transformation of languages. They are essentially independent of rational transductions, and can be used to represent other operations such as closure under union or rational closure. Full AFLs closed under substitution (like the family Alg) have quite remarkable properties. Here we give only some basic definitions and results on substitutions. More properties are proved in Chapters VI and VII.
First, we recall the following

**Definition** Let $\mathcal{M}$ be a family of languages. A $\mathcal{M}$-substitution is a substitution $\sigma: X^* \to Y^*$ such that $\sigma(x) \in \mathcal{M}$ for all $x \in X$.

Let $\mathcal{L}$ be a family of languages. Then we denote by $\mathcal{L} \square \mathcal{M}$ the set of all languages obtained from languages in $\mathcal{L}$ by applying $\mathcal{M}$-substitutions. Thus

$$\mathcal{L} \square \mathcal{M} = \{\sigma(L) \mid L \in \mathcal{L}, \sigma: X_L^* \to Y^* \text{ a } \mathcal{M}\text{-substitution}\}.$$

Clearly $\mathcal{L} \square \mathcal{M}$ is closed under copy, and thus $\square$ is a binary operation on families of languages. Sometimes, it is convenient to dispose of an unary notation, when one of the two operands $\mathcal{L}$ or $\mathcal{M}$ is some standard family of languages. We introduce the following equivalent notations:

$$\mathcal{L} \square \mathcal{M} = \mathcal{L}\bar{\mathcal{M}} = \mathcal{M}\underline{\mathcal{L}}.$$

Thus $\bar{\mathcal{M}}$ is an operator, parametrized by $\mathcal{M}$, and acting on the family $\mathcal{L}$; and $\underline{\mathcal{L}}$ is an operator acting on $\mathcal{M}$. These notations will be used in the next chapter, where an operator calculus is developed involving only unary operators.

**Example 5.1** The notation fits with the notation for rational closure of Section 3. Indeed, in view of Proposition 3.2,

$$\mathrm{Rat} \square \mathcal{L} = \mathrm{Rat}\,\bar{\mathcal{L}} = \mathcal{L}\,\underline{\mathrm{Rat}}$$

is the rational closure of $\mathcal{L}$.

**Example 5.2** For any family of languages $\mathcal{L}$,

$$\mathrm{Elm} \square \mathcal{L} = \mathrm{Elm}\,\bar{\mathcal{L}} = \mathcal{L}\,\underline{\mathrm{Elm}} \quad \text{and} \quad \mathrm{Fin} \square \mathcal{L} = \mathrm{Fin}\,\bar{\mathcal{L}} = \mathcal{L}\,\underline{\mathrm{Fin}}$$

are the closure under union and the closure under union and product of $\mathcal{L}$.

**Example 5.3** Substitution is not commutative. A Rat-substitution substitutes a rational set to each letter and thus is a rational substitution, a Elm-substitution substitutes a finite set of letters to each letter. In particular, any cone is closed under rational substitution:

$$\mathcal{L} \square \mathrm{Rat} = \mathcal{L}\,\overline{\mathrm{Rat}} = \mathrm{Rat}\,\underline{\mathcal{L}} \subset \mathcal{L}\Gamma,$$

but Rat $\square \mathcal{L}$ usually is not contained in $\mathcal{L}\Gamma$.

**Example 5.4** The family Rat is closed under rational substitution, thus

$$\mathrm{Rat} \square \mathrm{Rat} = \mathrm{Rat};$$

we also have $\mathrm{Alg} \square \mathrm{Alg} = \mathrm{Alg}$ since Alg is closed under context-free substitution.

The following formulas are clear; let $\mathcal{L}$, $\mathcal{L}'$, $\mathcal{M}$, $\mathcal{M}'$ be families of languages:

$$\mathcal{M} \subset \mathcal{M}' \Rightarrow \mathcal{L} \square \mathcal{M} \subset \mathcal{L} \square \mathcal{M}'; \tag{5.1}$$

$$\mathcal{L} \subset \mathcal{L}' \Rightarrow \mathcal{L} \square \mathcal{M} \subset \mathcal{L}' \square \mathcal{M}; \tag{5.2}$$

further, if $\{x\} \in \mathcal{L}$ for at least one letter $x \in \Sigma$ (and hence for all letters since $\mathcal{L}$ is a family of languages), then

$$\mathcal{M} \subset \mathcal{L} \square \mathcal{M}, \qquad \mathcal{M} \subset \mathcal{M} \square \mathcal{L}. \tag{5.3}$$

To see the first of the inclusions (5.3), consider $M \in \mathcal{M}$, $\{x\} \in \mathcal{L}$, and define a $\mathcal{M}$-substitution $\sigma : x^* \to X_M^*$ by $\sigma(x) = M$. Then $\sigma(x) = M \in \mathcal{L} \square \mathcal{M}$. For the second inclusion, define $\sigma' : X_M^* \to X_M^*$ to be the identity. Then $\sigma'(x) = \{x\} \in \mathcal{L}$ for all $x \in X_M$. Thus $\sigma'(M) = M \in \mathcal{M} \square \mathcal{L}$.
Relation (5.2) can be completed by

$$\left(\bigcup_{\lambda \in \Lambda} \mathcal{L}_\lambda\right) \square \mathcal{M} = \bigcup_{\lambda \in \Lambda} (\mathcal{L}_\lambda \square \mathcal{M}), \tag{5.4}$$

where $(\mathcal{L}_\lambda)_{\lambda \in \Lambda}$ is any set of families of languages.
The following theorem is important.

**Theorem 5.1** (Associativity) *Let $\mathcal{L}_1$, $\mathcal{L}_2$, $\mathcal{L}_3$ be families of languages; then $\mathcal{L}_1 \square (\mathcal{L}_2 \square \mathcal{L}_3) = (\mathcal{L}_1 \square \mathcal{L}_2) \square \mathcal{L}_3$.*

In unary notations, the associativity relation has the expressions:

$$\mathcal{L}_3(\underline{\mathcal{L}_2\underline{\mathcal{L}}_1}) = (\mathcal{L}_3\underline{\mathcal{L}}_2)\underline{\mathcal{L}}_1 \quad \text{or} \quad (\mathcal{L}_1\bar{\mathcal{L}}_2)\bar{\mathcal{L}}_3 = \mathcal{L}_1(\overline{\mathcal{L}_2\bar{\mathcal{L}}_3}).$$

Proof. We show first the inclusion $(\mathcal{L}_1 \square \mathcal{L}_2) \square \mathcal{L}_3 \subset \mathcal{L}_1 \square (\mathcal{L}_2 \square \mathcal{L}_3)$. Let $A \in (\mathcal{L}_1 \square \mathcal{L}_2) \square \mathcal{L}_3$. There exist a language $B \in \mathcal{L}_1 \square \mathcal{L}_2$, and a $\mathcal{L}_3$-substitution $\sigma : X_B^* \to X_A^*$ such that $\sigma(B) = A$. Since $B \in \mathcal{L}_1 \square \mathcal{L}_2$, there exist a language $C \in \mathcal{L}_1$, and a $\mathcal{L}_2$-substitution $\tau : X_C^* \to X_B^*$ such that $\tau(C) = B$. Let $\rho = \sigma \circ \tau : X_C^* \to X_A^*$. For each letter $z \in X_C$,

$$\rho(z) = \sigma(\tau(z)) \in \mathcal{L}_2 \square \mathcal{L}_3,$$

thus $\rho$ is a $\mathcal{L}_2 \square \mathcal{L}_3$-substitution. Since $A = \rho(C)$, we have $A \in \mathcal{L}_1 \square (\mathcal{L}_2 \square \mathcal{L}_3)$. This proves the inclusion.
We now establish the converse inclusion. Let $A \in \mathcal{L}_1 \square (\mathcal{L}_2 \square \mathcal{L}_3)$. There exist a language $C \in \mathcal{L}_1$ and a $\mathcal{L}_2 \square \mathcal{L}_3$-substitution $\rho : X_C^* \to X_A^*$ with $\rho(C) = A$. For each $z \in X_C$, $\rho(z) \in \mathcal{L}_2 \square \mathcal{L}_3$; consequently there is a language $B_z \in \mathcal{L}_2$ and a

$\mathscr{L}_3$-substitution

$$\sigma_z : X^*_{B_z} \to X^*_A$$

such that $\rho(z) = \sigma_z(B_z)$. By the closure under copy of $\mathscr{L}_2$, we may assume the alphabets $X_{B_z}$ $(z \in X_C)$ pairwise disjoint. Set .

$$Y = \bigcup_{z \in X_C} X_{B_z},$$

and define a substitution $\sigma : Y^* \to X^*_A$ by

$$\sigma(y) = \sigma_z(y) \quad \text{iff} \quad y \in X_{B_z}.$$

Then $\sigma$ is a $\mathscr{L}_3$-substitution, and $\rho(z) = \sigma(B_z)$ for all $z \in X_C$. Finally define a substitution $\tau : X^*_C \to Y^*$ by $\tau(z) = B_z$ for $z \in X_C$. Then $\tau$ is a $\mathscr{L}_2$-substitution, and

$$\rho(z) = \sigma(B_z) = \sigma \circ \tau(z) \qquad \text{for all} \qquad z \in X_C.$$

Now $\tau(C) \in \mathscr{L}_1 \square \mathscr{L}_2$, and

$$A = \rho(C) = \sigma(\tau(C)) \in (\mathscr{L}_1 \square \mathscr{L}_2) \square \mathscr{L}_3.$$ ■

By induction, we immediately obtain the following

**Corollary 5.2** *Let $\mathscr{L}$ and $\mathscr{M}$ be families of languages; then*

$$(\cdots((\underbrace{\mathscr{L} \square \mathscr{M}) \square \mathscr{M}) \square \ldots \square \mathscr{M}}_{k \text{ times}}) = \mathscr{L} \square (\cdots((\underbrace{\mathscr{M} \square \mathscr{M}) \cdots \square \mathscr{M}}_{k \text{ times}})$$

or equivalently $\mathscr{L}(\bar{\mathscr{M}})^k = \overline{\mathscr{L}\mathscr{M}(\bar{\mathscr{M}})^{k-1}}$. ■

Finally we note the following rule

**Corollary 5.3** *Let $\mathscr{L}, \mathscr{M}$ and $\mathscr{N}$ be families of languages: then*

$$\mathscr{N}\underline{\mathscr{L}}\bar{\mathscr{M}} = \mathscr{N}\bar{\mathscr{M}}\underline{\mathscr{L}}.$$

Proof. $\mathscr{N}\underline{\mathscr{L}}\bar{\mathscr{M}} = (\mathscr{L} \square \mathscr{N}) \square \mathscr{M} = \mathscr{L} \square (\mathscr{N} \square \mathscr{M}) = \mathscr{L} \square (\mathscr{N}\bar{\mathscr{M}}) = \mathscr{N}\bar{\mathscr{M}}\underline{\mathscr{L}}$. ■

The set of all families of languages equipped with the substitution $\square$ is a semigroup but not a monoid (Exercise 5.1). The operation $\square$ is not a binary operation between languages since the result $\sigma(L)$ depends of $L$ and of the $\text{Card}(X_L)$ languages $\sigma(x)(x \in X_L)$. There exists a special kind of substitution called syntactic substitution which depends only on two languages. This substitution is very important for two reasons: first, any substitution can be represented as the composition of a syntactic substitution and a rational transduction; in this sense, syntactic substitutions are "generic". Second, syntactic substitutions lead to "decompositions" of certain families of languages; results of this kind are the converse of closure properties. They will be exposed in Chapter VII.

**Definition** Let $L$ and $M$ be languages. The syntactic substitution of $M$ in $L$ is the substitution

$$\sigma_M : X_L^* \to (X_M \cup X_L)^*$$

defined by

$$\sigma_M(x) = Mx \qquad (x \in X_L).$$

We use also the symbol $\uparrow$ for syntactic substitution, defined by

$$L \uparrow M = \sigma_M(L).$$

By definition,

$$L \uparrow M = \{m_1 x_1 m_2 x_2 \cdots m_r x_r \mid r \geqslant 1, m_1, \ldots, m_r \in M, \\ x_1, \ldots, x_r \in X_L, x_1 \cdots x_r \in L\} \cup (1 \cap L). \tag{5.5}$$

Syntactic substitution is frequently used with the supplementary assumption that the alphabets $X_L$ and $X_M$ are disjoint; sometimes $1 \notin L, 1 \notin M$ are also required. The notation is justified by the following formulas which are similar to formulas for exponentiation:

$$\emptyset \uparrow M = \emptyset; \qquad 1 \uparrow M = 1, \qquad x \uparrow M = Mx \quad (x \in \Sigma); \qquad L \uparrow 1 = L;$$

$$(fg) \uparrow M = (f \uparrow M)(g \uparrow M); \qquad L \uparrow M = \bigcup_{f \in L} f \uparrow M;$$

$$(L_1 \cup L_2) \uparrow M = L_1 \uparrow M \cup L_2 \uparrow M; \qquad (L_1 L_2) \uparrow M = (L_1 \uparrow M)(L_2 \uparrow M).$$

Note also that

$$c^+ \uparrow M = \bigcup_{n \geqslant 1} c^n \uparrow M = \bigcup_{n \geqslant 1} (Mc)^n = (Mc)^+, \; c \in \Sigma.$$

Thus, Theorem 4.4 may be rewritten in the form

$$L\hat{\Gamma} = (c^+ \uparrow L)\Gamma, \qquad \text{for} \quad c \in \Sigma \backslash X_L.$$

Note that $\sigma_M$ is not a $\{M\}$-substitution, but if $L \in \mathscr{L}$ and $M \in \mathscr{M}$ and $\mathscr{M}$ is a cone, then $L \uparrow M \in \mathscr{L} \square \mathscr{M}$.

**Proposition 5.4** *Let $\mathscr{M}$ be a family of languages. Any $\mathscr{M}$-substitution $\sigma : X^* \to Y^*$ can be factorized into $\sigma = \tau \circ \sigma_N$ where $\sigma_N : X^* \to Z^*$ is a syntactic substitution, $N \in \mathscr{M}$ Elm and $\tau : Z^* \to Y^*$ is a rational transduction.*

Proof. For each $x \in X$, set $M_x = \sigma(x)$, and let $N_x$ be a copy of $M_x$ over an alphabet $Z_x$ in bijection with $Y$ by a mapping $\theta_x : Z_x \to Y$. We choose the alphabets $Z_x$, $(x \in X)$ pairwise disjoint and disjoint from $X$. Then set

$$Z = X \cup \bigcup_{x \in X} Z_x$$

and define a morphism $\theta: Z^* \to Y^*$ by

$$\theta(x)=1 \quad (x\in X) \qquad \theta(z)=\theta_x(z) \quad (z\in Z_x).$$

Set $$N=\bigcup_{x\in X} N_x. \tag{5.6}$$

Then the syntactic substitution $\sigma_N: X^* \to Z^*$ satisfies

$$\sigma_N(x)=Nx=\bigcup_{y\in X} N_y x \qquad (x\in X).$$

Define the transduction $\tau: Z^* \to Y^*$ by

$$\tau(f)=\theta(f\cap K) \qquad (f\in Z^*)$$

where $K=(\bigcup_{x\in X} Z_x^* x)^* \in \mathrm{Rat}(Z^*)$.

Clearly $\tau$ is rational. Further

$$\tau\circ\sigma_N(1)=\tau(1)=1=\sigma(1),$$

and for $w=x_1x_2\cdots x_r$, $r\geq 1$, $x_i\in X$, we have

$$\tau\circ\sigma_N(w)=\theta(Nx_1Nx_2\cdots Nx_r\cap K)$$
$$=\theta(N_{x_1}x_1N_{x_2}x_2\cdots N_{x_r}x_r)=M_{x_1}M_{x_2}\cdots M_{x_r}=\sigma(w).$$

Thus $\tau\circ\sigma_N=\sigma$. By (5.6), $N\in \mathcal{M}\,\underline{\mathrm{Elm}}$. This achieves the proof. ■

This proposition can be used to verify that a full semi-AFL is closed under substitution. First, we give a definition.

**Definition** Let $\mathcal{L}$ and $\mathcal{M}$ be two families of languages. Then $\mathcal{L}$ is closed under $\mathcal{M}$-substitution if $\mathcal{L}\,\Box\,\mathcal{M}\subset\mathcal{L}$, and $\mathcal{L}$ is closed under substitution if $\mathcal{L}\,\Box\,\mathcal{L}\subset\mathcal{L}$.

**Example 5.5** Any cone is closed under rational substitution:

$$\mathcal{L}=\mathcal{L}\Gamma \Rightarrow \mathcal{L}=\mathcal{L}\,\Box\,\mathrm{Rat}.$$

**Example 5.6** The families Rat and Alg are closed under substitution. The family Lin is a rational cone, but not a full AFL (see next section). Thus Lin is not closed under substitution in view of:

**Proposition 5.5** *A rational cone closed under substitution is a full* AFL.

Proof. Let $\mathcal{L}$ be closed under substitution. Then $\mathcal{L}\,\Box\,\mathcal{L}\subset\mathcal{L}$. If $\mathcal{L}$ is a rational cone, then $\mathrm{Rat}\subset\mathcal{L}$ by Proposition 2.1. Thus

$$\mathcal{L}\subset\mathcal{L}\,\underline{\mathrm{Rat}}=\mathrm{Rat}\,\Box\,\mathcal{L}\subset\mathcal{L}\,\Box\,\mathcal{L}\subset\mathcal{L},$$

and $\mathcal{L}=\mathcal{L}\,\underline{\mathrm{Rat}}$. ■

**Proposition 5.6** *Let $\mathscr{L}$ be a full semi-AFL. Then $\mathscr{L}$ is closed under substitution iff $\mathscr{L}$ is closed under syntactic substitution.*

Proof. Let $\mathscr{L}$ be closed under substitution, $M \in \mathscr{L}$, and let $\sigma_M : X^* \to (X_M \cup X)^*$ be a syntactic substitution. Then $\sigma_M(x) = Mx \in \mathscr{L}$ for $x \in X$ since $\mathscr{L}$ is a cone. Thus $\sigma_M$ is a $\mathscr{L}$-substitution, and $\mathscr{L}$ is closed under syntactic substitution.

Conversely, let $\sigma$ be a $\mathscr{L}$-substitution. In view of Proposition 5.4, there exists a factorization

$$\sigma = \tau \circ \sigma_N,$$

where $\sigma_N$ is a syntactic substitution, and $N \in \mathscr{L}\,\underline{\mathrm{Elm}} = \mathscr{L}$. Since $\mathscr{L}$ is closed under syntactic substitution, $\sigma_N(L) \in \mathscr{L}$ for $L \in \mathscr{L}$, and $\tau(\sigma_N(L)) = \sigma(L) \in \mathscr{L}\Gamma = \mathscr{L}$. ∎

**Exercises**

**5.1** Show that there is no neutral element for the operation $\square$.

**5.2** Let $L$ and $M$ be languages over disjoint alphabets and let $c$ be a new letter. Show that

$$L \uparrow (c^+ \uparrow M) \approx (L \uparrow c^+) \uparrow M.$$

**5.3** Let $L, M, K$ be languages over disjoint alphabets. Show that $L \uparrow (K \uparrow M) \leqslant (L \uparrow K) \uparrow M$. Discuss the converse inequality.

**5.4** Let $L, M, M'$ be languages over disjoint alphabets. Show that $M \approx M'$ implies $L \uparrow M \approx L \uparrow M'$. Show that the implication $M \approx M' \Rightarrow M \uparrow L \approx M' \uparrow L$ is generally false.

## V.6 Example: The Cone of Linear Languages

In this section, we study the family Lin of linear languages. We show that Lin is a principal cone, and we give generators of this cone. We also prove that Lin is not a full AFL.

**Definition** A context-free grammar $G = \langle V, X, P \rangle$ is linear if the right side of each production contains at most one nonterminal symbol:

$$\xi \to \alpha \in P \quad \Rightarrow \quad |\alpha|_V \leqslant 1.$$

A language $L$ is linear if there exists a linear grammar generating $L$.

We denote by Lin the set of all linear languages. Clearly, Lin is a family of languages.

**Example 6.1** The symmetric languages $S_n$ over $n$ letters. They are defined over the alphabet

$$Z_n = X_n \cup \bar{X}_n,$$

where $X_n = \{x_1, \ldots, x_n\}$, $\bar{X}_n = \{\bar{x}_1, ., \ldots, \bar{x}_n\}$ are two disjoint alphabets of $n$ letters, and they are generated by the grammars with productions

$$\xi \to \sum_{k=1}^{n} x_k \xi \bar{x}_k + 1.$$

Clearly $f \in S_n$ iff $f = g\bar{g}$ where $g = x_{i_1} \cdots x_{i_p} \in X_n^*$ and $\bar{g} = \bar{x}_{i_p} \cdots \bar{x}_{i_1}$.

**Example 6.2** Let $X = \{a, b\}$, and $\Theta = \{=, \neq, >, <, \geqslant, \leqslant\}$. The language $S_\theta$, $(\theta \in \Theta)$ is defined by

$$S_\theta = \{a^n b^m \mid n, m \geqslant 0 \quad \text{and} \quad n\theta m\}.$$

All these languages are linear. They are related by

$$S_< = S_= b^+ = S_\leqslant b; \qquad S_\leqslant = S_< \cup S_= = S_< b^{-1};$$
$$S_> = a^+ S_= = a S_\geqslant; \qquad S_\geqslant = S_> \cup S_= = a^{-1} S_>;$$
$$S_\neq = S_> \cup S_<; \qquad S_= \cup S_\neq = a^* b^*.$$

Of course, $S_=$ is equal to the symmetric language $S_1$.

The following lemma is useful.

**Lemma 6.1** *For any linear grammar $G = \langle V, X, P \rangle$ there exists a linear grammar $G' = \langle V', X, P' \rangle$ such that $V' \supset V$, further $L_{G'}(\sigma) = L_G(\sigma)$ for all $\sigma \in V$, and all productions of $G'$ are of one of the following forms ($\xi, \eta \in V', x \in V$):*

(i) $\xi \to x\eta$ *or* $\xi \to \eta x$;

(ii) $\xi \to x$ *or* $\xi \to 1$.

The proof of the lemma is left to the reader. A grammar $G'$ of the above form is said in canonical form.

**Theorem 6.2** *The family* Lin *is a rational cone:* Lin $\Gamma$ = Lin.

Proof. Let $G$ be a linear grammar, and consider the constructions of the grammars $\psi G$, $\varphi^{-1}G$, $G_K$ given in Section II.2. By inspection, $\psi G$ and $G_K$ are linear grammars and $\varphi^{-1}G$ is not a linear grammar. Thus Lin is closed under morphism and intersection with regular sets. In order to prove closure under inverse morphism, a special construction is necessary. Restricting ourselves to inverse projection, we show that the insertion of a new letter can be done just by moving sequentially in the right side of a production by remembering the number of the production in an auxiliary symbol.

Let $L \in \text{Lin}$ be generated by the linear grammar $G = \langle V, X, P \rangle$ in canonical form, let $Y \supset X$ and let $\pi$ be the projection of $Y^*$ onto $X^*$. We may assume $\text{Card}(Y \backslash X) = 1$, since the general case results by composition of such projections.

Let $z = Y \backslash X$, and number the productions from 1 to $n = \text{Card } P$. Define a grammar $\bar{G} = \langle \bar{V}, Y, \bar{P} \rangle$ as follows:

$$\bar{V} = V \cup \{\omega_1, \ldots, \omega_n\},$$

where the $\omega_i$ are new symbols. Further, if the $p$-th production of $G$ is

$$\xi \to x\eta \qquad (\text{or } \xi \to \eta x)$$

then put into $\bar{P}$ the productions

$$\xi \to \omega_p, \qquad \omega_p \to z\omega_p + x\eta;$$
$$(\text{or } \xi \to \omega_p, \qquad \omega_p \to \omega_p z + \eta x).$$

If the $p$-th production of $G$ is

$$\xi \to x \qquad (\text{or } \xi \to 1)$$

then put into $\bar{P}$ the productions

$$\xi \to \omega_p, \qquad \omega_p \to z\omega_p + \omega_p z + x;$$
$$(\text{or } \xi \to \omega_p, \qquad \omega_p \to z\omega_p + 1).$$

The grammar $\bar{G}$ obtained by this construction is linear, and it is easily seen that

$$L_{\bar{G}}(\xi) = \pi^{-1}(L_G(\xi)) \qquad \text{for} \quad \xi \in V.$$

Thus if $L = L_G(\sigma)$ for some $\sigma \in V$, then $\pi^{-1}(L)$ is in Lin. ∎

We shall prove that the symmetric languages $S_n$ are generators of the cone Lin for $n \geq 2$. First we prove

**Proposition 6.3** *The languages $S_n$ $(n \geq 2)$ are rationally equivalent; further $S_2 \geq S_1$.*

We shall prove later that $S_2 > S_1$. A direct proof is possible, but rather long; a simple proof will follow from general properties of iterative pairs developed in Chapter VIII.

Proof. Since

$$S_{n-1} = S_n \cap Z_{n-1}^* \qquad (n \geq 2),$$

it follows that $S_n \geq S_{n-1}$ for $n \geq 2$.

Conversely, let $n \geq 1$, and define a morphism

$$\alpha : Z_n^* \to Z_2^*$$

by $\quad \alpha(x_k) = x_1 x_2^k, \qquad \alpha(\bar{x}_k) = \bar{x}_2^k \bar{x}_1 \qquad 1 \leq k \leq n.$

Then $f \in S_2 \cap \alpha(Z_n^*)$ iff $f=1$ or

$$f = x_1 x_2^{k_1} x_1 x_2^{k_2} \cdots x_1 x_2^{k_p} \bar{x}_2^{k_p} \bar{x}_1 \cdots \bar{x}_2^{k_1} \bar{x}_1,$$

with $1 \leqslant k_i \leqslant n$ for $i = 1, \ldots, p$. Thus

$$\alpha^{-1}(S_2 \cap \alpha(Z_n^*)) = \alpha^{-1}(S_2) = S_n,$$

and $S_2 \geqslant S_n$ for $n \geqslant 1$. ∎

The method used to prove this proposition is an encoding of the alphabet $X_n$ in an alphabet of two letters. This is a quite general method which will be used several times.

**Theorem 6.4** Lin *is a principal cone, and any symmetric language* $S_n (n \geqslant 2)$ *is a generator of* Lin: Lin $= S_n \Gamma$.

Proof. Since $S_n \in$ Lin and Lin is a cone, we have $S_n \Gamma \subset$ Lin for all $n \geqslant 2$. Further $S_n \Gamma = S_{n'} \Gamma$ for $n, n' \geqslant 2$ by the preceding proposition. Thus to prove the theorem it suffices to show that any linear language $L$ belongs to $S_n \Gamma$ for some $n \geqslant 2$.

Let $L \subset X^*$ be a linear language, and let $G = \langle V, X, P \rangle$ be a linear grammar in canonical form such that $L = L_G(\sigma)$ for some $\sigma \in V$. Number the productions from 1 to $n = \text{Card } P$, and define a morphism $\alpha: Z_n^* \to X^*$ in the following way:

a) If the $p$-th production is

$$\xi \to x\eta \qquad (\text{or } \xi \to \eta x),$$

then set $\alpha(x_p) = x$, $\alpha(\bar{x}_p) = 1$; (or $\alpha(x_p) = 1$, $\alpha(\bar{x}_p) = x$).
b) If the $p$-th production is

$$\xi \to x \qquad (\text{or } \xi \to 1),$$

then set $\alpha(x_p) = x$, $\alpha(\bar{x}_p) = 1$ (or $\alpha(x_p) = \alpha(\bar{x}_p) = 1$).

Define the local rational language

$$K = (AX_n^* \cap X_n^* B) \backslash X_n^* C X_n^*$$

by

| | | |
|---|---|---|
| | $x_p \in A$ | iff the left side of the $p$-th production is $\sigma$ |
| | $x_p \in B$ | iff the right side of the $p$-th production is in $X \cup 1$ |
| | $x_p x_q \notin C$ | iff the left side of the $q$-th production is equal to the unique nonterminal in the right side of $p$-th production. |

We now prove

$$L = \alpha(S_n \cap K\bar{X}_n^*). \tag{6.1}$$

Indeed, consider a derivation

$$\sigma \to w_1 \to w_2 \to \cdots \to w_r = w \tag{6.2}$$

from $\sigma$ into $w \in L$ in the grammar $G$. This derivation is uniquely determined by the sequence $(p_1, p_2, \ldots, p_r)$ of the numbers of the productions used in the derivation. By construction, the word $x_{p_1}x_{p_2} \cdots x_{p_r}$ is in $K$, and $w = \alpha(x_{p_1}x_{p_2} \cdots x_{p_r}\bar{x}_{p_r} \cdots \bar{x}_{p_1})$.

Conversely, if $x_{p_1}x_{p_2} \cdots x_{p_r}\bar{x}_{p_r} \cdots \bar{x}_{p_1} \in S_n \cap K\bar{X}_n^*$, then there exists a derivation (6.2) for which $p_1, p_2, \ldots, p_r$ is the sequence of numbers of the productions. Moreover, $w = \alpha(x_{p_1}x_{p_2} \cdots x_{p_r}\bar{x}_{p_r} \cdots \bar{x}_{p_1})$. ■

Note that by formula (6.1), the above construction yields an analogue of the Chomsky–Schützenberger Theorem for linear languages. In fact, a weaker property can be shown for any principal cone (Exercise 6.1).

The rational cone Lin is principal, thus closed under union. We shall see that Lin is not closed under product and thus is not a full AFL. First we show how linear languages are linked to rational relations. The following proposition is in Rosenberg [1967].

**Proposition 6.5** *A language $L \subset X^*$ is linear iff there exists a rational relation $A \subset X^* \times X^*$ such that*

$$L = \{f\tilde{g} \mid (f, g) \in A\}.$$

Proof. Let $L \subset X^*$ be a linear language, and consider a linear grammar $G = \langle V, X, P\rangle$ in canonical form such that $L = L_G(\sigma)$ for some $\sigma \in V$.

Define a transducer $\boldsymbol{T} = \langle X, X, Q, \sigma, q_+, E\rangle$ by $Q = V \cup q_+$ with $q_+ \notin V$, and $E$ composed of the transitions

$$\begin{array}{llll} (\xi, x, 1, \eta) & \text{for} \quad \xi \to x\eta \in P & (\xi, x, 1, q_+) & \text{for} \quad \xi \to x \in P \\ (\xi, 1, x, \eta) & \text{for} \quad \xi \to \eta x \in P & (\xi, 1, 1, q_+) & \text{for} \quad \xi \to 1 \in P. \end{array}$$

It is easily seen that each computation $e = e_1e_2 \cdots e_n$ from $\xi$ to $\eta \in V$ (resp. to $q_+$) with label $|e| = (f, g)$ determines a derivation $\xi \xrightarrow{n} f\eta\tilde{g}$ (resp. $\xi \xrightarrow{n} f\tilde{g}$) and vice-versa. Thus $L = \{f\tilde{g} \mid (f, g) \in T(\sigma, q_+)\}$. Conversely, let $\tau: X^* \to X^*$ be the transduction with graph $A \in \mathrm{Rat}(X^* \times X^*)$ and let $\boldsymbol{T} = \langle X, X, Q, q_-, q_+, E\rangle$ be a transducer realizing $\tau$ and satisfying the conditions of Corollary III.6.2. Define $G = \langle V, X, P\rangle$ by $V = Q \backslash q_+$ and

$$P = \{q \to uq'\tilde{v} \mid (q, u, v, q') \in E, q' \neq q_+\} \cup \{q \to u\tilde{v} \mid (q, u, v, q_+) \in E\}.$$

Then $G$ is linear and $L_G(q_-) = \{f\tilde{g} \mid (f, g) \in A\}$. ■

Let us comment this proposition. First, we introduce a definition.

**Definition** For $M \subset X^*$ and $A \subset Y^* \times Z^*$, the bracket of $A$ and $M$ is the language over $X \cup Y \cup Z$ defined by

$$[A, M] = \{fw\tilde{g} \mid w \in M, (f, g) \in A\}.$$

Thus $L=\{f\tilde{g} \mid (f, g)\in A\}$ can be written as $L=[A, 1]$. Then, Proposition 6.5 states that for any $A\in \mathrm{Rat}(X^*\times X^*)$, the language $[A, 1]$ is linear, and conversely for any $L\in \mathrm{Lin}$, there exists a relation $A\in \mathrm{Rat}(X^*\times X^*)$ such that $L=[A, 1]$. However, if $L$ is linear, and $L=[A, 1]$ for some $A\subset X^*\times X^*$, then $A$ is not necessarily a rational relation (Exercise 6.4). Indeed, a factorization $w=f\tilde{g}$ is a "parsing" of $w$, and any parsing needs not to be linear.

We now prove an iteration lemma for linear languages.

**Proposition 6.6** (Iteration lemma for linear languages) *Let $L\subset X^*$ be a linear language; then there exists an integer $N\geqslant 1$ such that any word $f\in L$ of length $|f|\geqslant N$ admits a factorization*

$$f=aubvc \qquad (a, u, b, v, c\in X^*)$$

*satisfying*

$$au^nbv^nc\in L \quad \text{for} \quad n\geqslant 0, \quad |uv|\geqslant 1 \quad \text{and} \quad |auvc|\leqslant N.$$

The special feature of this iteration lemma is of course the very last condition which says that the factors $u$ and $v$ can be chosen near to the borders of the word $f$.

Proof. Let $A\subset X^*\times X^*$ be a rational relation such that $L=\{f\tilde{g}:(f, g)\in A\}$. Then the proposition is just a reformulation of the iteration lemma for rational relations together with the remark following Lemma III.3.3. ■

Proposition 6.6 can also be proved directly on a linear grammar generating $L$, without using Proposition 6.5 (Exercise 6.2).

**Example 6.3** The language $(S_=)^2=\{x^ny^nx^my^m \mid n, m\geqslant 0\}$ is not linear. Assume the contrary. Then by the iteration lemma, there would exist large enough integers $n,m$ such that $x^{n+\lambda}y^nx^my^{m+\mu}\in(S_=)^2$ for some integers $\lambda, \mu$ with $0<\lambda+\mu$, and this is impossible. This example shows that the full AFL generated by $S_=$ is not contained in Lin.

**Corollary 6.7** *The rational cone* Lin *is not a full* AFL. ■

We give some other examples of nonlinear languages.

**Example 6.4** The language $L=\{x^ny^{n+m}x^m \mid n, m\geqslant 0\}$ is not linear; this is shown as in Example 6.3.

**Example 6.5** The Dyck languages are not linear. Consider $D_1'^*, D_1^*\subset\{x, \bar{x}\}^*$. Then $D_1^*\cap x^*\bar{x}^*x^*=\{x^n\bar{x}^{n+m}x^n \mid n, m\geqslant 0\}$ thus $D_1^*\geqslant L$, where $L$ is the language of Example 6.4, and the linearity of $D_1^*$ would imply the linearity of $L$ since Lin is a cone. Next

$$D_1'^*\cap x^+\bar{x}^+x^+\bar{x}^+=\{x^{k+n}\bar{x}^nx^p\bar{x}^{p+k} \mid k\geqslant 0, n, p>0\}.$$

Let $C$ be the latter language, and assume $C \in \mathrm{Lin}$. Then $x^n\bar{x}^n x^p \bar{x}^p \in C$ for any $n, p > 0$, and by the iteration lemma, there exist $\lambda, \mu$, $0 < \lambda + \mu$ such that $x^{n-\lambda}\bar{x}^n x^p \bar{x}^{p-\mu} \in C$, provided $n, m$ has been chosen large enough. Thus $D_1'^* \notin \mathrm{Lin}$. We shall see later (Chapter VII) that any Dyck language over at least two letters is a generator of the cone Alg and therefore cannot be linear. Further, $D_1 \approx D_1' \approx D_1'^*$ (Proposition VII.1.3), thus $D_1$, $D_1' \notin \mathrm{Lin}$.

## Exercises

**6.1** Let $\mathscr{L} = A\Gamma$ be a principal cone. Show that there exist languages $(A_n)_{n \geq 1}$ in $\mathscr{L}$ such that any $L \in \mathscr{L}$ can be written in the form

$$L = \psi(A_n \cap K)$$

for some $n \geq 1$, where $\psi$ is an alphabetic morphism and $K$ is a regular language.

**6.2** Prove the iteration lemma 6.6 directly on a linear grammar for $L$.

**6.3** For $n \geq 1$, let $X_n = \{x_1, \ldots, x_n\}$ and define

$$M_n = \{f \in X_n^* \mid f = \tilde{f}\}.$$

$M_n$ is the language of palindromes over $n$ letters.

a) Show that $M_n$ is a linear context-free language first by exhibiting a grammar, then by showing that $S_n \geq M_n$.
b) Set $M_n' = \{f \in M_n \mid |f|_{x_n} = 1\}$. Show that $M_n \geq M_n' \geq S_{n-1}$ for $n \geq 2$.
c) Show that any $M_n (n \geq 2)$ is a generator of the cone Lin.

**6.4** Give an example of a linear language $L$ and an irrational relation $A$ such that $L = [A, 1]$. (For further discussion, see Rosenberg [1967].)

## V.7 Examples of Incomparable Languages

It is usually not difficult to prove that a language dominates rationally another language by providing a rational transduction mapping the first onto the second language. It is often much harder to prove nondomination or strict domination.

In some cases, known properties of cones give an easy answer. Thus the fact that the cone of linear languages is not closed under product implies that $\mathrm{Lin} \subsetneq \mathrm{Alg}$, and thus any generator of the cone of context-free languages strictly dominates all linear languages. On the other hand, there exists a general criterion based on so-called systems of iterative pairs which allows to prove strict domination in a great number of situations. This will be developed in Chapter VIII. For other classes of families of languages, the "syntactic lemma" proved in Chapter VII provides a tool to establish the existence of infinite chains of cones or full AFLs.

Nevertheless, there still remain languages for which strict domination or incomparability can only be shown by a direct proof. The proofs in these cases usually are by contradiction. Assuming the existence of a rational transduction, the pumping lemma is applied to the regular set given by Nivat's theorem in order to obtain a contradiction. In this section we illustrate this technique by two examples, proving the incomparability of the languages $S_<$, $S_>$, $S_{\neq}$.
We first fix some notations. Set

$$S_< = \{x^n y^m \mid 0 \leqslant n < m\}, \qquad X = \{x, y\};$$
$$S_{\neq} = \{a^n b^m \mid n, m \geqslant 0, n \neq m\}, \quad Y = \{a, b\};$$

we may assume $X \cap Y = \emptyset$, and set $Z = X \cup Y$. Finally, let $\alpha : Z^* \to X^*$ and $\beta : Z^* \to Y^*$ be the two projections from $Z^*$ onto $X^*$ and $Y^*$ respectively.

**Proposition 7.1** *The language $S_<$ is not a rational image of $S_{\neq}$.*

Proof. Assume the contrary. Then there exists a rational transduction $\tau : Y^* \to X^*$ such that $\tau(S_{\neq}) = S_<$. By Nivat's Theorem III.4.1, there exists a regular language $R \subset Z^*$ such that

$$\tau(f) = \alpha(\beta^{-1}(f) \cap R) \qquad f \in Y^*.$$

Let $N$ be the integer of the pumping lemmas associated with $R$ ($N$ is the number of states of a finite automaton recognizing $R$), let $r > 2N$, and let $w = x^r y^s \in S_<$ for some $s > r$. Then there exists $f = a^n b^m \in S_{\neq}$ such that

$$x^r y^s \in \alpha(\beta^{-1}(a^n b^m) \cap R),$$

and consequently there is a word $h \in R$ such that

$$\alpha h = x^r y^s, \qquad \beta h = a^n b^m.$$

Consider an arbitrary factorization

$$h = h_1' h_2' \qquad \text{with} \quad \alpha h_1' = x^r \quad \alpha h_2' = y^s.$$

Now $|h_1'|_x = r > 2N$. Thus, marking the letters $x$ in $h_1'$, we obtain by Ogden's iteration lemma applied twice a factorization $h_1' = v_1 u' v_2 u'' v_3$, with $0 < |u'|_x, |u''|_x \leqslant N$ and $v_1 u'^* v_2 u''^* v_3 h_2' \subset R$. Since $\beta h_1' \subset a^* b^*$, at least one of the two words $\beta(u')$, $\beta(u'')$ belongs to $a^* \cup b^*$. (This is the reason why we apply the iteration lemma twice!) Denote by $u$ one of the words $u'$, $u''$ whose image by $\beta$ is in $a^*$ or $b^*$. This provides a factorization

$$h = h_1 u h_2, \qquad \text{with} \quad 0 < |u|_x \leqslant N, \quad \beta(u) \in a^* \cup b^*, \quad h_1 u^* h_2 \subset R.$$

Now $\beta(h_1 u^p h_2) \in a^* b^*$ for $p \geqslant 0$, thus $\beta(h_1 u^p h_2) \in S_{\neq}$ for all $p \geqslant 0$ excepted at most one. This implies that

$$h_1 u^p h_2 \in R \cap \beta^{-1}(S_{\neq})$$

for all large enough $p$, and consequently

$$\alpha(h_1 u^p h_2) = x^{r+(p-1)|\alpha(u)|} y^s \in S_<$$

for all large enough $p$. Since $|\alpha(u)| = |u|_x > 0$, this is impossible. ■

By symmetry, we clearly obtain also the property: $S_{\neq} \not\geq S_>$. The next proposition, which is the converse of the previous one, is much more difficult to prove.

**Proposition 7.2** *The language $S_{\neq}$ is not a rational image of $S_<$.*

There are several consequences of these propositions:

**Theorem 7.3** *The languages $S_<$, $S_>$, $S_{\neq}$ are pairwise incomparable; the language $S_=$ strictly dominates the languages $S_<$, $S_>$, $S_{\neq}$.*

Proof. In view of Propositions 7.1 and 7.2, the languages $S_{\neq}$ and $S_<$ are incomparable. By symmetry, the languages $S_{\neq}$ and $S_>$ are incomparable. Assume $S_> \geq S_<$; then $S_> \geq S_> \cup S_< = S_{\neq}$. Thus $S_>$ and $S_<$ are incomparable. By Example 6.2, $S_=$ dominates $S_>$, $S_<$ and $S_{\neq}$. Assume $S_{\neq} \geq S_=$; then $S_= \geq S_<$ would imply the relation $S_{\neq} \geq S_<$, in contradiction to Proposition 7.1. Thus $S_= > S_{\neq}$. In the same manner, if $S_> \geq S_=$, then since $S_= \geq S_{\neq}$, it would follow that $S_> \geq S_{\neq}$. Thus $S_= > S_>$ and $S_= > S_<$. ■

The situation described by Theorem 7.3 can be pictorially represented as in Fig. V. 2.

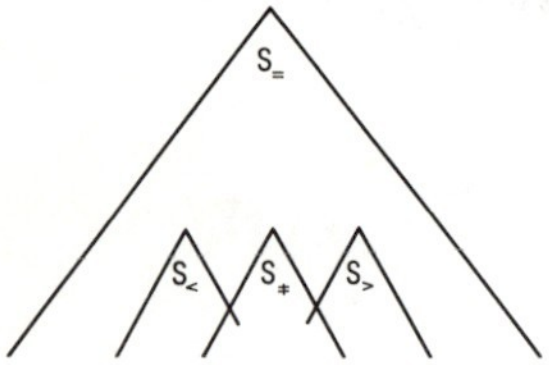


Fig. V.2

The proof of Proposition 7.2 is similar to the proof of 7.1: Assuming the existence of a rational transduction $\tau$ such that $\tau(S_<) = S_{\neq}$, we obtain a contradiction by showing that in fact $\tau(S_<) \not\supseteq S_{\neq}$. The difficulty of the proof comes from the scarceness of words in $a^*b^* \backslash S_{\neq}$. This implies some complications reflected by the use of factorials.

For $k \geq 2$, define

$$A_k = \{a^{n!k} b^{n!} \mid n \geq 1\}.$$

We first prove a simple lemma.

**Lemma 7.4** *There is no regular language $K$ such that $A_k \subset K \subset S_{\neq}$.*

Proof. Assume the contrary. Then $K$ is infinite, and by the pumping lemma, there is an integer $p$ such that $n \geqslant p$ implies

$$a^{k \cdot n!} b^{n!+\lambda p} \in K \qquad \text{for all } \lambda \geqslant 0.$$

Since $p \leqslant n$, $p$ divides $n!$, and $\mu = \frac{n!}{p}(k-1)$ is an integer. Thus $a^{kn!} b^{n!+\mu p} \in K$; since $n! + \mu p = kn!$, the language $K$ is not contained in $S_{\neq}$. ■

Proof of Proposition 7.2. We first sketch the proof. Assuming the existence of a rational transduction mapping $S_<$ onto $S_{\neq}$, we first show that there are always arbitrarily long words in $S_{\neq}$ which are images, through the transduction, only of very long words in $S_<$. This allows us to assume that words $h$ in the intermediate regular language $R$ contain always enough letters of each kind. In a second step, we apply the pumping lemma in a naive fashion. This gives the result excepted in one case. Then we use the pumping lemma in exactly the same way than in the proof of Proposition 7.1. This also gives the result excepted in one special case. Combining now the two factors supplied by the pumping lemma in these two reserved situations yields the contradiction by the use of the factorials. We now proceed to the formal proof.

a) Assume the result is false. Then there exists a rational transduction $\tau: X^* \to Y^*$ such that $\tau(S_<) = S_{\neq}$. Then this transduction verifies:

$$\begin{gathered}\text{for each } N \text{ there exists a } n \geqslant N \text{ such that any word} \\ x^r y^s \in S_< \text{ for which } a^{kn!} b^{n!} \in \tau(x^r y^s) \text{ satisfies } r \geqslant N.\end{gathered} \tag{7.1}$$

Indeed, assume the contrary; then there is an integer $N$ such that, for any $n \geqslant N$, there exists some word $x^r y^s \in S_<$ with $r < N$ and $a^{kn!} b^{n!} \in \tau(x^r y^s)$. This is equivalent to

$$\{a^{kn!} b^{n!} \mid n \geqslant N\} \subset \tau(B_N),$$

where $B_N = \{x^r y^s \mid r < N, r < s\}$. The language $B_N$ is regular. For $n = 1, \ldots, N-1$, let $w_n \in S_<$ be a word such that $a^{kn!} b^{n!} \in \tau(w_n)$; then $B = B_N \cup \{w_1, \ldots, w_{N-1}\}$ is a regular language contained in $S_<$, and

$$A_k \subset \tau(B) \subset \tau(S_<) = S_{\neq}. \tag{7.2}$$

The language $\tau(B)$ is regular, and (7.2) is in contradiction with Lemma 7.4. This proves (7.1).

b) By Nivat's Theorem, there is a regular language $R \subset Z^*$ such that

$$\tau(f) = \beta(\alpha^{-1}(f) \cap R) \qquad \text{for} \quad f \in X^*.$$

Let $q$ be the number of states of a finite automaton recognizing $R$, and fix $k$ ($k$ is the index of the language $A_k$) to be $k = q! + 1$. Finally take $N$ in (7.1) to satisfy $N \geq 3q$. In view of (7.1), there is an integer $n \geq N$ such that $a^{kn!}b^{n!} \in \tau(x^r y^s)$ with $N \leq r < s$. Let $h \in R$ be a word such that

$$\alpha h = x^r y^s, \qquad \beta h = a^{kn!}b^{n!}.$$

c) Let $h_1$ be the right factor of length $q$ of $h$. Then $h_1 \in \{y, b\}^q$ since both $s$ and $n!$ are greater that $q$. By the pumping lemma, $h_1$ has a factor $u$ with $0 < |u| \leq q$ and such that, by setting

$$h = guh', \tag{7.3}$$

$gu^*h' \subset R$, and $|uh'| \leq q$. From $u \in \{y, b\}^*$, it follows that $\alpha(u) \in y^*$, $\beta(u) \in b^*$, thus $\alpha(gu^\lambda h') \in S_<$ for all $\lambda \geq 1$, and thus $\beta(gu^\lambda h') \in S_{\neq}$ for $\lambda \geq 1$. We shall see that $|\beta(u)| = 0$. Assume $t = |\beta(u)| > 0$; then

$$\beta(gu^\lambda h') = a^{kn!}b^{n!+(\lambda-1)t} \in S_{\neq} \qquad \lambda \geq 1,$$

and $t$ divides $n!$ (since $1 \leq t \leq |u| \leq q \leq N \leq n$); thus $\mu = \dfrac{n!}{t}(k-1)+1$ is an integer, and $n!+(\mu-1)t = kn!$; this implies $\beta(gu^\mu h) \notin S_{\neq}$. (Note that this argument was already used for the proof of Lemma 7.4.) Thus $|\beta(u)| = 0$, and we set $p_u = |\alpha(u)| = |u|_y = |u|$. Then $0 < p_u \leq q$.

d) Consider now the left factor $g$ of $h$ given by (7.3). We have $|g|_b \geq |h|_b - q \geq 2q$. Arguing exactly as in the proof of Proposition 7.1 by a double application of Ogden's iteration lemma to the marked $b$'s of $g$, there exists a factorization $g = g'vg''$, and thus a factorization

$$h = g'vg''uh', \qquad g'v^*g''u^*h' \subset R,$$

with $0 < |v|_b \leq q$, $\alpha(v) \in x^* \cup y^*$, $\beta(v) \in b^*$.

e) Set $p_v = |\alpha(v)|$, $q_v = |v|_b = |\beta(v)|$, and define

$$h_{\lambda\mu} = g'v^{1+\lambda}g''u^{1+\mu}h' \qquad (\lambda, \mu \geq 0).$$

If $\alpha(v) \in y^*$, then $\alpha(h_{\lambda,\mu}) \in S_<$ for all $\lambda, \mu \geq 0$, thus $\beta(h_{\lambda,0}) = a^{kn!}b^{n!+\lambda q_v} \in S_{\neq}$ for $\lambda \geq 0$. By $0 < q_v \leq q$, this leads to a contradiction by the same factorial argument as in c). Thus $\alpha(v) \in x^*$, and

$$\alpha(h_{\lambda,\mu}) = x^{r+\lambda p_v}y^{s+\mu p_u}.$$

Consequently, $h_{\lambda p_u, \mu p_v} \in \alpha^{-1}(S_<) \cap R$ for all $\lambda, \mu$ such that $0 \leq \lambda \leq \mu$. On the other hand,

$$\beta(h_{\lambda,\mu}) = a^{kn!}b^{n!+\lambda q_v}$$

since $|\beta(u)| = 0$. Thus

$$\beta(h_{\lambda p_u, \mu p_v}) = a^{kn!}b^{n!+\lambda p_u q_v} \qquad (\lambda, \mu \geq 0).$$

We now use once more the factorial argument: since $p_u \leq q$, $p_u$ divides $k-1=q!$; since $q_v \leq q \leq n$, $q_v$ divides $n!$. Thus the number $t=\dfrac{(k-1)n!}{p_u q_v}$ is an integer, and

$$\beta(h_{tp_u, tp_v}) = a^{kn!}b^{kn!} \notin S_{\neq}.$$

Since $\alpha(h_{tp_u, tp_v}) \in S_<$, this gives the final contradiction. ■

**Exercises**

**7.1** (Ginsburg and Greibach [1969]) Assume that the full AFL generated by a language $A$ is equal to the rational closure of the cone generated by $A$ (Theorem VI.4.10): $A\hat{\Gamma} = (A\Gamma)\underline{\mathrm{Rat}}$. Prove that the full AFLs $S_>\hat{\Gamma}$ and $S_<\hat{\Gamma}$ are incomparable.

**7.2** For rational numbers $0 < r \leq s$ define $M(r, s) \subset \{x, y\}^*$ by

$$M(r,s) = 1 \cup \left\{x^n y^m \mid n, m \in \mathbb{N}\backslash 0,\ r \leq \frac{m}{n} \leq s\right\}.$$

a) Show that $M(r, s) \approx M(kr, ks)$ for any integer $k > 0$.

b) Show that $M(r, s) \approx M(qr, qs)$ for any rational number $q > 0$.

c) Show that for rational numbers $p, q$, $0 < q \leq p < 1$ implies $M(q, 1) \leq M(p, 1)$. (Hint. Use the fact that $M(p, 1) \approx M(p^{n+1}, p^n)$ for integers $n \geq 0$, and that $p^{n+1} < q \leq p^n$ for some integer $n \geq 1$.)

d) Show that $M(1, 1) \geq M\left(\dfrac{k}{k+1}, 1\right)$ for any integer $k > 0$.

e) Conclude that $M(q, 1) \leq M(p, 1)$ for all rational numbers $p, q$ such that $0 < q \leq p \leq 1$, and that all languages $M(r, s)$ are linear languages.

f) Prove that for $0 < q < 1$, $M(q, 1) > S_>, S_{\neq}, S_<$.

# VI Operators

Operators on families of languages are defined as mappings of families of languages satisfying an additional property which links together the languages in the families. All transformations of families of languages introduced in the preceding chapter are operators in this sense. We develop some general properties of operators. In the last two sections of this chapter, operator calculus is used to derive some fundamental results on cones and full AFLs. By this way, we can separate parts of proofs which are only formal calculations from other parts which rely on a few, more or less standard constructions. The exposition follows Lewis [1970] and Nivat [1975].

## VI.1 Operators

In the preceding chapter, we introduced several operations on families of languages, such as closure under rational transduction, substitution and so on. These transformations can be considered as mappings which associate a new family of languages to any given family. Then these mappings can be composed, and the result defines a new operation on families of languages. On the other hand, some of the properties of families of languages proved in the previous chapter are in fact results on transformations. Thus, the equality $\mathscr{L}\Gamma\Gamma = \mathscr{L}\Gamma$ (Proposition V.2.3) means that $\Gamma$ is idempotent: $\Gamma\Gamma = \Gamma$.

These considerations show that it is natural to investigate functions over the set of all families of languages (over some fixed alphabet $\Sigma$), with the aim to find relations between these functions. It is to be expected that the consideration of the totality of these mappings will give no information, since all transformations of families of languages defined up to now are related to transformations of the languages themselves. This relation must be expressed on the level of families of languages. Therefore we consider only transformations $\omega$ on families which satisfy the following additional property: any language $A$ in the image of the family $\mathscr{L}$ by $\omega$ depends only on a finite number of languages in $\mathscr{L}$.

In the first three sections of this chapter, we investigate transformations of this kind. A concrete use of these operators will be made in the last two sections.

Let $\Sigma$ be a fixed infinite alphabet. We denote by $\mathbb{F}$ the set of all families of languages over $\Sigma$ in the sense defined in the preceding chapter: any language is contained in the free monoid $X^*$ generated by some finite $X \subset \Sigma$. Let $\omega$ be a mapping from $\mathbb{F}$ into $\mathbb{F}$. The image of a family $\mathscr{L} \in \mathbb{F}$ under $\omega$ is denoted by $\mathscr{L}\omega$.

It is a family of languages, and in particular closed under copy. If $\omega_1$ and $\omega_2$ are two mappings from $\mathbb{F}$ into itself, the composition is denoted by juxtaposition and is defined by

$$\mathcal{L}(\omega_1\omega_2)=(\mathcal{L}\omega_1)\omega_2 \qquad (\mathcal{L}\in\mathbb{F}).$$

This composition is associative, and the identity mapping is a neutral element. We denote it by id.

**Definition** A mapping $\omega:\mathbb{F}\to\mathbb{F}$ is called monotone if it satisfies the following condition:

$$\text{for any } \mathcal{L}, \mathcal{M}\in\mathbb{F}, \quad \mathcal{L}\subset\mathcal{M}\Rightarrow\mathcal{L}\omega\subset\mathcal{M}\omega \tag{1.1}$$

Further, $\omega$ is locally finite if it verifies:

for any $\mathcal{L}\in\mathbb{F}$, and for each $A\in\mathcal{L}\omega$, there is a finite family $\mathcal{L}_A\subset\mathcal{L}$ such that $A\in\mathcal{L}_A\omega$. (1.2)

Recall that a family of languages $\mathcal{M}$ is finite (see Section V.1) if there is a finite number of languages $M_1,\ldots,M_n\in\mathcal{M}$ such that each $M\in\mathcal{M}$ is a copy of one of the $M_1,\ldots,M_n$.

If in (1.2), the family $\mathcal{L}_A$ is always composed of a single language (and its copies), then $\omega$ is called unitary.

**Definition** A mapping $\omega$ from $\mathbb{F}$ into itself is an operator on families of languages if it is both monotone and locally finite. The set of all operators over $\mathbb{F}$ is denoted by $\Omega$.

We first show that $\Omega$ is a monoid, and that the set of unitary operators is a submonoid of $\Omega$.

**Proposition 1.1** id *is an unitary operator; if $\omega_1, \omega_2$ are operators, then $\omega_1\omega_2$ is an operator; further, if $\omega_1, \omega_2$ are unitary operators, then $\omega_1\omega_2$ is unitary.*

Proof. Clearly, the identity mapping is monotone and unitary. Let $\omega_1, \omega_2$ be operators, and let $\mathcal{L}$, $\mathcal{M}$ be families of languages. If $\mathcal{L}\subset\mathcal{M}$, then $\mathcal{L}\omega_1\subset\mathcal{M}\omega_1$, and thus $(\mathcal{L}\omega_1)\omega_2\subset(\mathcal{M}\omega_1)\omega_2$; thus $\omega_1\omega_2$ is monotone. Next set $\mathcal{M}=\mathcal{L}\omega_1$. Let $A\in\mathcal{M}\omega_2$. Since $\omega_2$ is locally finite, there is a finite family $\mathcal{M}_A=\{M_1,\ldots,M_n\}\subset\mathcal{M}$ such that $A\in\mathcal{M}_A\omega_2$. For each $M_i\in\mathcal{L}\omega_1$, there exists a finite family $\mathcal{L}_i\subset\mathcal{L}$ such that $M_i\in\mathcal{L}_i\omega_i$. Let $\mathcal{L}'=\mathcal{L}_1\cup\mathcal{L}_2\cup\cdots\cup\mathcal{L}_n$. Then $\mathcal{L}'$ is a finite family of languages, and $\mathcal{L}_i\omega_1\subset\mathcal{L}'\omega_1$ for $i=1,\ldots,n$, by the monotonicity of $\omega_1$. Thus $M_i\in\mathcal{L}'\omega_1$, $(i=1,\ldots,n)$, and $\mathcal{M}_A\subset\mathcal{L}'\omega_1$. Since $\omega_2$ is monotone, this implies $\mathcal{M}_A\omega_2\subset\mathcal{L}'\omega_1\omega_2$; consequently, $A\in\mathcal{L}'\omega_1\omega_2$. Thus $\omega_1\omega_2$ is locally finite. A similar proof shows that if $\omega_1$ and $\omega_2$ are unitary, then $\omega_1\omega_2$ is also unitary. ■

We now define an order on operators by setting:

$$\omega_1\leqslant\omega_2 \quad\Leftrightarrow\quad \mathcal{L}\omega_1\subset\mathcal{L}\omega_2 \quad \text{for all} \quad \mathcal{L}\in\mathbb{F}.$$

Clearly, $\leqslant$ is an order relation. No confusion seems possible between this order on operators, and the preorder on languages defined by rational transductions in Section V.2.

We note the formula:

$$\omega_1 \leqslant \omega_2 \Rightarrow \omega_1\omega \leqslant \omega_2\omega \quad \text{and} \quad \omega\omega_1 \leqslant \omega\omega_2.$$

Indeed, if $\mathscr{L}\omega_1 \subset \mathscr{L}\omega_2$, then $\mathscr{L}\omega_1\omega \subset \mathscr{L}\omega_2\omega$ by the monotonicity of $\omega$. The inclusion $(\mathscr{L}\omega)\omega_1 \subset (\mathscr{L}\omega)\omega_2$ is true by definition.

The most interesting operators are closure operators:

**Definition** An operator $\omega$ is called
(i) extensive if $\mathrm{id} \leqslant \omega$, thus if $\mathscr{L} \subset \mathscr{L}\omega$ for $\mathscr{L} \in \mathbb{F}$
(ii) idempotent if $\omega = \omega\omega$, thus if $(\mathscr{L}\omega)\omega = \mathscr{L}\omega$ for $\mathscr{L} \in \mathbb{F}$
(iii) a closure operator if it is both extensive and idempotent.

Usually, a closure operator with respect to some order is a mapping which is extensive, monotone and idempotent. In the present case, monotonicity is supplied by definition.

### Exercises

**1.1** Let $\mathscr{L}_0 = \{A_0\}$ be the family consisting of the single language $A_0$. Let $\omega$ be defined by $\mathscr{L}\omega = \mathscr{L}_0$ for $\mathscr{L} \in \mathbb{F}$. Show that $\omega$ is an idempotent, unitary operator which is not extensive.

**1.2** Let $\omega$ be a closure operator, $\mathscr{L}$, $\mathscr{M}$ be families of languages. Show that $(\mathscr{L} \cup \mathscr{M})\omega = (\mathscr{L}\omega \cup \mathscr{M}\omega)\omega$.

**1.3** Let $\omega$ be an unitary operator. Show that for all sets $(\mathscr{L}_i)_{i\in I}$ of families,

$$\left(\bigcup_{i\in I} \mathscr{L}_i\right)\omega = \bigcup_{i\in I} (\mathscr{L}_i\omega).$$

## VI.2 Examples of Operators

We list here the principal operators and some simple relations between operators.

**2.1** $\Phi$ is the "morphism" operator defined by

$$\mathscr{L}\Phi = \{\varphi(L) \mid L \in \mathscr{L},\ \varphi : X_L^* \to Y^* \text{ a morphism}\}.$$

$\Phi$ is a unitary closure operator.

**2.2** $\Phi^{-1}$ is the "inverse morphism" operator defined by

$$\mathscr{L}\Phi^{-1} = \{\varphi^{-1}(L) \mid L \in \mathscr{L},\ \varphi : Y^* \to X_L^* \text{ a morphism}\}.$$

$\Phi^{-1}$ also is a unitary closure operator.

**2.3** The operators

$\Phi_a$ : alphabetic morphism; $\Phi_a^{-1}$ : inverse alphabetic morphism;
$\Phi_\pi$ : projection; $\Phi_\pi^{-1}$ : inverse projection

are defined as $\Phi$ and $\Phi^{-1}$, by replacing morphism by alphabetic morphism and projection. They are all unitary closure operators.

**2.4** $I_R$ is the operator "intersection with regular sets" defined by

$$\mathscr{L}I_R = \{L \cap K \mid L \in \mathscr{L}, K \in \mathrm{Rat}\}.$$

Clearly $I_R$ is monotone and unitary. Since the intersection of two regular languages is regular, $I_R$ is a closure operator.

**2.5** $\Gamma$ is the operator of closure under rational transduction:

$$\mathscr{L}\Gamma = \{L' \mid \exists L \in \mathscr{L}, L \geqslant L'\}.$$

Clearly, $\Gamma$ is an unitary operator, and $\Gamma$ is a closure operator by Proposition V.2.3.

Recall (see Section V.5) that if $\mathscr{L}, \mathscr{M}$ are families of languages, we introduced two unary notations for the substitution $\mathscr{L} \square \mathscr{M}$ by setting

$$\mathscr{L} \square \mathscr{M} = \mathscr{L}\bar{\mathscr{M}} = \mathscr{M}\underline{\mathscr{L}}.$$

This yields two mappings $\bar{\mathscr{M}}$ and $\underline{\mathscr{L}}$.

**2.6** $\bar{\mathscr{M}}$ is an operator for any family $\mathscr{M}$.
Indeed, $\bar{\mathscr{M}}$ is monotone by formula (V.5.2); further, if $A \in \mathscr{L}\bar{\mathscr{M}}$, then $A = \sigma(L)$ for some language $L \in \mathscr{L}$ and some $\mathscr{M}$-substitution $\sigma$. Thus, $A \in \{L\}\bar{\mathscr{M}}$ and $\bar{\mathscr{M}}$ is unitary. In general, $\bar{\mathscr{M}}$ is not a closure operator.

**2.7** $\underline{\mathscr{L}}$ is an operator for any family $\mathscr{L}$.
Indeed, $\underline{\mathscr{L}}$ is monotone by formula (V.5.1); and if $A \in \mathscr{M}\underline{\mathscr{L}} = \mathscr{L}\bar{\mathscr{M}}$, then $A = \sigma(L)$, where $L \in \mathscr{L}$ and $\sigma : X_L^* \to Y^*$ is some $\mathscr{M}$-substitution. Thus $A \in \mathscr{M}_A\underline{\mathscr{L}}$, where $\mathscr{M}_A = \{\sigma(x) : x \in X_L\}$. In general, $\underline{\mathscr{L}}$ is not unitary. This is the reason why we cannot restrict to unitary operators.

**2.8** $\underline{\mathrm{Rat}}$, $\underline{\mathrm{Fin}}$, $\underline{\mathrm{Elm}}$ are closure operators.

**2.9** $\hat{\Gamma}$ is the "full AFL" operator.

Clearly, $\hat{\Gamma}$ is monotone. To show that $\hat{\Gamma}$ is locally finite, consider a family of languages $\mathscr{L}$, and a language $A \in \mathscr{L}\hat{\Gamma}$. Then $A$ has be obtained from languages in $\mathscr{L}$ by a finite number of rational transductions and rational operations. Thus there is a finite family $\mathscr{L}_A \subset \mathscr{L}$ such that $A \in \mathscr{L}_A\hat{\Gamma}$. Thus $\hat{\Gamma}$ is locally finite; moreover $\hat{\Gamma}$ is a closure operator. Note that we shall prove in Section 4 the equality $\hat{\Gamma} = \Gamma\underline{\mathrm{Rat}}$; this shows a posteriori that $\hat{\Gamma}$ is an operator.

We now give some examples of relations between operators.

**1.** Nivat's Theorem gives

$$\Gamma = \Phi^{-1} I_R \Phi = \Phi_a^{-1} I_R \Phi_a = \Phi_\pi^{-1} I_R \Phi_\pi.$$

**2.** The inequalities $\Phi^{-1}$, $I_R\Phi$, $\overline{\mathrm{Rat}} \leq \Gamma$ hold since inverse morphisms, morphisms, intersection with regular sets, rational substitutions are special cases of rational transductions.

**3.** The associativity of the substitution $\mathscr{L} \square \mathscr{M}$ gives the commutativity formula:

$$\underline{\mathscr{L}}\bar{\mathscr{M}} = \bar{\mathscr{M}}\underline{\mathscr{L}}. \tag{2.1}$$

This is precisely Corollary V.5.3.

**4.** If $\mathscr{L}_i$, $(i \in I)$ are cones, then $\bigcup_{i \in I} \mathscr{L}_i$ is a cone. Indeed, $\Gamma$ is unitary, and by Exercise 1.3,

$$\left(\bigcup_{i \in I} \mathscr{L}_i\right)\Gamma = \bigcup_{i \in I} (\mathscr{L}_i \Gamma) = \bigcup_{i \in I} \mathscr{L}_i.$$

(This is not true for the full AFL operator $\hat{\Gamma}$, see Exercise V.4.2: $\hat{\Gamma}$ is not unitary.)

## Exercises

**2.1** Let $\mathscr{M}$ be a family of languages. Show that if $\{x\} \in \mathscr{M}$ for $x \in \Sigma$, then $\underline{\mathscr{M}}$ and $\bar{\mathscr{M}}$ are extensive.

**2.2** Give an example of two closure operators $\omega_1$, $\omega_2$ such that $\omega_1\omega_2$ is not idempotent, and thus is not a closure operator (this shows that the theorem of Elgot and Mezei about the idempotency of $\Gamma$ cannot be obtained as a formal consequence of Nivat's Theorem).

## VI.3 Closure Operators

Let $\mathscr{L}$ be a family of languages, and let $\omega$ be an operator. Suppose $\mathscr{L}$ is not closed under $\omega$ (i.e. $\mathscr{L}\omega$ is not contained in $\mathscr{L}$). In some cases, there exists a least family $\mathscr{L}'$ containing $\mathscr{L}$ and closed under $\omega$. This defines a new operator $\omega^*$ which associates $\mathscr{L}'$ to $\mathscr{L}$, and $\omega^*$ is a closure operator.

In this section, we give an explicit construction of the closure operator $\omega^*$, and more generally of the operator $\{\omega_1, \ldots, \omega_k\}^*$ which associates to a family $\mathscr{L}$ the least family of languages containing $\mathscr{L}$ and closed under $\omega_1, \ldots, \omega_k$. The main result of this section is a particularly simple expression of $\{\omega_1, \ldots, \omega_k\}^*$ which holds provided $\omega_1, \ldots, \omega_k$ are closure operators satisfying some relations of commutation.

The reader will observe that the results of this section do not use the fact that the objects under consideration are families of languages. Similarily, the local finiteness of operators is only used to derive a "continuity" property. In fact, all computations are valid in the much more general framework of the lattice of continuous functions over a complete lattice, and lead to rather well-known results in this algebraic structure. We prefer to restrict ourselves to the special case of families of languages and to prove directly the few results necessary in the sequel. (The reader interested in the general theory may consult, e.g. Maeda and Maeda [1974]).

The set $\mathbb{F}$ of all families of languages is a complete semi-lattice for the union. It would be a complete lattice if the empty family would be added. The first lemma shows that operators are upper-continuous mappings.

**Lemma 3.1** (Continuity) *Let $\mathscr{L}_1 \subset \mathscr{L}_2 \subset \cdots \subset \mathscr{L}_n \subset \cdots$ be an increasing sequence of families of languages, and let $\omega$ be an operator. Then*

$$\left(\bigcup_{i \geq 1} \mathscr{L}_i\right)\omega = \bigcup_{i \geq 1} (\mathscr{L}_i\omega) \tag{3.1}$$

Proof. Set $\mathscr{L} = \bigcup_{i \geq 1} \mathscr{L}_i$. For any $i \geq 1$, $\mathscr{L}_i \subset \mathscr{L}$; consequently $\mathscr{L}_i\omega \subset \mathscr{L}\omega$ since $\omega$ is monotone. Thus $\bigcup_{i \geq 1} \mathscr{L}_i\omega \subset \mathscr{L}\omega$.

Conversely, let $A \in \mathscr{L}\omega$. Since $\omega$ is locally finite, there is a finite family $\mathscr{L}_A \subset \mathscr{L}$ such that $A \in \mathscr{L}_A\omega$. Since the sequence $(\mathscr{L}_i)_{i \geq 1}$ is increasing, there is an integer $k$ such that $\mathscr{L}_A \subset \mathscr{L}_k$. Using monotonicity once more, this yields $A \in \mathscr{L}_k\omega$. Thus $\mathscr{L}\omega \subset \bigcup_{i \geq 1} \mathscr{L}_i\omega$. ∎

Note that if $\omega$ is unitary, then (3.1) holds without the assumption that the sequence is increasing (Exercise 1.3).

Let $\omega$ be an operator. A family of languages $\mathscr{L}$ is closed under $\omega$ if $\mathscr{L}\omega \subset \mathscr{L}$. If $\omega$ is extensive, this is equivalent to $\mathscr{L}\omega = \mathscr{L}$. If $\omega$ is a closure operator, then $\mathscr{L}\omega$ is always closed under $\omega$. Our aim is to define the operator of closure under $\omega$.

To any set $(\omega_i)_{i \in I}$ of operators, we associate a mapping denoted by $\bigcup_{i \in I} \omega_i$ and defined by

$$\mathscr{L}\left(\bigcup_{i \in I} \omega_i\right) = \bigcup_{i \in I} (\mathscr{L}\omega_i) \qquad \mathscr{L} \in \mathbb{F}.$$

To verify that $\omega = \bigcup_{i \in I} \omega_i$ is an operator, we first observe that $\omega$ is clearly monotone. $\omega$ is also locally finite. Indeed, let $A \in \mathscr{L}\omega = \bigcup_{i \in I} (\mathscr{L}\omega_i)$. Then there exists an index $k \in I$ such that $A \in \mathcal{L}\omega_k$. Since $\omega_k$ is locally finite, $A \in \mathcal{L}_A\omega_k$ for some finite family $\mathcal{L}_A \subset \mathcal{L}$. Then $A \in \bigcup_{i \in I} (\mathcal{L}_A\omega_i) = \mathcal{L}_A\omega$.

If $I$ is finite, $I = \{i_1, \ldots, i_r\}$, we write

$$\bigcup_{i \in I} \omega_i = \omega_{i_1} \cup \omega_{i_2} \cup \cdots \cup \omega_{i_r}.$$

In particular, $\omega \leqslant \omega'$ iff $\omega' = \omega \cup \omega'$.

**Definition** Let $\omega$ be an operator; then $\omega^*$ is the operator defined by

$$\omega^* = \bigcup_{n \geqslant 0} \omega^n,$$

where $\omega^0 = \mathrm{id}$.

**Proposition 3.2** *Let $\omega$ be an extensive operator; then $\omega^*$ is a closure operator. For any family of languages $\mathcal{L}$, $\mathcal{L}\omega^*$ is the least family of languages containing $\mathcal{L}$ and closed under $\omega$; moreover, $\mathcal{L}\omega^*$ is closed under $\omega^*$.*

Proof. Let $\mathcal{L}$ be a family of languages. Then

$$\mathcal{L} \subset \bigcup_{n \geqslant 0} \mathcal{L}\omega^n = \mathcal{L}\omega^*,$$

showing that $\omega^*$ is extensive. We show that $\mathcal{L}\omega^*$ is closed under $\omega$. Since $\omega$ is extensive, the sequence $(\mathcal{L}\omega^n)_{n \geqslant 0}$ is increasing. Thus by the continuity lemma,

$$\mathcal{L}\omega^*\omega = \Big(\bigcup_{n \geqslant 0} \mathcal{L}\omega^n\Big)\omega = \bigcup_{n \geqslant 0} \mathcal{L}\omega^n\omega = \bigcup_{n \geqslant 1} \mathcal{L}\omega^n = \bigcup_{n \geqslant 0} \mathcal{L}\omega^n = \mathcal{L}\omega^*. \tag{3.2}$$

Next, let $\mathcal{M}$ be a family of languages containing $\mathcal{L}$ and closed under $\omega$. Since $\omega$ is monotone, $\mathcal{L}\omega^n \subset \mathcal{M}\omega^n = \mathcal{M}$ for $n \geqslant 0$, and $\mathcal{L}\omega^* \subset \mathcal{M}$, showing that $\mathcal{L}\omega^*$ is the least family containing $\mathcal{L}$ and closed under $\omega$. Finally, we verify that $\omega^*$ is idempotent. By (3.2), $\omega^*\omega = \omega^*$. Therefore, $\omega^*\omega^n = \omega^*$ for any $n \geqslant 0$. Thus

$$\mathcal{L}\omega^*\omega^* = \bigcup_{n \geqslant 0} (\mathcal{L}\omega^*)\omega^n = \mathcal{L}\omega^*. \qquad \blacksquare$$

**Example 3.1** Let $\mathcal{M}$ be a family of languages such that $\{x\} \in \mathcal{M}$ for $x \in \Sigma$. Then $\bar{\mathcal{M}}$ is extensive (Exercise 2.1). For any family of languages $\mathcal{L}$, the family

$$\mathcal{L}(\bar{\mathcal{M}})^* = \bigcup_{n \geqslant 0} \mathcal{L}(\bar{\mathcal{M}})^n$$

is the least family of languages containing $\mathcal{L}$ and closed under $\mathcal{M}$-substitution.

In particular, $\mathcal{M}\bar{\mathcal{M}}^*$ is the least family containing $\mathcal{M}$ and closed under $\mathcal{M}$-substitution. We prove that $\mathcal{M}\bar{\mathcal{M}}^*$ is closed under substitution, i.e. under $\mathcal{M}\bar{\mathcal{M}}^*$-substitution. For this, we first extend formula (2.1) and show

$$\bar{\mathcal{M}}^*\underline{\mathcal{N}} = \underline{\mathcal{N}}\bar{\mathcal{M}}^* \qquad (\bar{\mathcal{M}} \text{ extensive}). \tag{3.3}$$

Indeed, the formula $\bar{\mathcal{M}}\underline{\mathcal{N}} = \underline{\mathcal{N}}\bar{\mathcal{M}}$ implies, by induction,

$$\bar{\mathcal{M}}^n\underline{\mathcal{N}} = \underline{\mathcal{N}}\bar{\mathcal{M}}^n \qquad (n \geq 0).$$

For any family $\mathcal{L}$, $(\mathcal{L}\bar{\mathcal{M}}^n)_{n\geq 0}$ is an increasing sequence since $\bar{\mathcal{M}}$ is extensive; by continuity

$$\mathcal{L}\bar{\mathcal{M}}^*\underline{\mathcal{N}} = \bigcup_{n\geq 0} \mathcal{L}(\bar{\mathcal{M}}^n\underline{\mathcal{N}}) = \bigcup_{n\geq 0} \mathcal{L}\underline{\mathcal{N}}\bar{\mathcal{M}}^n = \mathcal{L}\underline{\mathcal{N}}\bar{\mathcal{M}}^*.$$

This proves (3.3).

**Corollary 3.3** *Let $\mathcal{M}$ be a family of languages such that $\{x\} \in \mathcal{M}$ for $x \in \Sigma$; then $\mathcal{M}\bar{\mathcal{M}}^*$ is closed under substitution.*

Proof. We have

$$\mathcal{M}\bar{\mathcal{M}}^*\overline{\mathcal{M}\bar{\mathcal{M}}^*} = \mathcal{M}\bar{\mathcal{M}}^*\underline{\mathcal{M}\bar{\mathcal{M}}}^* = \mathcal{M}\underline{\mathcal{M}\bar{\mathcal{M}}}^*\bar{\mathcal{M}}^*$$

by (3.3). Further,

$$\mathcal{M}\underline{\mathcal{M}\bar{\mathcal{M}}}^*\bar{\mathcal{M}}^* = \mathcal{M}\bar{\mathcal{M}}^*\bar{\mathcal{M}}\bar{\mathcal{M}}^*.$$

Now $\bar{\mathcal{M}}^*\bar{\mathcal{M}} = \bar{\mathcal{M}}^*\bar{\mathcal{M}}^* = \bar{\mathcal{M}}^*$ by Proposition 3.2, thus

$$\mathcal{M}\bar{\mathcal{M}}^*\overline{\mathcal{M}\bar{\mathcal{M}}^*} = \mathcal{M}\bar{\mathcal{M}}^*.$$ ■

We now extend Proposition 3.2 to the case of several operators. Let $\omega_1, \ldots, \omega_k$, $(k \geq 1)$ be operators. Define $\Bbbk = \{1, 2, \ldots, k\}$, and for a word

$$f = i_1 i_2 \cdots i_n \in \Bbbk^+ \qquad (i_1, \ldots, i_n \in \Bbbk),$$

set $\qquad \omega_f = \omega_{i_1}\omega_{i_2} \cdots \omega_{i_n}.$

Finally, let $\varepsilon$ be the empty word of $\Bbbk^*$ and set $\omega_\varepsilon = \mathrm{id}$.

**Definition** Let $\omega_1, \omega_2, \ldots, \omega_k$ be operators; then $\{\omega_1, \omega_2, \ldots, \omega_k\}^*$ is the operator defined by

$$\{\omega_1, \omega_2, \ldots, \omega_k\}^* = \bigcup_{f\in\Bbbk^*} \omega_f.$$

**Proposition 3.4** *Let $\omega_1, \ldots, \omega_k$ be extensive operators; then $\{\omega_1, \ldots, \omega_k\}^*$ is a closure operator. For any family of languages $\mathcal{L}$, $\mathcal{L}\{\omega_1, \ldots, \omega_k\}^*$ is the least family of languages containing $\mathcal{L}$ and closed under $\omega_1, \ldots, \omega_k$; moreover, $\mathcal{L}\{\omega_1, \ldots, \omega_k\}^*$ is closed under $\{\omega_1, \ldots, \omega_k\}^*$.*

Proof. Let $\omega = \{\omega_1, \ldots, \omega_k\}^*$. Clearly $\omega$ is extensive. We show that $\omega\omega_i = \omega$ for $i = 1, \ldots, k$. First note that, since each $\omega_i$ is extensive, $\omega_f \leq \omega_g$ for any subword $f$ of $g$, $(f, g \in \Bbbk^*)$.

Let $\mathcal{L}$ be a family of languages and define

$$\mathcal{L}^{(n)} = \bigcup_{f\in\Bbbk^n} \mathcal{L}\omega_f.$$

The sequence $(\mathscr{L}^{(n)})_{n\geq 0}$ is increasing; thus by continuity

$$\mathscr{L}\omega\omega_i = \left(\bigcup_{n\geq 0} \mathscr{L}^{(n)}\right)\omega_i = \bigcup_{n\geq 0} \mathscr{L}^{(n)}\omega_i \qquad i=1,\ldots,k. \tag{3.4}$$

Note that the inclusion $\mathscr{L}^{(n)}\omega_i \subset \mathscr{L}^{(n+1)}$ is not necessarily true except if $\omega_i$ is unitary, since in that case $\omega_i$ can be passed through the union. Let $g$ be the word of length $nk^n$ obtained by concatenating all words of $\mathbb{k}^n$. Then $\mathscr{L}\omega_f \subset \mathscr{L}\omega_g$ for all $f\in\mathbb{k}^n$, therefore

$$\mathscr{L}^{(n)} \subset \mathscr{L}\omega_g$$

and $$\mathscr{L}^{(n)}\omega_i \subset \mathscr{L}\omega_g\omega_i = \mathscr{L}\omega_{gi} \subset \mathscr{L}\omega. \tag{3.5}$$

By (3.4) and (3.5), we have $\mathscr{L}\omega\omega_i = \mathscr{L}\omega$ for $i=1,\ldots,k$, showing that $\mathscr{L}\omega$ is closed under $\omega_1,\ldots,\omega_k$. Consequently, $\omega\omega_f=\omega$ for all $f\in\mathbb{k}^*$, and

$$\mathscr{L}\omega\omega = \bigcup_{f\in\mathbb{k}^*} \mathscr{L}\omega\omega_f = \mathscr{L}\omega.$$

Thus $\omega$ is idempotent. Finally, let $\mathscr{M}$ be a family of languages containing $\mathscr{L}$ and closed under $\omega_1,\ldots,\omega_k$. Then $\mathscr{L}\omega_f \subset \mathscr{M}\omega_f = \mathscr{M}$ for $f\in\mathbb{k}^*$, and therefore $\mathscr{L}\omega\subset\mathscr{M}$. This achieves the proof. ∎

**Example 3.2** Let $\omega_1=\Gamma$, $\omega_2=\underline{\mathrm{Rat}}$. Then $\mathscr{L}\{\Gamma,\underline{\mathrm{Rat}}\}^*$ is the least family containing $\mathscr{L}$ and closed under $\Gamma$ and $\underline{\mathrm{Rat}}$. Thus $\{\Gamma,\underline{\mathrm{Rat}}\}^*=\hat{\Gamma}$, and

$$\hat{\Gamma} = \bigcup_{n\geq 1} \omega_{i_1}\omega_{i_2}\cdots\omega_{i_n}, \quad \text{with } \omega_{i_j}=\Gamma \text{ or } \omega_{i_j}=\underline{\mathrm{Rat}}.$$

We now show that the operator $\{\omega_1,\ldots,\omega_k\}^*$ has a very simple form, provided certain conditions are verified.

**Theorem 3.5** *Let $\omega_1,\omega_2,\ldots,\omega_k$ be closure operators; if $\omega_j\omega_i\leq\omega_i\omega_j$ for $1\leq i<j\leq k$, then*

$$\{\omega_1,\omega_2,\ldots,\omega_k\}^* = \omega_1\omega_2\cdots\omega_k.$$

Proof. Clearly $\omega_1\omega_2\cdots\omega_k \leq \bigcup_{f\in\mathbb{k}^*}\omega_f$. Conversely, let $f=i_1i_2\cdots i_n\in\mathbb{k}^n$. If $i_1>i_2$, then $\omega_{i_1}\omega_{i_2}\leq\omega_{i_2}\omega_{i_1}$. Thus $\omega_f\leq\omega_{i_2}\omega_{i_1}\cdots\omega_{i_n}$. Comparing then $i_3$ to $i_1$ and so on yields finally

$$\omega_f \leq \omega_{j_1}\omega_{j_2}\cdots\omega_{j_n},$$

where $1\leq j_1\leq j_2\leq\cdots\leq j_n\leq k$ and $(j_1,\ldots,j_n)$ is a permutation of $(i_1,\ldots,i_n)$. Using the idempotency and the extensivity of $\omega_1,\ldots,\omega_k$, we obtain

$$\omega_{j_1}\omega_{j_2}\cdots\omega_{j_n} \leq \omega_1\cdots\omega_k.$$

Thus $\omega_f\leq\omega_1\cdots\omega_k$ for any $f\in\mathbb{k}^*$; therefore

$$\{\omega_1,\ldots,\omega_k\}^* \leq \omega_1\omega_2\cdots\omega_k.$$ ∎

**Example 3.2** (continued). In order to prove the equality $\hat{\Gamma} = \Gamma\,\underline{\text{Rat}}$, it suffices, in view of Theorem 3.5, to show the inequality $\underline{\text{Rat}}\,\Gamma \leqslant \Gamma\,\underline{\text{Rat}}$. This will be done in the next section.

**Exercises**

**3.1** Let $\omega_1, \omega_2, \ldots, \omega_k$ be extensive operators. Show that

$$\{\omega_1, \ldots, \omega_k\}^* = \bar{\omega}^*$$

where $\bar{\omega} = \omega_1 \cup \omega_2 \cup \cdots \cup \omega_k$.

**3.2** Let Word be the family of all words of $\Sigma^*$. Show that $\underline{\text{Word}}$ is the operator of closure under product. Show that $\underline{\text{Elm}}\,\underline{\text{Word}} \leqslant \underline{\text{Word}}\,\underline{\text{Elm}}$, and prove that $\underline{\text{Fin}} = \underline{\text{Word}}\,\underline{\text{Elm}}$.

## VI.4 Subcommutative Relations

In this section, we use the results of the preceding section, and especially Theorem 3.5 to derive relations between some frequently appearing operators. As a result, we obtain that the operations defining cones and full AFLs can always be applied in some specific order. For this, we use standard techniques from formal language theory to prove some inequalities between fundamental operators; from these, other relations can then be deduced in a completely formal way.

As an illustration, we first reformulate several results of Chapter III as relations between operators.

**Proposition 4.1** $\Gamma = \Phi^{-1} I_R \Phi = \Phi_a^{-1} I_R \Phi_a = \Phi_\pi^{-1} I_R \Phi_\pi$.

This is the theorem of Nivat (Theorem III.4.1).

**Proposition 4.2** $\Gamma\Gamma = \Gamma$.

This is the theorem of Elgot and Mezei (Theorem III.4.4). We give the proof of the theorem once more, but using now operators. We need the following three lemmas.

**Lemma 4.3** $\Phi_\pi \Phi_\pi^{-1} \leqslant \Phi_\pi^{-1} \Phi_\pi$.

This is just Lemma III.4.5.

**Lemma 4.4** $I_R \Phi^{-1} \leqslant \Phi^{-1} I_R$, $\quad I_R \Phi_a^{-1} \leqslant \Phi_a^{-1} I_R$, $\quad I_R \Phi_\pi^{-1} \leqslant \Phi_\pi^{-1} I_R$.

Proof. We prove the first inequality. Let $\mathcal{L}$ be a family of languages, and let $A \in \mathcal{L} I_R \Phi^{-1}$. Thus there exist a language $L \in \mathcal{L}$, a regular language $K \subset X_L^*$, and a morphism $\varphi: X_A^* \to X_L^*$ such that $A = \varphi^{-1}(L \cap K)$. Thus $A = \varphi^{-1}(L) \cap \varphi^{-1}(K)$, and since $\varphi^{-1}(L) \in \mathcal{L}\Phi^{-1}$ and $\varphi^{-1}(K) \in \text{Rat}$, we have $A \in \mathcal{L}\Phi^{-1} I_R$, showing that $\mathcal{L} I_R \Phi^{-1} \subset \mathcal{L}\Phi^{-1} I_R$. ■

**Lemma 4.5** $\Phi I_R \leqslant I_R \Phi$; $\Phi_a I_R \leqslant I_R \Phi_a$; $\Phi_\pi I_R \leqslant I_R \Phi_\pi$.

Proof. We prove the first inequality. Let $\mathscr{L}$ be a family of languages, and let $A \in \mathscr{L}\Phi I_R$. Thus there exist a language $L \in \mathscr{L}$, a morphism $\varphi : X_L^* \to X_A^*$ and $K \in \mathrm{Rat}(X_A^*)$ such that $A = \varphi(L) \cap K$. Now

$$A = \varphi(L \cap \varphi^{-1}(K)),$$

and since $\varphi^{-1}(K) \in \mathrm{Rat}(X_L^*)$, we get $A \in \mathscr{L} I_R \Phi$. This shows that $\mathscr{L}\Phi I_R \subset \mathscr{L} I_R \Phi$. ∎

The proof of Elgot and Mezei's theorem can now be rewritten as follows:

$$\begin{aligned} \Gamma\Gamma &= \Phi_\pi^{-1} I_R \Phi_\pi \Phi_\pi^{-1} I_R \Phi_\pi && \text{(Nivat's Theorem)} \\ &\leqslant \Phi_\pi^{-1} I_R \Phi_\pi^{-1} \Phi_\pi I_R \Phi_\pi && \text{(Lemma 4.3)} \\ &\leqslant \Phi_\pi^{-1} \Phi_\pi^{-1} I_R I_R \Phi_\pi \Phi_\pi && \text{(Lemma 4.4 and 4.5)} \\ &= \Phi_\pi^{-1} I_R \Phi_\pi = \Gamma. && \text{(idempotency)} \end{aligned}$$

Since clearly $\Gamma\Gamma \geqslant \Gamma$, we finally get $\Gamma = \Gamma\Gamma$.

Remark. The three Lemmas 4.3, 4.4 and 4.5 imply, in view of Theorem 3.5, the equality

$$\{\Phi_\pi^{-1}, I_R, \Phi_\pi\}^* = \Phi_\pi^{-1} I_R \Phi_\pi.$$

Since $\{\Phi_\pi^{-1}, I_R, \Phi_\pi\}^*$ is a closure operator, it is idempotent, and this shows directly that $\Gamma\Gamma = \Gamma$. On the other hand, by the equality $\Gamma = \{\Phi_\pi^{-1}, I_R, \Phi_\pi\}^*$ we see directly that for any family $\mathscr{L}$, $\mathscr{L}\Gamma$ is the least family containing and closed by inverse projection, intersection with rational languages, and projection.

We now investigate similar relations involving substitutions. First, we prove

**Lemma 4.6** $\underline{\mathrm{Elm}}\, \Gamma \leqslant \Gamma\, \underline{\mathrm{Elm}}$.

Proof. Let $\mathscr{L}$ be a family of languages, and $A \in \mathscr{L}\, \underline{\mathrm{Elm}}\, \Gamma$. Then there exist a language $B \in \mathscr{L}\, \underline{\mathrm{Elm}}$ and a rational transduction $\tau : X_B^* \to X_A^*$ such that $A = \tau(B)$. Since $B \in \mathscr{L}\, \underline{\mathrm{Elm}}$, there is an integer $n \geqslant 0$ such that

$$B = \bigcup_{1 \leqslant i \leqslant n} L_i \qquad L_i \in \mathscr{L}, \quad i = 1, \ldots, n.$$

This implies $\tau(B) = \bigcup_{1 \leqslant i \leqslant n} \tau(L_i)$, and since $\tau(L_i) \in \mathscr{L}\Gamma$ for $i = 1, \ldots, n$, we finally get $A \in \mathscr{L}\Gamma\, \underline{\mathrm{Elm}}$. ∎

**Theorem 4.7** $\{\Gamma, \underline{\mathrm{Elm}}\}^* = \Gamma\, \underline{\mathrm{Elm}}$, *or equivalently: The full semi-*AFL *generated by a family of languages* $\mathscr{L}$ *is equal to the closure under union of the rational cone generated by* $\mathscr{L}$.

Proof. This follows from Theorem 3.5 in view of the preceding lemma. ∎

We now want to show that Theorem 4.7 remains true when Elm is replaced by Rat, and more precisely that the equality $\mathscr{L}\hat{\Gamma} = \mathscr{L}\Gamma\, \underline{\mathrm{Rat}}$ holds for any family of languages $\mathscr{L}$. We need only to prove that $\mathscr{L}\Gamma\,\underline{\mathrm{Rat}}$ is a cone, and for this, it suffices to show the inequality $\underline{\mathrm{Rat}}\,\Gamma \leq \Gamma\,\underline{\mathrm{Rat}}$. Writing $\Gamma=\Phi^{-1}I_R\Phi$, we have to "pass" the operator $\underline{\mathrm{Rat}}$ through the three operators $\Phi^{-1}$, $I_R$, $\Phi$. This will be done by proving three inequalities, each one involving one of these three operators. It appears that the operator $\underline{\mathrm{Rat}}$ can be replaced, with only minor changes, by any operator of the form $\underline{\mathscr{L}}$; this gives, as a corollary, another important property of cones. Therefore, we show the more general

**Proposition 4.8** *Let $\mathscr{L}$ be a family of languages; then*

$$\underline{\mathscr{L}}\Gamma \leq \Gamma\underline{\mathscr{L}\Gamma} \quad \text{or equivalently} \quad \bar{\mathscr{L}}\Gamma \leq \Gamma\overline{\mathscr{L}\Gamma}.$$

First, we note the following

**Corollary 4.9** $\underline{\mathrm{Rat}}\,\Gamma \leq \Gamma\,\underline{\mathrm{Rat}}$.

Proof. By Proposition 4.8, we have $\underline{\mathrm{Rat}}\,\Gamma \leq \Gamma\,\underline{\mathrm{Rat}\,\Gamma}$, and since Rat is a cone, $\mathrm{Rat}\,\Gamma=\mathrm{Rat}$. ∎

Using Theorem 3.5, this corollary implies the following theorem

**Theorem 4.10** (Ginsburg and Greibach [1969]) $\hat{\Gamma}=\{\Gamma,\underline{\mathrm{Rat}}\}^{*}=\Gamma\,\underline{\mathrm{Rat}}$ *or equivalently: The full* AFL *generated by a family of languages $\mathscr{L}$ is equal to the rational closure of the cone generated by $\mathscr{L}$.* ∎

By this result, AFL-properties can be separated in properties of rational cones, and in properties of rational closure.

For the proof of Proposition 4.8 we need three lemmas. The first lemma involves morphisms.

**Lemma 4.11** *Let $\mathscr{L}$ be a family of languages; then*

$$\underline{\mathscr{L}}\Phi_\pi=\Phi_\pi\underline{\mathscr{L}};\quad \underline{\mathscr{L}}\Phi_a=\Phi_a\underline{\mathscr{L}};\quad \underline{\mathscr{L}}\Phi=\Phi\underline{\mathscr{L}}$$

*or equivalently*

$$\bar{\mathscr{L}}\Phi_\pi=\overline{\mathscr{L}\Phi_\pi};\quad \bar{\mathscr{L}}\Phi_a=\overline{\mathscr{L}\Phi_a};\quad \bar{\mathscr{L}}\Phi=\overline{\mathscr{L}\Phi}.$$

Proof. Let $\mathscr{M}$ be a family of languages; then

$$\mathscr{M}\underline{\mathscr{L}}\Phi=\mathscr{M}\Phi\underline{\mathscr{L}} \quad\Leftrightarrow\quad \mathscr{L}\bar{\mathscr{M}}\Phi=\mathscr{L}\overline{\mathscr{M}\Phi}.$$

Thus the two sequences of formulas are equivalent.

Let $A\in\mathscr{L}\bar{\mathscr{M}}\Phi$. Then there exist a language $B\in\mathscr{L}\bar{\mathscr{M}}$, and a morphism $\varphi: X_B^*\to X_A^*$ such that $A=\varphi(B)$. Since $B\in\mathscr{L}\bar{\mathscr{M}}$, there are a language $L\in\mathscr{L}$, and a $\mathscr{M}$-substitution $\sigma: X_L^*\to X_B^*$ with $B=\sigma(L)$. Thus $A=(\varphi\circ\sigma)(L)$. Now $\varphi\circ\sigma: X_L^*\to X_A^*$ is a $\mathscr{M}\Phi$-substitution since $\varphi\sigma(z)\in\mathscr{M}\Phi$ for each $z\in X_L$. This shows that $A\in\mathscr{L}\overline{\mathscr{M}\Phi}$. Thus $\mathscr{L}\bar{\mathscr{M}}\Phi\subset\mathscr{L}\overline{\mathscr{M}\Phi}$.

Conversely, let $A\in\mathscr{L}\overline{\mathscr{M}\Phi}$. Then there exist a language $L\in\mathscr{L}$, and a $\mathscr{M}\Phi$-substitution $\sigma: X_L^*\to X_A^*$ with $A=\sigma(L)$. For each $z\in X_L$, there exist an alphabet $Y_z$, a language $M_z \subset Y_z^*$, $M_z \in \mathcal{M}$, and a morphism $\varphi_z : Y_z^* \to X_A^*$ such that $\sigma(z) = \varphi_z(M_z)$. By copying if necessary, we may suppose the alphabets $Y_z$, $(z \in X_L)$ pairwise disjoint. Set

$$Y = \bigcup_{z \in X_L} Y_z,$$

and define a morphism $\varphi : Y^* \to X_A^*$ by $\varphi(y) = \varphi_z(y)$ iff $y \in Y_z$. Finally, define a substitution

$$\sigma' : X_L^* \to Y^*$$

by $\sigma'(z) = M_z$ for $z \in X_L$. Then $\sigma = \varphi \circ \sigma'$. Since $\sigma'$ is a $\mathcal{M}$-substitution, it follows that $A = \varphi(\sigma'(L)) \in \mathcal{L}\bar{\mathcal{M}}\Phi$, and thus $\mathcal{L}\mathcal{M}\Phi \subset \mathcal{L}\bar{\mathcal{M}}\Phi$. The proofs of the two other identities are similar. ■

The second lemma handles inverse morphism. The situation is complicated by the fact that inverse morphisms are not substitutions.

**Lemma 4.12** *Let $\mathcal{L}$ be a family of languages; then*

$$\underline{\mathcal{L}}\Phi_\pi^{-1} \leqslant \underline{\mathcal{L}}\Phi_a^{-1} \leqslant \Gamma\underline{\mathcal{L}\Phi_\pi^{-1}} \quad \textit{or equivalently} \quad \bar{\mathcal{L}}\Phi_\pi^{-1} \leqslant \bar{\mathcal{L}}\Phi_a^{-1} \leqslant \Phi_\pi^{-1}\overline{\mathcal{L}\Gamma}.$$

Proof. Let $\mathcal{M}$ be a family of languages. Then

$$\mathcal{M}\underline{\mathcal{L}}\Phi_a^{-1} \subset \mathcal{M}\Gamma\underline{\mathcal{L}\Phi_\pi^{-1}} \quad \Leftrightarrow \quad \mathcal{L}\bar{\mathcal{M}}\Phi_a^{-1} \subset \mathcal{L}\Phi_\pi^{-1}\overline{\mathcal{M}\Gamma}.$$

This shows that the two sets of inequalities are equivalent. The inequality $\underline{\mathcal{L}}\Phi_\pi^{-1} \leqslant \underline{\mathcal{L}}\Phi_a^{-1}$ is clear, and only the last inequality has to be shown. Let $A \in \mathcal{L}\bar{\mathcal{M}}\Phi_a^{-1}$. Thus there exist a language $B \in \mathcal{L}\bar{\mathcal{M}}$, and an alphabetic morphism $\varphi : X_A^* \to X_B^*$ such that $A = \varphi^{-1}(B)$. Further there are a language $L \in \mathcal{L}$, and a $\mathcal{M}$-substitution $\sigma : X_L^* \to X_B^*$ such that $B = \sigma(L)$; consequently, we have $A = \varphi^{-1}(\sigma(L))$. Note that $\varphi^{-1} \circ \sigma$ is not a substitution in general, since $\varphi^{-1}(1) \neq \{1\}$. Let $y$ be a new letter not in $X_L$, set $Y = X_L \cup \{y\}$, and let $\theta : Y^* \to X_L^*$ be the projection from $Y^*$ onto $X_L^*$. Set $L' = \theta^{-1}(L)$; then $L' \in \mathcal{L}\Phi_\pi^{-1}$. Define a substitution $\sigma' : Y^* \to X_A^*$ by

$$\sigma'(z) = \varphi^{-1}(\sigma(z)) \qquad \text{for} \quad z \in X_L; \qquad \sigma'(y) = \varphi^{-1}(1).$$

Then $\sigma'$ is a $\mathcal{M}\Gamma$-substitution since $\varphi^{-1}(1) \in \mathrm{Rat} \subset \mathcal{M}\Gamma$. Finally, we have

$$\sigma'(\theta^{-1}(w)) = \varphi^{-1}(\sigma(w)) \qquad \text{for all} \quad w \in X_L^*,$$

and thus $\sigma'(L') = \sigma'(\theta^{-1}(L)) = \varphi^{-1}(\sigma(L)) = A$, thus $A \in \mathcal{L}\Phi_\pi^{-1}\overline{\mathcal{M}\Gamma}$. Consequently $\mathcal{L}\bar{\mathcal{M}}\Phi_a^{-1} \subset \mathcal{L}\Phi_\pi^{-1}\overline{\mathcal{M}\Gamma}$. ■

Finally, we investigate intersection with regular sets.

**Lemma 4.13** *Let $\mathcal{L}$ be a family of languages; then*

$$\underline{\mathcal{L}}I_R \leqslant I_R\underline{\mathcal{L}\Gamma} \quad \textit{or equivalently} \quad \bar{\mathcal{L}}I_R \leqslant \overline{\Gamma\mathcal{L}I_R}.$$

Proof. Let $\mathcal{M}$ be a family of languages. Then

$$\mathcal{M}\underline{\mathcal{L}}I_R \subset \mathcal{M}I_R\underline{\mathcal{L}\Gamma} \quad \Leftrightarrow \quad \mathcal{L}\bar{\mathcal{M}}I_R \subset \mathcal{L}\Gamma\overline{\mathcal{M}I_R}.$$

Therefore, the two inequalities are equivalent. Let $A \in \mathcal{L}\bar{\mathcal{M}}I_R$. Then there exist $K \in \mathrm{Rat}(X_A^*)$ and $B \in \mathcal{L}\bar{\mathcal{M}}$ such that $A = B \cap K$. Next, there exist $L \in \mathcal{L}$, and a $\mathcal{M}$-substitution $\sigma : X_L^* \to X_B^*$ such that

$$A = B \cap K = \sigma(L) \cap K.$$

Let $\mathbf{A} = \langle X_A, Q, q_-, Q_+\rangle$ be a finite automaton accepting $K : K = |\mathbf{A}|$. For each pair $q$, $q' \in Q$, define

$$K_{q,q'} = \{f \in X_A^* \mid q \cdot f = q'\}.$$

Clearly, these languages are regular. Next, set

$$Y = Q \times X_A \times Q,$$

and define a transduction $\tau : X_L^* \to Y^*$ by

$$\tau(1) = \begin{cases} \{1\} & \text{if } \ 1 \in K; \\ \emptyset & \text{if } \ 1 \notin K, \end{cases}$$

$\tau(z_1 z_2 \cdots z_n) = \{(q_-, z_1, q_1)(q_1, z_2, q_2) \cdots (q_{n-1}, z_n, q_n) \mid q_i \in Q, q_n \in Q_+\}$ for $n \geq 1$, $z_1, \ldots, z_n \in X_L$. Clearly this transduction is rational, and $\tau(L) \in \mathcal{L}\Gamma$. Finally, define a substitution $\sigma' : Y^* \to X_A^*$ by

$$\sigma'(q, z, q') = \sigma(z) \cap K_{q,q'} \qquad (q, z, q') \in Y.$$

It follows that $\sigma'$ is a $\mathcal{M}I_R$-substitution. For any word

$$w = z_1 z_2 \cdots z_n \in X_L^+ \qquad (z_i \in X_L)$$

we have

$$\sigma'(\tau(w)) = \bigcup (\sigma(z_1) \cap K_{q_-,q_1})(\sigma(z_2) \cap K_{q_1,q_2}) \cdots (\sigma(z_n) \cap K_{q_{n-1},q_n}),$$

where the union is over all $(q_1, \ldots, q_n) \in Q^n$ with $q_n \in Q_+$. Therefore $\sigma'(\tau(\underline{w})) = \sigma(w) \cap K$. Since $\sigma'(\tau(1)) = K \cap \sigma(1)$, we have $A = \sigma'(\tau(L))$, and $A \in \mathcal{L}\Gamma\mathcal{M}I_R$. ■

Now we are ready to prove Proposition 4.8.

Proof of Proposition 4.8. We have

$$\begin{aligned} \bar{\mathcal{L}}\Gamma = \bar{\mathcal{L}}\Phi_a^{-1}I_R\Phi_a &\leq \Phi_\pi^{-1}\overline{\mathcal{L}\Gamma}I_R\Phi_a && \text{(Lemma 4.12)} \\ &\leq \Phi_\pi^{-1}\Gamma\overline{\mathcal{L}\Gamma I_R}\Phi_a = \Gamma\overline{\mathcal{L}\Gamma}\Phi_a && \text{(Lemma 4.13)} \\ &= \Gamma\overline{\mathcal{L}\Gamma\Phi_a} = \Gamma\overline{\mathcal{L}}\Gamma. && \text{(Lemma 4.11).} \end{aligned}$$

■

Remark. A more symmetrical statement of this inequality is the following: $\overline{\mathscr{L}\Gamma}\Gamma \leqslant \Gamma\overline{\mathscr{L}\Gamma}$. The converse statement: $\Gamma\overline{\mathscr{L}\Gamma} \leqslant \overline{\mathscr{L}\Gamma}\Gamma$ is not true in general (Exercise 4.1).

The following result also is obtained by a simple application of Proposition 4.8.

**Theorem 4.14** (Ginsburg and Spanier [1970]). *Let $\mathscr{L}$ be a rational cone (resp. a full semi-*AFL*, a full* AFL) *and let $\mathscr{M}$ be a rational cone; then $\mathscr{L}\bar{\mathscr{M}} = \mathscr{L} \,\square\, \mathscr{M}$ is a cone (resp. a full semi-*AFL*, a full* AFL).

Proof. In view of Proposition 4.8,

$$\mathscr{L}\bar{\mathscr{M}}\Gamma \subset \mathscr{L}\Gamma\overline{\mathscr{M}\Gamma}.$$

Since $\mathscr{L} = \mathscr{L}\Gamma$, $\mathscr{M} = \mathscr{M}\Gamma$, it follows $\mathscr{L}\bar{\mathscr{M}}\Gamma = \mathscr{L}\bar{\mathscr{M}}$, and therefore $\mathscr{L} \,\square\, \mathscr{M}$ is a cone. Next assume $\mathscr{L}$ is a full semi-AFL. Then by (2.1)

$$\mathscr{L}\bar{\mathscr{M}}\,\underline{\mathrm{Elm}} = \mathscr{L}\,\underline{\mathrm{Elm}}\,\bar{\mathscr{M}} = \mathscr{L}\bar{\mathscr{M}},$$

and thus $\mathscr{L}\bar{\mathscr{M}}$ is closed under union. Similarly, if $\mathscr{L}$ is a full AFL,

$$\mathscr{L}\bar{\mathscr{M}}\,\underline{\mathrm{Rat}} = \mathscr{L}\,\underline{\mathrm{Rat}}\,\bar{\mathscr{M}} = \mathscr{L}\bar{\mathscr{M}}.$$

■

We show now that substitution by full AFLs instead of cones do not change the resulting family. We deduce this fact from the following equality.

**Lemma 4.15** *Let $\mathscr{L}$ be a family of languages; then*

$$\Gamma\overline{\mathscr{L}\Gamma} = \Gamma\overline{\mathscr{L}\hat{\Gamma}} \quad \textit{or equivalently} \quad \Gamma\underline{\mathscr{L}\Gamma} = \hat{\Gamma}\underline{\mathscr{L}\Gamma}.$$

Proof. We have $\Gamma\overline{\mathscr{L}\hat{\Gamma}} = \Gamma\overline{\mathscr{L}\Gamma\,\underline{\mathrm{Rat}}} = \Gamma\overline{\mathrm{Rat}}\,\overline{\mathscr{L}\Gamma}$.
Now $\overline{\mathrm{Rat}} \leqslant \Gamma$ since each rational substitution is a rational transduction; therefore $\Gamma\overline{\mathrm{Rat}}\,\overline{\mathscr{L}\Gamma} \leqslant \Gamma\Gamma\overline{\mathscr{L}\Gamma} = \Gamma\overline{\mathscr{L}\Gamma}$. Thus $\Gamma\overline{\mathscr{L}\hat{\Gamma}} \leqslant \Gamma\overline{\mathscr{L}\Gamma}$. The converse inequality is clear. ■

**Corollary 4.16** *Let $\mathscr{L}$ and $\mathscr{M}$ be cones; then $\mathscr{L} \,\square\, \mathscr{M} = \mathscr{L} \,\square\, \mathscr{M}\hat{\Gamma}$.*

Proof. Indeed $\mathscr{L}\bar{\mathscr{M}} = \mathscr{L}\Gamma\overline{\mathscr{M}\Gamma} = \mathscr{L}\Gamma\overline{\mathscr{M}\hat{\Gamma}}$ by the preceding lemma. ■

Finally we show that $\mathscr{L}\bar{\mathscr{M}}^*$, the least family containing $\mathscr{L}$ and closed under $\mathscr{M}$-substitution, is a cone provided $\mathscr{L}$ and $\mathscr{M}$ are cones.

**Proposition 4.17** *Let $\mathscr{L}$ and $\mathscr{M}$ be cones; then $\mathscr{L}\bar{\mathscr{M}}^*$ is a cone. Further, if $\mathscr{L}$ is a full* AFL, *then $\mathscr{L}\bar{\mathscr{M}}^*$ is a full* AFL. For any cone $\mathscr{L}$, *$\mathscr{L}\bar{\mathscr{L}}^*$ is a full* AFL *closed under substitution.*

Proof. Assume $\mathscr{L}$ and $\mathscr{M}$ are cones. In view of Theorem 4.14, and arguing by induction, $\mathscr{L}\bar{\mathscr{M}}^k$ is a cone for any integer $k \geqslant 0$ (with $\bar{\mathscr{M}}^0 = id$). Thus $\mathscr{L}\bar{\mathscr{M}}^* = \bigcup_{k \geqslant 0} \mathscr{L}\bar{\mathscr{M}}^k$ is a cone, since the union of cones is a cone. Assume $\mathscr{L}$ is a full AFL.

Then by formula (3.3),

$$\mathscr{L}\bar{\mathscr{M}}^* \underline{\mathrm{Rat}} = \mathscr{L}\, \underline{\mathrm{Rat}}\, \bar{\mathscr{M}}^* = \mathscr{L}\bar{\mathscr{M}}^*,$$

and thus $\mathscr{L}\bar{\mathscr{M}}^*$ is a full AFL.

Finally, $\mathscr{L}\bar{\mathscr{L}}^*$ is a cone and is closed under substitution by Corollary 3.3, thus is a full AFL by Proposition V.5.5. ■

**Exercises**

**4.1** Show that $\Gamma\overline{\mathscr{L}\Gamma} \neq \overline{\mathscr{L}\Gamma}\Gamma$. (Hint. Apply the operators to Elm.)

**4.2** Show that $\Phi_\pi^{-1}\Phi_\pi \leqslant \Phi_\pi\Phi_\pi^{-1}$, and thus $\Phi_\pi^{-1}\Phi_\pi = \Phi_\pi\Phi_\pi^{-1}$. Discuss the situation for the operators $\Phi_a$ and $\Phi$.

**4.3** Show that $\Phi^{-1}I_R \neq I_R\Phi^{-1}$.

**4.4** (Ginsburg, Greibach and Hopcroft [1969]). Let $\mathscr{L}_1$ and $\mathscr{L}_2$ be two families of languages. The wedge of $\mathscr{L}_1$ and $\mathscr{L}_2$ is the family $\mathscr{L}$ defined by

$$\mathscr{L} = \mathscr{L}_1 \wedge \mathscr{L}_2 = \{L_1 \cap L_2 \mid L_1 \in \mathscr{L}_1, L_2 \in \mathscr{L}_2\}.$$

a) show that if $\mathscr{L}_1$ and $\mathscr{L}_2$ are closed under inverse morphism and intersection with regular sets, then the same holds for $\mathscr{L}$.

b) show that if $\mathscr{L}_1$ and $\mathscr{L}_2$ are rational cones, then $\mathscr{L}\Phi$ is a rational cone (Hint. Apply the operator $\Phi\Gamma$ to $\mathscr{L} = \mathscr{L}\Phi^{-1}I_R$.)

c) show that if $\mathscr{L}_1$ and $\mathscr{L}_2$ are full AFLs, then $\mathscr{L}\Phi$ is a full AFL.

d) show that if $\mathscr{L}_1$ is a cone and $\mathscr{L}_2$ is a full AFL, then $\mathscr{L}_1 \square \mathscr{L}_2 \subset \mathscr{L}\Phi$ (Hint. Show that $L_1 \uparrow L_2 \in \mathscr{L}_1 \wedge \mathscr{L}_2$ if $L_1 \in \mathscr{L}_1$, $L_2 \in \mathscr{L}_2$ are over disjoint alphabets, and use Proposition V.5.4.)

**4.5** Show that $\Gamma\, \underline{\mathrm{Elm}} = \underline{\mathrm{Elm}}\, \Gamma$.

## VI.5 Marked Substitution

In the preceding section, we showed that $\mathscr{L} \square \mathscr{M}$ is a rational cone provided $\mathscr{L}$ and $\mathscr{M}$ are cones. In fact, a much more precise result holds: if $\mathscr{L}$ and $\mathscr{M}$ are principal cones (or full AFLs), then $\mathscr{L} \square \mathscr{M}$ is also principal. Moreover, a generator of $\mathscr{L} \square \mathscr{M}$ can be given which is explicitly expressed in terms of generators of $\mathscr{L}$ and $\mathscr{M}$ by means of the syntactic substitution. The question of the converse property, namely whether the principality of $\mathscr{L} \square \mathscr{M}$ implies the principality of $\mathscr{L}$ and/or $\mathscr{M}$ will be considered in the next chapter.

It will appear in the sequel that the proof of the principality of $\mathscr{L} \square \mathscr{M}$ requires inequalities in operators which are precisely the converses of the inequalities proved in Section 4. Since some of these inequalities are sharp, we have to go through a series of inequalities involving a special type of substitution called marked substitutions. They differ from the general substitutions by the condition that the substituted languages have endmarkers. Marked substitutions are therefore closely related to, and in fact a generalization of syntactic substitutions.

A language $A$ is said **marked** if it is of the form $A = Bc$, with $c$ a letter, $c \in X_A \backslash X_B$. A marked language never contains the empty word.

**Definition** The marking operator $\#$ is the operator which associates, to any family of languages $\mathcal{L}$, the family

$$\mathcal{L}\# = \{Lc \mid L \in \mathcal{L}, c \in \Sigma \backslash X_L\}.$$

Clearly, $\mathcal{L}\#$ is closed under copy and thus is a family of languages. It is also clear that $\#$ is monotone and unitary, and therefore is an operator. Note that $\#$ is neither extensive nor idempotent.

**Definition** Let $\mathcal{M}$ be a family of languages. A marked $\mathcal{M}$-substitution is a $\mathcal{M}\#$-substitution.

**Example 5.1** A syntactic substitution is a marked substitution. More precisely, let $A$, $B$ be languages over disjoint alphabets. Then

$$A \uparrow B = \sigma_B(A)$$

where $\sigma_B(x) = Bx$ for $x \in X_A$. Thus $\sigma_B(x) \in \{B\}\#$, and $\sigma_B$ is a marked $\{B\}$-substitution.

Any marked substitution involving only two languages can be reduced to syntactic substitution in the following sense.

**Proposition 5.1** *Let A and B be languages over disjoint alphabets. Then*

$$(\{A\} \,\square\, (\{B\}\#))\Gamma = (A \uparrow B)\Gamma.$$

Proof. We just have seen that $A \uparrow B \in \{A\} \,\square\, (\{B\}\#)$; thus $(A \uparrow B)\Gamma \subset (\{A\} \,\square\, (\{B\}\#))\Gamma$. Conversely, let $L \in \{A\} \,\square\, (\{B\}\#)$. Then there exists a $\{B\}\#$-substitution $\sigma : X_A^* \to X_L^*$ such that $L = \sigma(A)$. Hence for each $x \in X_A$,

$$\sigma(x) = B_x c_x,$$

where $B_x$ is a copy of $B$, and $c_x \in X_L \backslash X_{B_x}$. Let $\theta_x : X_B^* \to X_{B_x}^*$ be the copy isomorphism. Then $\theta_x(B) = B_x$ for each $x \in X_A$. Consider the relation

$$R = \Big( \bigcup_{x \in X_A} \{(y, \theta_x(y)) : y \in X_B\}^*(x, c_x) \Big)^* \subset (X_A \cup X_B)^* \times X_L^*.$$

Clearly, $R$ is rational, and the transduction $\tau : (X_A \cup X_B)^* \to X_L^*$ defined by $R$ is rational. Next

$$\mathrm{dom}(\tau) = \Big( \bigcup_{x \in X_A} X_B^* x \Big)^*,$$

and for $w_1, w_2, \ldots, w_n \in X_B^*$, $x_1, x_2, \ldots, x_n \in X_A$,

$$\tau(w_1x_1w_2x_2\cdots w_nx_n) = \theta_{x_1}(w_1)c_{x_1}\theta_{x_2}(w_2)c_{x_2}\cdots\theta_{x_n}(w_n)c_{x_n}.$$

Thus $\quad \tau(A \uparrow B) = \sigma(A) = L.$

Consequently, $L \in (A \uparrow B)\Gamma$, and $(\{A\} \square (\{B\}\#))\Gamma \subset (A \uparrow B)\Gamma$. ■

This proposition will be helpful to construct generators of the cone $\mathcal{L} \square \mathcal{M}$ from generators of cones $\mathcal{L}$ and $\mathcal{M}$. We now prove some simple facts on the operator $\#$. First note that $\mathcal{L}\#$ never is a cone, even if $\mathcal{L}$ is a cone, since $\{1\} \notin \mathcal{L}\#$.

**Lemma 5.2** $\# \leqslant \Gamma$, $\Gamma\# \leqslant \Gamma$, $\#\Gamma = \Gamma$, $\hat{\Gamma}\# \leqslant \hat{\Gamma}$, $\#\hat{\Gamma} = \hat{\Gamma}$.

Proof. Let $A$ and $B$ be languages such that $A = Bc$, with $c \notin X_B$. Then $A \approx B$ by Corollary V.2.5. Consequently, $\mathcal{L}\#\Gamma = \mathcal{L}\Gamma$ for any family of languages $\mathcal{L}$. Thus $\#\Gamma = \Gamma$, which implies $\# \leqslant \#\Gamma = \Gamma$, and $\Gamma\# \leqslant \Gamma\Gamma = \Gamma$. Next $\hat{\Gamma}\# \leqslant \hat{\Gamma}\Gamma = \hat{\Gamma}$ and $\#\hat{\Gamma} = \#\Gamma\,\underline{\mathrm{Rat}} = \Gamma\,\underline{\mathrm{Rat}}$. ■

As already noted several times, an inverse alphabetic morphism is not a substitution. By means of the marking operator, inverse alphabetic morphisms can be expressed using rational substitutions.

**Lemma 5.3** $\Phi_a^{-1} \leqslant \#\overline{\mathrm{Rat}}$.

Proof. Let $\mathcal{L}$ be a family of languages, and let $A \in \mathcal{L}\Phi_a^{-1}$. Then $A = \varphi^{-1}(B)$, where $B \in \mathcal{L}$ and $\varphi: X_A^* \to X_B^*$ is an alphabetic morphism. Define

$$X_z = \{x \in X_A : \varphi x = z\} \quad (z \in X_B); \qquad X_0 = \{x \in X_A : \varphi x = 1\}.$$

Then $\varphi^{-1}(z) = X_0^*X_zX_0^*$ for $z \in X_B$ and $\varphi^{-1}(1) = X_0^*$. Let $c \in \Sigma \backslash X_B$, and define a rational substitution

$$\sigma: (X_B \cup c)^* \to X_A^* \quad \text{by} \quad \sigma(z) = X_0^*X_z \quad (z \in X_B); \qquad \sigma(c) = X_0^*.$$

Then $\sigma(fc) = \varphi^{-1}(f)$ for all $f \in X_B^*$. Hence $A = \sigma(Bc)$. Since $Bc \in \mathcal{L}\#$, $A \in \mathcal{L}\#\overline{\mathrm{Rat}}$, proving thus the inequality. ■

We now turn to the construction of a set of generators of the cone $\mathcal{L}\Gamma \square \mathcal{M}\Gamma$ in terms of $\mathcal{L}$ and $\mathcal{M}$. We shall prove that $\mathcal{L}\Gamma \square \mathcal{M}\Gamma$ is generated by $\mathcal{L}\# \square \mathcal{M}\hat{\Gamma}$. The inclusion $(\mathcal{L}\# \square \mathcal{M}\hat{\Gamma}) \subset \mathcal{L}\Gamma \square \mathcal{M}\Gamma$ is easy to show. The converse inclusion can be written as an inequality between operators:

$$\Gamma\overline{\mathcal{M}\Gamma} \leqslant \#\overline{\mathcal{M}\hat{\Gamma}}\Gamma. \tag{5.1}$$

Note that (5.1) is a weak converse of Proposition 4.8. To prove (5.1), we proceed as in the previous section and write

$$\Gamma\overline{\mathcal{M}\Gamma} = \Phi_a^{-1}I_R\Phi_a\overline{\mathcal{M}\Gamma}.$$

We thus have to prove three inequalities involving $\Phi_a$, $I_R$, $\Phi_a^{-1}$ and the operator $\overline{\mathcal{M}\Gamma}$. Alphabetic morphisms are easy to handle. For intersection with regular sets, we have the following "converse" of Lemma 4.13.

**Lemma 5.4** *Let $\mathcal{L}$ be a family of languages; then*

$$I_R\bar{\mathcal{L}} \leqslant \overline{\mathcal{L}\#\Gamma} \quad \textit{or equivalently} \quad \underline{\mathcal{L}I_R} \leqslant \#\underline{\mathcal{L}\Gamma}.$$

Proof. Let $\mathcal{M}$ be a family of languages. First note that

$$\mathcal{M}I_R\bar{\mathcal{L}} \subset \mathcal{M}\overline{\mathcal{L}\#\Gamma} \quad \Leftrightarrow \quad \mathcal{L}\underline{\mathcal{M}I_R} \subset \mathcal{L}\#\underline{\mathcal{M}}\Gamma.$$

Thus the two inequalities are equivalent.
Next, let $A \in \mathcal{M}I_R\bar{\mathcal{L}}$. Then there exist $M \in \mathcal{M}$, $K \in \mathrm{Rat}(X_M^*)$ and a $\mathcal{L}$-substitution $\sigma: X_M^* \to X_A^*$ such that $A = \sigma(M \cap K)$. We may assume $X_M \cap X_A = \emptyset$. Define a $\mathcal{L}\#$-substitution

$$\sigma': X_M^* \to (X_M \cup X_A)^*$$

by $\qquad \sigma'(x) = \sigma(x)x, \qquad x \in X_M.$

Next, consider the rational substitution

$$\sigma'': X_M^* \to (X_M \cup X_A)^*$$

defined by

$$\sigma''(x) = X_A^* x, \qquad x \in X_M.$$

Then $K' = \sigma''(K)$ is a rational subset of $(X_M \cup X_A)^*$. Further, for $f \in X_M^+$, we have

$$w \in \sigma'(f) \cap K'$$

iff $\qquad w = w_1x_1w_2x_2 \cdots w_nx_n,$

with $w_i \in \sigma(x_i)$, $x_i \in X_M$, $f = x_1x_2 \cdots x_n \in K$. Clearly $\sigma'(1) \cap K' = 1 \cap K$. Let $\pi$ be the projection from $(X_M \cup X_A)^*$ onto $X_A^*$. It follows that

$$A = \sigma(M \cap K) = \pi(\sigma'(M) \cap K').$$

Thus $A \in \mathcal{M}\overline{\mathcal{L}\#\Gamma}$, and the inclusion is proved. ■

Inverse morphisms will be treated by Lemma 5.3. Thus we are ready to prove following

**Proposition 5.5** *Let $\mathcal{L}$ and $\mathcal{M}$ be families of languages. Then*

(i) $\qquad \mathcal{L}\Gamma \square \mathcal{M}\Gamma = \mathcal{L}\Gamma \square \mathcal{M}\hat{\Gamma} = (\mathcal{L}\# \square \mathcal{M}\hat{\Gamma})\Gamma;$

(ii) $\qquad \mathcal{L}\Gamma\,\underline{\mathrm{Elm}} \square \mathcal{M}\Gamma = \mathcal{L}\Gamma\,\underline{\mathrm{Elm}} \square \mathcal{M}\hat{\Gamma} = (\mathcal{L}\# \square \mathcal{M}\hat{\Gamma})\Gamma\,\underline{\mathrm{Elm}};$

(iii) $\qquad \mathcal{L}\hat{\Gamma} \square \mathcal{M}\Gamma = \mathcal{L}\hat{\Gamma} \square \mathcal{M}\hat{\Gamma} = (\mathcal{L}\# \square \mathcal{M}\hat{\Gamma})\hat{\Gamma}.$

Proof. We first prove

$$\Gamma\overline{\mathcal{M}\Gamma} = \#\overline{\mathcal{M}\hat{\Gamma}\Gamma}. \tag{5.2}$$

Indeed, $\Gamma\overline{\mathcal{M}\Gamma} = \Phi_a^{-1} I_R \Phi_a \overline{\mathcal{M}\Gamma} \leqslant \Phi_a^{-1} I_R \,\overline{\mathrm{Rat}}\,\overline{\mathcal{M}\Gamma}$

since a morphism is a rational substitution. In view of $\overline{\mathrm{Rat}}\,\overline{\mathcal{M}\Gamma} = \overline{\mathcal{M}\hat{\Gamma}}$, we obtain

$$\begin{aligned} \Gamma\overline{\mathcal{M}\Gamma} &\leqslant \Phi_a^{-1} I_R \overline{\mathcal{M}\hat{\Gamma}} \\ &\leqslant \Phi_a^{-1} \overline{\mathcal{M}\hat{\Gamma}\#\Gamma} && \text{(by Lemma 5.4)} \\ &\leqslant \Phi_a^{-1} \overline{\mathcal{M}\hat{\Gamma}\Gamma} && \text{(by Lemma 5.2)} \\ &\leqslant \#\overline{\mathrm{Rat}}\,\overline{\mathcal{M}\hat{\Gamma}\Gamma} && \text{(by Lemma 5.3)} \\ &= \#\overline{\mathcal{M}\hat{\Gamma}\Gamma}. \end{aligned}$$

Conversely, in view of Proposition 4.8,

$$\#\overline{\mathcal{M}\hat{\Gamma}\Gamma} \leqslant \#\Gamma\overline{\mathcal{M}\hat{\Gamma}\Gamma} = \#\Gamma\overline{\mathcal{M}\hat{\Gamma}} = \Gamma\overline{\mathcal{M}\hat{\Gamma}}$$

by Lemma 5.2, and in view of Lemma 4.15, $\Gamma\overline{\mathcal{M}\hat{\Gamma}} = \Gamma\overline{\mathcal{M}\Gamma}$. This proves (5.2). Next, since $\underline{\mathcal{N}}\bar{\mathcal{M}} = \bar{\mathcal{M}}\underline{\mathcal{N}}$ for any families $\mathcal{N}$ and $\mathcal{M}$, it suffices to prove (i). Now (i) rewrites as

$$\mathcal{L}\Gamma\overline{\mathcal{M}\Gamma} = \mathcal{L}\Gamma\overline{\mathcal{M}\hat{\Gamma}} = \mathcal{L}\#\overline{\mathcal{M}\hat{\Gamma}\Gamma},$$

and thus follows directly from (5.2) and Lemma 4.15. ∎

Proposition 5.5 is not satisfactory, since in the case where $\mathcal{L}$ and $\mathcal{M}$ are singletons, it does not give a single generator for the family $\mathcal{L}\Gamma \,\square\, \mathcal{M}\Gamma$. To get this generator, we prove

**Lemma 5.6** *Let $\mathcal{L}$ be a family of languages. Then*

$$\overline{\mathcal{L}\Gamma} \leqslant \overline{\mathcal{L}\#\Gamma} \quad \textit{or equivalently} \quad \Gamma\underline{\mathcal{L}} \leqslant \#\underline{\mathcal{L}}\Gamma.$$

Proof. Let $\mathcal{M}$ be a family of languages. First, observe that

$$\mathcal{M}\overline{\mathcal{L}\Gamma} \subset \mathcal{M}\overline{\mathcal{L}\#\Gamma} \quad \Leftrightarrow \quad \mathcal{L}\Gamma\underline{\mathcal{M}} \subset \mathcal{L}\#\underline{\mathcal{M}}\Gamma,$$

thus the inequalities are equivalent.
Next, let $A \in \mathcal{M}\overline{\mathcal{L}\Gamma}$. Then $A = \sigma(B)$, where $B \in \mathcal{M}$ and $\sigma: X_B^* \to X_A^*$ is a $\mathcal{L}\Gamma$-substitution. Thus, for each letter $z \in X_B$, $\sigma(z) \in \mathcal{L}\Gamma$, and there exist an alphabet $Y_z$, a rational transduction $\tau_z: Y_z^* \to X_A^*$, and a language $L_z \in \mathcal{L}$, $L_z \subset Y_z^*$ such that

$$\sigma(z) = \tau_z(L_z) \qquad z \in X_B.$$

We may assume that the alphabets $Y_z$ are pairwise disjoint and disjoint from $X_B$. Set

$$Y = X_B \cup \bigcup_{z \in X_B} Y_z,$$

and define a substitution $\sigma' : X_B^* \to Y^*$ by

$$\sigma'(z) = L_z z \qquad z \in X_B.$$

Thus $\sigma'$ is a $\mathcal{L}\#$-substitution. Next, let $\tau : Y^* \to X_A^*$ be the transduction defined by

$$\tau(1) = 1; \qquad \tau(f) = \emptyset \quad \text{if } f \notin (\bigcup Y_z^* z)^*;$$
$$\tau(f_1 z_1 \cdots f_n z_n) = \tau_{z_1}(f_1) \cdots \tau_{z_n}(f_n), \quad \text{if } f_i \in Y_{z_i}^*,\ z_i \in B.$$

Then
$$\tau(\sigma'(z)) = \tau(L_z z) = \tau_z(L_z) = \sigma(z) \qquad (z \in X_B),$$
$$\tau(\sigma'(1)) = \tau(1) = \sigma(1)$$
$$\tau(\sigma'(fg)) = \tau(\sigma'(f)\sigma'(g)) = \sigma(f)\sigma(g) \qquad \text{for} \quad f, g \in X_B^*.$$

It follows that $\tau \circ \sigma'$ is a substitution, and $\tau \circ \sigma' = \sigma$. Consequently,

$$A = \sigma(B) = \tau(\sigma'(B)).$$

Now $\sigma'(B) \in \mathcal{M}\overline{\mathcal{L}\#}$, and in order to achieve the proof, it suffices to show that $\tau$ is rational. For this, let

$$R_z \subset Y_z^* \times X_A^*$$

be the graph of $\tau_z (z \in X_B)$. Then

$$S = \Big( \bigcup_{z \in X_B} R_z \cdot (z, 1) \Big)^*$$

is the graph of $\tau$ and since each $R_z$ is rational, $S$ is a rational relation. ■

**Theorem 5.7** (Ginsburg and Greibach [1970]) *Let $\mathcal{L}$ and $\mathcal{M}$ be principal cones (full AFLs). Then $\mathcal{L} \square \mathcal{M}$ is a principal cone (full AFL). For any cone (full AFL) generators $L$ of $\mathcal{L}$ and $M$ of $\mathcal{M}$ and for any new letters $c, d \notin X_L \cup X_M$, $Ld \uparrow (Mc)^+$ is a cone (full AFL) generators of $\mathcal{L} \square \mathcal{M}$:*

$$L\Gamma \square M\Gamma = (Ld \uparrow (Mc)^+)\Gamma \tag{5.3}$$
$$L\hat{\Gamma} \square M\hat{\Gamma} = (Ld \uparrow (Mc)^+)\hat{\Gamma} \tag{5.4}$$

*Further if $L \neq \{1\}$, then*

$$L\Gamma \square M\Gamma = (L \uparrow (Mc)^+)\Gamma \tag{5.5}$$
$$L\hat{\Gamma} \square M\hat{\Gamma} = (L \uparrow (Mc)^+)\hat{\Gamma}. \tag{5.6}$$

Proof. By (2.1) and Corollary 4.16, $L\hat{\Gamma} \,\square\, M\hat{\Gamma} = (L\Gamma \,\square\, M\Gamma)\ \underline{\mathrm{Rat}}$. Thus (5.4) and (5.6) are direct consequences of (5.3) and (5.5) respectively. To show (5.3), we first note that in view of (5.2)

$$\Gamma\overline{M\Gamma} = \#\overline{M\hat{\Gamma}}\Gamma.$$

Next by Theorem V.4.4, $M\hat{\Gamma} = M_1\Gamma$, where $M_1 = (Mc)^+$. Consequently

$$\begin{aligned}\Gamma\overline{M\hat{\Gamma}}\# \overline{M_1\Gamma}\Gamma &\leqslant \#\overline{M_1\#}\Gamma && \text{(by Lemma 5.6)}\\ &\leqslant \#\Gamma\overline{M_1\#\Gamma} && \text{(by Proposition 4.8)}\\ &\leqslant \Gamma\overline{M_1\Gamma} && \text{(by Lemma 5.2)}\\ &= \Gamma\overline{M\hat{\Gamma}}.\end{aligned}$$

Thus $\Gamma\overline{M\hat{\Gamma}} = \#\overline{M_1\#}\Gamma$ and by Lemma 4.15, $\Gamma\overline{M\Gamma} = \#\overline{M_1\#\Gamma}$. Consequently

$$L\Gamma \,\square\, M\Gamma = (\{L\}\# \,\square\, \{M_1\}\#)\Gamma = (\{Ld\} \,\square\, \{M_1\}\#)\Gamma.$$

By Proposition 5.1, we obtain

$$L\Gamma \,\square\, M\Gamma = (Ld \uparrow (Mc)^+)\Gamma.$$

Suppose now $L \neq \{1\}$. To prove (5.5), it suffices to show

$$Ld \uparrow (Mc)^+ \approx L \uparrow (Mc)^+.$$

Set $X = X_L$, $Y = X_M$, $Z = X \cup Y \cup \{c, d\}$. Consider the following rational relations of $Z^* \times Z^*$:

$$R = \{(y, y) : y \in Y\}^*(c, c); \qquad I = \Big(\bigcup_{x \in X} R^+(x, x)\Big)^*; \qquad E = Y^*c \times \{1\}.$$

The transduction $Z^* \to Z^*$ with graph $I$ is the identity on $((Y^*c)^+X)^*$. Let $\tau_1 : Z^* \to Z^*$ be the rational transduction with graph

$$A = I \cdot E^+ \cdot \{(d, 1)\}.$$

Then for $f_1, \ldots, f_{n+1} \in (Y^*c)^+$, $x_1, \ldots, x_n \in X$,

$$\tau_1(f_1x_1 \cdots f_nx_nf_{n+1}d) = f_1x_1 \cdots f_nx_n.$$

Consequently $\tau_1(Ld \uparrow (Mc)^+) = L \uparrow (Mc)^+$. (The hypothesis $L \neq \{1\}$ was not used.) Conversely, consider the transductions $Z^* \to Z^*$ with graph

$$S_x = R^+(1, x)R^+(x, d) \qquad (x \in X).$$

These transductions map a word $m_1c \cdots m_kcx$, ($m_i \in Y^*$, $k \geqslant 2$) into the words $m_1c \cdots m_icxm_{i+1}c \cdots m_kcd$, ($1 \leqslant i < k$). Define a transduction $\tau_2 : Z^* \to Z^*$ with graph

$$B = I \cdot \Big(\bigcup_{x \in X} S_x\Big).$$

Then $\tau_2((Mc)^+x_1 \cdots (Mc)^+x_n) = (Mc)^+x_1 \cdots (Mc)^+x_n(Mc)^+d$ for $x_1, \ldots, x_n \in X$, $n \geq 1$. Consequently, if $1 \notin L$ then

$$\tau_2(L \uparrow (Mc)^+) = Ld \uparrow (Mc)^+.$$

If $1 \in L$, then $Ld \uparrow (Mc)^+ = \tau_2(L \uparrow (Mc)^+) \cup (Mc)^+d$. In this case, consider the rational transduction $\tau_3: Z^* \to Z^*$ with graph

$$C = \Big(\bigcup_{x \in X} E^+(x, 1)\Big)^* \Big(\bigcup_{x \in X} R^+(x, d)\Big).$$

Then for $k \geq 1$, $f_1, \ldots, f_k \in (Y^*c)^+$, $x_1, \ldots, x_k \in X$,

$$\tau_3(f_1x_1 \cdots f_kx_k) = f_kd.$$

Thus if $L \neq \{1\}$, $\tau_3(L \uparrow (Mc)^+) = (Mc)^+d$. Hence

$$Ld \uparrow (Mc)^+ = (\tau_2 \cup \tau_3)(L \uparrow (Mc)^+). \qquad \blacksquare$$

## Exercises

**5.1** Compute $(Ld \uparrow (Mc)^+)\Gamma$ and $(L \uparrow (Mc)^+)\Gamma$ for $L = \{1\}$.

**5.2** Show that for $L \neq \{1\}$, $L\hat{\Gamma} \,\square\, M\hat{\Gamma} = ((Ld)^+ \uparrow (Mc)^+)\Gamma$.

**5.3** Let $\mathscr{L}$ and $\mathscr{M}$ be principal cones. Show that there exist $L \in \mathscr{L}$, $M \in \mathscr{M}$ such that $\mathscr{L} \,\square\, \mathscr{M} = (L \uparrow M)\Gamma$. Give an example of languages $L$, $M$ such that $(L \uparrow M)\Gamma \subsetneq L\Gamma \,\square\, M\Gamma$ (see also Exercises 5.6, 5.7 below).

**5.4** (Lewis [1970]) For any family of languages $\mathscr{L}$ define $\mathring{\mathscr{L}}$ by

$$\mathring{\mathscr{L}} = \{A \mid \{A\} \,\square\, \mathscr{L} \subset \mathscr{L}\}.$$

a) Show that $\mathring{\mathscr{L}} \,\square\, \mathscr{L} = \mathscr{L}$, and that $\mathring{\mathscr{L}}$ is closed under substitution.
b) Show that if $\mathscr{L}$ contains at least one language $\{x\}$ with $x \in \Sigma$, then $\mathring{\mathscr{L}} \subset \mathscr{L}$.
c) Show that if $\mathscr{L}$ is a full AFL, then $\mathring{\mathscr{L}}$ is a full AFL.

**5.5** (Greibach and Hopcroft [1969]) Let $\mathscr{L}$ be a family of languages closed under rational substitution and intersection with regular sets. Show that if $\mathscr{L}$ contains at least one nonempty language $\neq \{1\}$, then $\mathscr{L}$ is a rational cone. (Hint. Show that $\mathscr{L}\# \subset \mathscr{L}$ and use Lemma 5.3.)

The two following exercises are difficult:

**5.6** Show that $S_2 \uparrow (Mc)^+ \approx S_2 \uparrow M$, where $S_2$ is the symmetric language, $S_2$ and $M$ are over disjoint alphabets, and $c$ is a new letter. (Hint. Show first that $S_2 \uparrow M \approx S_3 \uparrow M$, and use the third letter of the alphabet of $S_3$ to create the letter $c$.)

**5.7** Show that $D_1'^* \uparrow (Mc)^+ \approx D_1'^* \uparrow M$, where $D_1'^*$ is the restricted Dyck language, and $M$, $c$ are as above. Same problem for $D_n'^*$, $(n \geq 2)$. (Hint. For $D_1'^*$, consider first a rational transduction which replaces, in words of $D_1'^* \uparrow M$, factors of the form $mxm'\bar{x}$ by $mc$. For $D_n'^*$, $(n \geq 2)$, Proposition VII.1.1 may be useful.)

# VII Generators

The first section contains examples of generators of the cone Alg of context-free languages. The Dyck languages over at least two pairs of parentheses and the language $E$ of completely parenthesized arithmetic expressions are shown to be generators. Sections 2 and 3 are concerned with S. Greibach's Syntactic Lemma and its applications. In particular, we prove that in a substitution closed principal full AFL, the nongenerators form a substitution closed full AFL. Next the Syntactic Lemma is used to exhibit infinite ascending chains of cones, and thus nonprincipal cones of context-free languages. In Section 4, we study the family of languages recognized by one counter pushdown automata and we prove this family to be the full AFL generated by $D_1'^*$. The last section deals with the family of quasi-rational or nonexpansive languages. Several characterizations of this family are given.

## VII.1 Generators of the Context-Free Languages

The family Alg of context-free or algebraic languages is a full AFL closed under substitution (see Theorem II.2.1.). In this section, we shall see that Alg is a principal cone, by describing some of its generators. Each cone generator is also a full AFL generator of Alg. As a consequence of the Syntactic Lemma, we shall see that conversely, each full AFL generator generates the cone Alg, i.e. $L\hat{\Gamma} = \text{Alg}$ iff $L\Gamma = \text{Alg}$. The family of (full AFL or cone) generators of Alg is the most "powerful" class of context-free languages.
No characterization of the generators of Alg is known. The Chomsky-Schützenberger Theorem implies that the Dyck languages $D_n'$, $D_n'^*$, $D_n$, $D_n^*$ are generators for $n \geq 2$. Other examples of generators are related to the language $E$ of completely parenthesized arithmetic expressions introduced in Section II.4. Autebert and Beauquier [1974] observed that these languages are all rationally equivalent to languages of the form $D_n'^* \cap K$, with $K$ rational, and gave a characterization of generators of this form. Let us mention a remarkable property of the generators of Alg proved by Beauquier [1976a, b][1978a]: For any generator $L$ of Alg, there is a regular language $K$ such that $L \cap K$ is still a generator and is an unambiguous language.
Recall that the Dyck languages

$$D_n^*, D_n'^*, D_n, D_n' \qquad (n \geq 1)$$

are defined over an alphabet $Z_n = X_n \cup \bar{X}_n$, with $X_n = \{x_1, \ldots, x_n\}$, $\bar{X}_n = \{\bar{x}_1, \ldots, \bar{x}_n\}$. We first prove:

**Proposition 1.1** *For* $n \geq 2$, $D_n^* \approx D_n'^* \approx D_n \approx D_n' \approx D_2'^*$; *further* $D_2'^* \geq D_1^*$, $D_1'^*$, $D_1$, $D_1'$.

The Dyck languages over one pair of parentheses are not generators. The full AFL generated by $D_1'^*$, $D_1$, $D_1'$ is the family of one counter languages studied in section 4.

Proof. For $n \geq 1$ we have

$$D_n^* = D_{n+1}^* \cap Z_n^*, \quad D_n'^* = D_{n+1}'^* \cap Z_n^*, \quad D_n = D_{n+1} \cap Z_n^*,$$
$$D_n' = D_{n+1}' \cap Z_n^*,$$

thus

$$D_{n+1}^* \geq D_n^*, \quad D_{n+1}'^* \geq D_n'^*, \quad D_{n+1} \geq D_n, \quad D_{n+1}' \geq D_n'.$$

Conversely, consider the morphism $\alpha: Z_n^* \to Z_2^*$ defined by

$$\alpha(x_i) = x_1^i x_2 x_1^i, \qquad \alpha(\bar{x}_i) = \bar{x}_1^i \bar{x}_2 \bar{x}_1^i \qquad i = 1, \ldots, n.$$

Then $U = \alpha(Z_n)$ is a bifix code. We shall verify that

$$D_n^* = \alpha^{-1}(D_2^* \cap U^*), \quad D_n'^* = \alpha^{-1}(D_2'^* \cap U^*),$$
$$D_n = \alpha^{-1}(D_2 \cap U^*), \quad D_n' = \alpha^{-1}(D_2' \cap U^*),$$

which imply

$$D_n^* \approx D_2^*, \quad D_n'^* \approx D_2'^*, \quad D_n \approx D_2, \quad D_n' \approx D_2'. \tag{1.1}$$

The inclusion $\alpha(D_n^*) \subset D_2^* \cap U^*$ (and similarly for $\alpha(D_n'^*) \subset D_2'^* \cap U^*$, $\alpha(D_n) \subset D_2 \cap U^*$, $\alpha(D_n') \subset D_2' \cap U^*$) is shown by a straightforward induction. To prove conversely $D_n^* \supset \alpha^{-1}(D_2^* \cap U^*)$ we argue by induction on the length of words. Let $f \in \alpha^{-1}(D_2^* \cap U^*)$. If $|f| = 0$, then $f = 1 \in D_n^*$. Assume $|f| > 0$. Then $w = \alpha(f) \neq 1$. Since $w \in D_2^*$, there exists a Dyck reduction

$$w = w_0 \vdash w_1 \vdash w_2 \vdash \cdots \vdash w_p = 1.$$

Consider the smallest index $i$ such that $|w_i|_{x_2} < |w|_{x_2}$. Then since $w \overset{*}{\vdash} w_i$, it follows immediately from Lemma II.3.6 that $w$ admits a factorization

$$w = u_0 \bar{x}_2 v x_2 u_1 \qquad \text{or} \qquad w = u_0 x_2 v \bar{x}_2 u_1$$

with $v \equiv 1 (\text{mod } \delta_2)$ and $v \in \{x_1, \bar{x}_1\}^*$. Suppose for instance that $w$ has the first of the above factorizations. Since $w \in U^*$, there are integers $i, j$, $(1 \leq i, j \leq n)$ such that

$$w = u' \bar{x}_1^i \bar{x}_2 \bar{x}_1^i x_1^j x_2 x_1^j u''$$

with $u'\bar{x}_1^i = u_0$, $v = \bar{x}_1^i x_1^j$ and $x_1^j u'' = u_1$. Since $v \in D_2^*$, we have $i = j$, and $u', u'' \in U^*$. Consequently there are $f', f'' \in Z_n^*$ such that $\alpha(f') = u'$, $\alpha(f'') = u''$ and $f = f'\bar{x}_i x_i f''$. Next $\alpha(f'f'') = u'u'' \in D_2^* \cap U^*$. By induction, $f'f'' \in D_n^*$, whence $f \in D_n^*$. To prove the inclusion $D_n \supset \alpha^{-1}(D_2 \cap U^*)$, assume that there is a word $f \in \alpha^{-1}(D_2 \cap U^*)$ such that $f \notin D_n$. By the above proof, $f \in D_n^*$, and consequently $f = f_1 f_2$, with $f_1, f_2 \in D_n^* \backslash 1$. But then $\alpha(f_1), \alpha(f_2) \in D_2^* \backslash 1$, and $\alpha(f) \notin D_2$. The proofs for $D_n'^*$ and $D_n'$ are identical.

In view of the Chomsky-Schützenberger Theorem II.3.10 and by (1.1), there are integers $n, m, p, q \geq 2$ such that

$$D_n^* \geq D_2'^* \approx D_m'^* \geq D_2 \approx D_p \geq D_2' \approx D_q' \geq D_2^* \approx D_n^*.$$

This concludes the proof. ■

Proposition 1.1 implies immediately

**Theorem 1.2** *The cone* Alg *of algebraic languages is principal; any of the languages* $D_n^*$, $D_n'^*$, $D_n$, $D_n'(n \geq 2)$ *is a generator of* Alg.

Proof. Let $L$ be an algebraic language; then in view of the Chomsky-Schützenberger Theorem, $D_n'^* \geq L$ for some $n \geq 2$. Since $D_n'^* \approx D_2'^*$ by the previous proposition, $D_2'^* \geq L$, whence $\text{Alg} \subset D_2'^*\Gamma$. The converse implication is obvious. ■

**Proposition 1.3** $D_1 \approx D_1' \approx D_1'^*$.

We shall see later (Proposition VIII.7.1) that none of these languages is a generator of Alg. The situation of $D_1^*$ is particular: Clearly Proposition 1.3 implies

$$D_1^*\hat{\Gamma} \subset D_1\hat{\Gamma} = D_1'\hat{\Gamma} = D_1'^*\hat{\Gamma}.$$

Thus $D_1^*$ is not a generator of Alg (since otherwise $D_1'^*$ would be a generator). In fact, it can be proved (Boasson [1973a]) that the above inclusion is strict, and that further the cones $D_1^*\Gamma$ and $D_1'^*\Gamma$ are incomparable.

Proof. The equalities

$$D_1' = x_1 D_1'^* \bar{x}_1, \qquad D_1'^* = x_1^{-1} D_1' \bar{x}_1^{-1}$$

show that $D_1'$ and $D_1'^*$ are rationally equivalent. Define an isomorphism $\beta : Z_1^* \to Z_1^*$ by

$$\beta(x_1) = \bar{x}_1, \qquad \beta(\bar{x}_1) = x_1,$$

and set $\bar{D}_1' = \beta(D_1')$. Then $\bar{D}_1' \approx D_1'$, whence $D_1' \geq D_1' \cup \bar{D}_1'$.

Next we claim that

$$D_1 = D_1' \cup \bar{D}_1'. \tag{1.2}$$

We shall see indeed that, with the notations of Section II.3,

$$D'_1 = D_{1,x_1} = x_1 Z_1^* \bar{x}_1 \cap D_1, \qquad \bar{D}'_1 = D_{1,\bar{x}_1} \tag{1.3}$$

The inclusions $D'_1 \subset D_{1,x_1}$ and $\bar{D}'_1 \subset D_{1,\bar{x}_1}$ are clear. Conversely, consider a word $w$ in $D_{1,x_1}$. Then by formula (II.(3.5), (3.6)), $w = x_1 w_1 w_2 \cdots w_n \bar{x}_1$, where $w_1, \ldots, w_n \in D_{1,x_1}$. Thus the conclusion follows by induction. The same proof holds for $\bar{D}'_1$.

By (1.2), $D'_1 \geq D_1$, and in view of (1.3), we have conversely $D_1 \geq D'_1$. ■

There are other generators of Alg, related to the language $E$ of completely parenthesized arithmetic expressions. The first languages of this type have been described by Schützenberger [1973]; other languages were found by Autebert and Beauquier [1974]. A rather complete list was given in Autebert [1973].

For $n \geq 1$, consider the alphabets

$$A_n = \{a_1, \ldots, a_n\}, \qquad C_n = \{c_1, \ldots, c_n\}, \qquad Y_n = A_n \cup C_n \cup \{b, d\}.$$

Let $E_n$ be the language generated by the grammar $G_n$ with productions:

$$\xi \to \sum_{k=1}^{n} a_k \xi b \xi c_k + d.$$

For $n = 1$ we set $a = a_1$, $c = c_1$. Thus $E_1 = E$ is the language introduced in Section II.4. The languages $E_n$ have properties very similar to those of $E$.

**Lemma 1.4** *The language $E_n$ is bifix; more precisely, if $w \in E_n$, then*

$$|w|_{A_n} = |w|_b = |w|_{C_n} = |w|_d - 1,$$

*and any nonempty proper left (right) factor $w'$ of $w$ satisfies*

$$|w'|_{A_n} > |w'|_{C_n} \qquad (|w'|_{A_n} < |w'|_{C_n}).$$

The simple proof is left to the reader.

**Lemma 1.5** *The language $E_n$ is equal to the class of $d$ in the congruence $\eta_n$ generated by the relations*

$$a_k d b d c_k \sim d \qquad k = 1, \ldots, n.$$

The proof is the same as for Theorem II.4.5 and is also left to the reader.

**Lemma 1.6** *Let $w \in E_n$, and $w = u a_k u'$, with $u, u' \in Y_n^*$. Then there exists a unique factorization $u' = w_1 b w_2 c_k u''$ with $w_1, w_2 \in E_n$, $u'' \in Y_n^*$.*

Proof. Let $w = u a_k u'$. Since $w \in E_n$, and $w \neq d$, there exist $w', w'' \in E_n$ such that $w = a_l w' b w'' c_l$ for some $l, (1 \leq l \leq n)$. If $u = 1$, then $u' = w' b w'' c_l$ is the desired factorization. If $u \neq 1$, then two possibilities arise. Either $|u a_k| \leq |a_l w'|$, and then $w'$ is of the form $w' = u_1 a_k u'_1$ with $a_l u_1 = u$ and $u'_1$ a left factor of $u'$. Otherwise $|a_k u'| \leq |w'' c_l|$, and $w''$ has the form $w'' = u_1 a_k u'_1$, with $u'_1 c_l = u'$ and

$u_1$ a right factor of $u$. In both cases, the existence of a factorization follows by induction. The unicity is obvious since $E_n$ is prefix. ∎

**Theorem 1.7** *The languages $E_n (n \geq 1)$ are rationally equivalent.*

Thus, in opposition to the situation for Dyck languages, we also have $E_1 \approx E_n$.

Proof. The proof is in two parts. In the first part we prove the relations $E_n \approx E_2 \geq E_1$ by a standard encoding argument similar to that employed for Proposition 1.1. In the second part we prove $E_1 \geq E_2$ by a more complicated encoding.

First we observe that

$$E_n = E_{n+1} \cap Y_n^*,$$

hence $E_{n+1} \geq E_n$ for $n \geq 1$. Next, we show that $E_2 \geq E_n$. For simplicity, set $a = a_1$, $c = c_1$, $a' = a_2$, $c' = c_2$. Define a morphism

$$\theta : Y_n^* \to Y_2^*$$

by $$\theta(a_k) = a(dba')^k; \qquad \theta(c_k) = c'^k c \qquad k = 1, \ldots, n$$

$$\theta(b) = b, \qquad \theta(d) = d.$$

The set $U = \theta(Y_n)$ is a code. Further we have the derivations

$$\xi \xrightarrow{*} a(dba')^k \xi b \xi c'^k c \qquad k = 1, \ldots, n$$

in the grammar $G_n$. This shows that

$$\theta(E_n) \subset E_2 \cap U^*.$$

To show the converse inclusion $E_n \supset \theta^{-1}(E_2)$ and thereby the relation $E_2 \geq E_n$, we argue by induction on the length of a word $f \in \theta^{-1}(E_2)$. If $|f| = 1$, then $f = d$ and $f \in E_n$. Assume $|f| > 1$ and set $w = \theta(f)$. Then $w \in E_2 \cap U^*$, $|w| > 1$ and consequently $w = a(dba')^k w'$ for some $k$, $(1 \leq k \leq n)$ and some $w' \in U^*$. Applying Lemma 1.6 $k$ times shows the existence of words $u_0, u_1 \in E_2$ such that

$$w = a(dba')^k u_0 b u_1 c'^k c.$$

We verify that $u_0, u_1 \in U^*$. Assume $u_0 \notin U^*$. Then $u_0 b$ is a left factor of a word $v \in U^*$, and by the form of the elements of $U$, $v \in U^* \theta(a_i)$ for some $i \in \{1, \ldots, n\}$. But then either $adb$ or $a'db$ is a right factor of $u_0 b$, and $d$ is a proper right factor of $u_0$ which is impossible since $u_0 \in E_2$. Thus $u_0 \in U^*$ and similarly $u_1 \in U^*$. Consequently $f$ factorizes into $f = a_k f_0 b f_1 c_k$ with $\theta(f_0) = u_0$, $\theta(f_1) = u_1$. Since $u_0, u_1 \in E_2 \cap U^*$, it follows by induction that $f_0, f_1 \in E_n$, whence $f \in E_n$. This completes the first part of the proof.

In order to prove the relation $E = E_1 \geq E_2$, consider the morphism $\psi : Y_2^* \to Y^*$

with $Y = Y_1$ defined by

$$\psi(a_1) = \alpha_1 = adba \qquad \psi(c_1) = \gamma_1 = cc$$
$$\psi(a_2) = \alpha_2 = aa \qquad \psi(c_2) = \gamma_2 = cbdc$$
$$\psi(b) = b \qquad \psi(d) = d.$$

The set $V = \psi(Y_2)$ is a prefix code. Next

$$\xi \xrightarrow{*} \alpha_i \xi b \xi \gamma_i \qquad i = 1, 2$$

in the grammar $G = G_1$ generating $E$, and therefore

$$\psi(E_2) \subset E \cap V^*.$$

The converse inclusion $\psi^{-1}(E \cap V^*) \subset E_2$ is proved by induction. Let $f \in \psi^{-1}(E \cap V^*)$. If $|f| = 1$, then $f = d \in E_2$. Suppose $|f| > 1$, and set $w = \psi(f)$. Since $w \neq d$, $w$ contains occurrences of $\alpha_1$ or $\alpha_2$, thus $f$ contains occurrences of $a_1$, $a_2$. We consider the right most occurrence of a letter $a_1$ or $a_2$ in $f$. Then $f$ factorizes into

$$f = g a_k h' \qquad k \in \{1, 2\},$$

and $h' \in (Y_2 \backslash \{a_1, a_2\})^*$. Define $u = \psi(g)$, $v' = \psi(h')$. Then $v' \in \{b, d, c\}^*$.
If $k = 1$, then $w = uadbav'$, and in view of Lemma 1.6, there exist words $w_0, w_1 \in E$, such that

$$w = uadbaw_0bw_1ccv, \qquad \text{with } v' = w_0bw_1ccv.$$

Since $v'$ contains no occurrence of the letter $a$, we have $w_0 = w_1 = d$. Thus $v' = dbdccv$, and $v \in V^*$. Therefore

$$f = ga_1dbdc_1h,$$

with $\psi(h) = v$. Consider the word $f' = gdh$. Then $\psi(f') = udv$, and since $w \equiv udv (\mathrm{mod}\ \eta_1)$, Lemma 1.5 implies that $udv \in E$. Since also $udv \in V^*$, we have $f' \in E_2$ by induction. Next $f' \equiv f (\mathrm{mod}\ \eta_2)$, whence $f \in E_2$.
If $k = 2$, then $w = uaav'$, and in view of Lemma 1.6 there are words $w_0, w_1, w_2 \in E$ such that

$$w = uaaw_0bw_1cbw_2cv, \qquad v' = w_0bw_1cbw_2cv.$$

As above, $w_0 = w_1 = w_2 = d$, which implies that $v \in V^*$ and

$$f = ga_2dbdc_2h$$

with $\psi(h) = v$. We conclude as above that $f \in E_2$. Thus $E_2 = \psi^{-1}(E \cap V^*)$ and therefore $E \geqslant E_2$. ∎

**Corollary 1.8** *The languages $E_n (n \geq 1)$ are generators of the cone of algebraic languages.*

Proof. We prove that $E_3 \geq D_2'^*$. Let indeed $\varphi: Y_3^* \to Z_2^*$ be the morphism defined by

$$\varphi(a_i) = x_i, \qquad \varphi(c_i) = \bar{x}_i \qquad i = 1, 2$$
$$\varphi(a_3) = \varphi(c_3) = \varphi(b) = \varphi(d) = 1.$$

The language $\varphi(E_3)$ is generated by the grammar $\varphi(G_3)$ with productions:

$$\xi \to x_1 \xi\xi \bar{x}_1 + x_2 \xi\xi \bar{x}_2 + \xi\xi + 1.$$

It is straightforward (see Exercise II.3.2) that this grammar generates $D_2'^*$. Thus $\varphi(E_3) = D_2'^*$. ∎

We mention the following consequence of Theorem 1.7:

**Corollary 1.9** *Let $X$ be an alphabet, and let $A = \{\alpha, \beta, \gamma, \delta\} \subset X^+$ be a code. Then the language $E_A$ generated by the grammar with productions*

$$\xi \to \alpha\xi\beta\xi\gamma + \delta$$

*is a generator of the cones of algebraic languages.*

Proof. Let $Y = \{a, b, c, d\}$ and $E \subset Y^*$ as above. The morphism $\varphi: Y^* \to X^*$ defined by $\varphi(a) = \alpha$, $\varphi(b) = \beta$, $\varphi(c) = \gamma$, $\varphi(d) = \delta$ clearly satisfies $\varphi(E) = E_A$. Since $A$ is a code, $\varphi$ is injective. Consequently $E = \varphi^{-1}\varphi(E) = \varphi^{-1}(E_A)$ and $E_A \approx E$. Thus the corollary follows from Theorem 1.7. ∎

We now examine some variations of the language $E$. For a complete classification, the reader is referred to the paper of Autebert [1973] already mentioned.

**Proposition 1.10** *The language $E' \subset \{a, c\}^*$ generated by the grammar with productions*

$$\xi \to a\xi\xi c + 1$$

*is a generator of the cone* Alg.

**Corollary 1.11** *The languages $E_1'$ and $E_2'$ defined by the equations*

$$E_1' = aE_1' bE_1' c \cup 1; \qquad E_2' = aE_2' E_2' c \cup d$$

*are generators of the cone* Alg.

Proof. The language $E'$ is an obvious homomorphic image of both $E_1'$ and $E_2'$. ∎

Proof of Proposition 1.10. Set $X=\{a, c\}$ and define

$$\alpha = aaac, \qquad \beta = ca, \qquad \gamma = accc, \qquad \delta = ac,$$

$$A=\{\alpha, \beta, \gamma, \delta\}, \qquad K=(\alpha^* \cup \delta\gamma^*\beta)^*\delta\gamma^*.$$

Clearly $K$ is a regular language, and $A$ is a code.

We shall prove the equality

$$E' \cap K = E_A$$

which proves the assertion in view of Corollary 1.9. For this, define $L=E' \setminus \{1\}$. Since $L \cap K = E' \cap K$, it suffices to show

$$L \cap K = E_A.$$

The language $L$ is generated by the grammar $H$ with productions

$$\xi \to a\xi\xi c + a\xi c + ac.$$

This implies that $L$ is contained in $D_1'^*$ (with $x_1 = a$, $\bar{x}_1 = c$) by a straightforward induction. Thus $L$ is bifix. The following formulas are easily verified:

$$\xi \xrightarrow[H]{*} \alpha\xi\beta\xi\gamma; \qquad \xi \xrightarrow[H]{*} \delta; \tag{1.3}$$

$$\alpha K \beta K \gamma \subset K, \ \delta \in K; \tag{1.4}$$

$$\alpha w \gamma \in K \Rightarrow w \in K; \tag{1.5}$$

$$awc \in L \Rightarrow w \in 1 \cup L \cup L^2. \tag{1.6}$$

By (1.3) $\alpha L \beta L \gamma \cup \delta \subset L$, whence $L \supset E_A$. By (1.4) we have $K \supset E_A$, and thus $E_A \subset L \cap K$. The converse inclusion $L \cap K \subset E_A$ is proved by induction. Let $w \in L \cap K$. The shortest word in $K$ is $\delta$, and $\delta \in E_A$. Assume $w \neq \delta$. Then $w$ starts with a letter $a$, and ends with a letter $c$. Since $w \in A^*$, $w$ starts with $\alpha$, $\gamma$ or $\delta$ and ends with $\alpha$, $\gamma$ or $\delta$. Since $w \in L$ and $L$ is bifix, $w$ cannot start or end with $\delta$. Next, by the form of $K$, $w$ cannot start with $\gamma$ nor end with $\alpha$. Thus

$$w = \alpha g \gamma = aaacgaccc$$

for some $g \in X^*$. By (1.5), $g \in K$. By (1.6),

$$w' = aacgacc \in L \cup L^2.$$

Assume $w' \in L$. Then $w'' = acgac \in L \cup L^2$ by (1.6), and since $L$ is bifix and $ac \in L$, one has $g = 1$ and $w = \alpha\gamma$. But since $\alpha\gamma \notin K$ this is impossible. Thus $w' \in L^2$, and there are words $u, v \in 1 \cup L \cup L^2$ such that

$$w' = aacgacc = aucavc, \quad \text{thus} \quad acgac = ucav.$$

Since $u \neq 1$ and $u \neq a$, $ac$ is a left factor of $u$ and similarly $ac$ is a right factor of $v$. Thus $u = ach_1$ and $v = h_2ac$ for some words $h_1$, $h_2 \in 1 \cup L$. Thus

$$w = \alpha h_1 \beta h_2 \gamma.$$

Next $h_1 \neq 1$ since $\alpha\beta$ is not a left factor of a word of $K$. Similarly, $h_2 \neq 1$, and $h_1, h_2 \in L$. We claim that

$$h_1, h_2 \in L, \qquad h_1\beta h_2 \in K \quad \Rightarrow \quad h_1, h_2 \in K. \tag{1.7}$$

Taking (1.7) for granted for a moment, we recall that $g = h_1\beta h_2 \in K$. Thus $h_1, h_2 \in L \cap K$ and $h_1\, h_2 \in E_A$ by the induction hypothesis. Consequently $w \in E_A$ which is the desired conclusion.

To prove (1.7), consider the unique factorization

$$g = h_1\beta h_2 = u_1 u_2 \cdots u_s$$

of $g$ with $u_1, \ldots, u_s \in A$. Let $n$ be the least integer such that $h_1$ is a left factor of $u_1u_2 \cdots u_n$. If $h_1 = u_1 \cdots u_n$, then $\beta = u_{n+1}$ since $\beta$ is the only word in $A$ starting with the letter $c$. Consequently $h_2 = u_{n+2} \cdots u_s \in A^*$, and since $h_1\beta h_2 \in K$, we have $h_1, h_2 \in K$.

Assume $h_1 \neq u_1 \cdots u_n$. Then $h_1 v = u_1 \cdots u_n$, where $v \neq 1$ is a proper right factor of $u_n$. Since $\beta = ca$ is a proper factor of none of the words in $A$, $\beta$ cannot be a left factor of $v$. Thus necessarily $v = c$. Since the words in $L$ have even length, $|h_1c|$ is odd. The length of the words in $A^*$ are even, and consequently $|h_1c| = |u_1 \cdots u_n|$ is even. This is a contradiction. Thus (1.7) is proved. ■

## Exercises

**1.1** Define a morphism $\beta : Z_n^* \to Z_2^*$ by

$$\beta(x_i) = x_2 x_1^i, \qquad \beta(\bar{x}_i) = \bar{x}_1^i \bar{x}_2 \qquad i = 1, \ldots, n.$$

Show that $D_n'^* = \beta^{-1}(D_2'^* \cap \beta(Z_n^*))$, and $D_n^* \subsetneq \beta^{-1}(D_2^* \cap \beta(Z_n^*))$.

**1.2** Show that the language $L \subset \{a, b, c\}^*$ defined by $L = aLbLcL \cup \{1\}$ is a generator of Alg.

**1.3** (Schützenberger [1973]) Show that the language $M \subset \{a, \bar{a}\}^*$ defined by $M = aMM\bar{a} \cup a\bar{a}$ is a generator of Alg. (Hint. Show that $M \geqslant M'$, where $M'$ is given by $M' = aM'M'\bar{a} \cup b\bar{b}$.)

**1.4** Let $L$ be defined by $L = aLbL \cup d$. Show that $L \approx D_1'^*$. (As already mentioned $D_1'^*$ is not a generator of Alg. Thus the letter $c$ cannot be erased in $E$ without loosing the generator property of $E$.)

**1.5** Let $N$ be defined by $N = aNNa \cup d$. Show that $N$ is a generator of Alg. (Hint. Proceed as in the proof of Proposition 1.10, with $\alpha = aada$, $\beta = daaaad$, $\gamma = adaa$, $\delta = d$.)

**1.6** Show that $[w]_{\delta_n} \approx [w']_{\delta_n}$ for any $w, w' \in Z_n^*$, and that $[w]_{\delta_n}$ is a generator of Alg for any $w \in Z_n^*$. Same exercise for $\delta_n'$.

**1.7** Prove a Chomsky-Schützenberger like theorem for the family $E_n$, $(n \geq 1)$: For any $L \in \mathrm{Alg}$, there exist an integer $n$, a morphism $\varphi$ and a regular language $K$ such that $L = \varphi(E_n \cap K)$.

## VII.2 The Syntactic Lemma

For the investigation of families of languages with respect to substitution, the "Syntactic Lemma" due to S. A. Greibach is a basic tool. By means of this lemma, some information about the families $\mathscr{L}$ and $\mathscr{M}$ can be derived from properties of the substitution $\mathscr{L} \square \mathscr{M}$. Further the lemma gives a method to construct infinite ascending chains of cones and full AFLs, and thus of nonprincipal full AFLs. These applications will be given in the next section. The material for this and the next section is from Greibach [1970]. The present statement of the Syntactic Lemma is more precise than the original one. It is due to Beauquier [1978b]. For related topics and extensions, see Boasson, Crestin and Nivat [1973] and Greibach [1972].

**Theorem 2.1** (Syntactic Lemma) *Let $L$ and $M$ be languages over disjoint alphabets, and let $\mathscr{L}$, $\mathscr{M}$ be cones. If $L \uparrow M \in \mathscr{L} \square \mathscr{M}$, then $L \in \mathscr{L}$ or $M \in \mathscr{M}$.*

If $A = L \uparrow M \in \mathscr{L} \square \mathscr{M}$, then $A = \theta(B)$ for some $B \in \mathscr{L}$ and for some $\mathscr{M}$-substitution $\theta$. Thus $L \uparrow M$ and $\theta(B)$ represent alternative parsings of words in $A$. The theorem claims that the form of $A$ forces such constraints on the second parsing that either $L$ can be recovered from $B$ or $M$ can be obtained from one of the languages $\theta(x)$, $(x \in X_B)$ by a rational transduction.

Note that in the theorem, the conclusion cannot be replaced by: $L \in \mathscr{L}$ and $M \in \mathscr{M}$ (Exercise 2.1).

Proof. Let $L \subset X^*$ and $M \subset Y^*$. First we handle some trivial cases. If $L = \emptyset$ or $M = \emptyset$, then $L \in \mathscr{L}$ or $M \in \mathscr{M}$. If $L = \{1\}$, or $M = \{1\}$, then $L \in \mathscr{L}$ or $M \in \mathscr{M}$. Next $(L \backslash 1) \uparrow (M \backslash 1) = L \uparrow M \cap (Y^+X)^+$, thus $L \uparrow M \in \mathscr{L} \square \mathscr{M}$ implies $(L \backslash 1) \uparrow (M \backslash 1) \in \mathscr{L} \square \mathscr{M}$. Further, if $L \backslash 1 \in \mathscr{L}$ then $L \in \mathscr{L}$ and similarly for $M$. Thus it suffices to prove the theorem in the case where $1 \notin L$, $1 \notin M$.

Assume $\emptyset \neq L \subset X^+$, $\emptyset \neq M \subset X^+$. Since $A = L \uparrow M \in \mathscr{L} \square \mathscr{M}$, there exist a language $B \in \mathscr{L}$, and a $\mathscr{M}$-substitution $\theta : X_B^* \to (X \cup Y)^*$ such that $A = L \uparrow M = \theta(B)$. Thus any word $f \in A$ has at least two factorizations, a first one of the form

$$f = m_1 x_1 m_2 x_2 \cdots m_p x_p \tag{2.1}$$

with $m_1, \ldots, m_p \in M$, $x_1, \ldots, x_p \in X$, $x_1 \cdots x_p \in L$, and a second one of the form

$$f = d_1 d_2 \cdots d_q, \tag{2.2}$$

where $d_j \in \theta(z_j)$, $z_j \in X_B$, and $z_1 \cdots z_q \in B$. A factorization (2.2) is a $\theta$-factorization of $f$. Note that a word $f \in A$ has just one factorization (2.1), but may admit several $\theta$-factorizations.

Let $k \geq 1$ be an integer. We say that the $\theta$-factorization (2.2) splits at order $k$ or $k$-splits if each $d_j$, $(1 \leq j \leq q)$ contains at most $k$ letters $x \in X$. We now distinguish two cases:

a) There exists an integer $k \geq 1$ such that for each $u = x_1 \cdots x_p \in L$, there are words $m_1, \ldots, m_p \in M$ such that $f = m_1 x_1 \cdots m_p x_p$ has a $\theta$-factorization that splits at order $k$.

b) For all integers $k \geq 1$, there is a word $u = x_1 \cdots x_p \in L$ such that for all $m_1, \ldots, m_p \in M$, no $\theta$-factorization of $f = m_1 x_1 \cdots m_p x_p$ splits at order $k$.

Assume that a) holds and define a substitution $\bar{\theta}: X_B^* \to (X \cup Y)^*$ by

$$\bar{\theta}(z) = \theta(z) \cap T_k \qquad z \in X_B,$$

where $T_k = \{g \in (X \cup Y)^* \mid |g|_X \leq k\}$. Clearly $\bar{\theta}(z) \in \mathcal{M}$ since $\theta(z) \in \mathcal{M}$ and $\mathcal{M}$ is a cone. Further $\bar{\theta}(B) \subset \theta(B) = A$. In fact, $\bar{\theta}(B)$ is precisely the set of words in $A$ that have a $k$-splitting $\theta$-factorization. Thus by the assumption $\bar{\theta}(B) = A$. Next let $\pi: (X \cup Y)^* \to X^*$ be the projection. Then for $z \in X_B$, $\pi(\bar{\theta}(z)) \subset 1 \cup X \cup \cdots \cup X^k$. Thus $\sigma = \pi \circ \bar{\theta}: X_B^* \to X^*$ is a finite substitution, hence a rational transduction, and $\sigma(B) \in \mathcal{L}$. Since

$$\sigma(B) = \pi(\bar{\theta}(B)) = \pi(A) = L,$$

we have $L \in \mathcal{L}$.

Assume now that b) holds. Let $n \geq 1$ be an integer, and define

$$M_n = \{w \in M : |w| \leq n\}, \qquad k = k_n = \text{Card}(M_n).$$

Let $u = x_1 \cdots x_p \in L$ be a word such that the conclusion of b) holds for the fixed integer $k = k_n$. Then clearly $p \geq k$. Define $q, s$ by $p = qk + s$, $(0 \leq s < k)$. Consider any enumeration of the words in $M_n$:

$$M_n = \{w_1, w_2, \ldots, w_k\}$$

and define a word $f = w_1 x_1 \cdots w_p x_p$ by setting $w_j = w_i$ whenever $j \equiv i \pmod{k}$ and $0 < i \leq k$, i.e.

$$f = w_1 x_1 \cdots w_k x_k w_1 x_{k+1} \cdots w_k x_{2k} \cdots w_k x_{qk} w_1 x_{qk+1} \cdots w_s x_p.$$

Next consider a $\theta$-factorization (2.2) of this word $f$. By assumption, there is a factor $d_j$ that contains at least $k+1$ letters in $X$, thus $d_j$ has the form

$$d_j = u' x_r w_t x_{r+1} \cdots w_k x_{jk} w_1 x_{jk+1} \cdots w_{t-1} x_{r+k} v' \tag{2.3}$$

for some integers $r, t, j$, and some words $u', v'$.

Now define a transduction $\tau: (X \cup Y)^* \to Y^*$ by

$$\tau(g) = \{m \in Y^* : g \in (X \cup Y)^* X m X (X \cup Y)^*\} \qquad g \in (X \cup Y)^*.$$

Thus $\tau(g)$ is the set of all factors $m \in Y^*$ of $g$ such that $xmx'$ is a factor of $g$ for some letters $x, x' \in X$. The transduction $\tau$ is rational since its graph

$$R = [(X \cup Y)^* X \times \{1\}]\{(y, y) : y \in Y\}^* [X(X \cup Y)^* \times \{1\}]$$

is a rational relation. It follows immediately from (2.3) that $\tau(d_j) \supset M_n$. Next obviously $\tau(\theta(z)) \subset M$ for all $z \in X_B$. Combining both inclusions, there exists a letter $z \in X_B$ such that

$$M_n \subset \tau(\theta(z)) \subset M.$$

Thus if b) holds, then for any integer $n \geq 1$, there is a letter $z = z_n \in X_B$ such that

$$M_n \subset \tau(\theta(z_n)). \tag{2.4}$$

This implies that $M = \tau(\theta(\hat{z}))$ for some $\hat{z} \in X_B$. Indeed, assume the contrary. Then there is a word $w \in M$ which is in none of the languages $\tau(\theta(z))$, $(z \in X_B)$. Setting $n = |w|$, this contradicts (2.4). Since $\theta$ is an $\mathcal{M}$-substitution, $\theta(\hat{z}) \in \mathcal{M}$. Further, since $\mathcal{M}$ is a cone and $\tau$ is rational, $M \in \mathcal{M}$.

Thus we have proved that if a) holds, then $L \in \mathcal{L}$, and if b) holds then $M \in \mathcal{M}$. Since one of these conditions is always satisfied, the proof is complete. ■

There are several useful formulations of the Syntactic Lemma in terms of families of languages.

**Corollary 2.2** *Let $\mathcal{L}, \mathcal{L}', \mathcal{M}, \mathcal{M}'$ be cones. If $\mathcal{L}' \backslash \mathcal{L} \neq \emptyset$ and $\mathcal{M}' \backslash \mathcal{M} \neq \emptyset$, then $\mathcal{L}' \square \mathcal{M}' \backslash \mathcal{L} \square \mathcal{M} \neq \emptyset$.*

Proof. Let $L \in \mathcal{L}' \backslash \mathcal{L}$, $M \in \mathcal{M}' \backslash \mathcal{M}$. Since the families are closed under copy, we may assume $X_L \cap X_M = \emptyset$. Next $A = L \uparrow M \in \mathcal{L}' \square \mathcal{M}'$. Suppose that $A \in \mathcal{L} \square \mathcal{M}$. Then by the Syntactic Lemma $L \in \mathcal{L}$ or $M \in \mathcal{M}$. Thus $A \notin \mathcal{L} \square \mathcal{M}$. ■

The property of "factorization" of the Syntactic Lemma appears clearly in the following corollary which is just a reformulation of the above statement.

**Corollary 2.3** *Let $\mathcal{L}, \mathcal{L}', \mathcal{M}, \mathcal{M}'$ be cones. If $\mathcal{L} \square \mathcal{M} \subset \mathcal{L}' \square \mathcal{M}'$, then $\mathcal{L} \subset \mathcal{L}'$ or $\mathcal{M} \subset \mathcal{M}'$.*

Proof. Suppose the conclusion is false. Then $\mathcal{L} \backslash \mathcal{L}' \neq \emptyset$ and $\mathcal{M} \backslash \mathcal{M}' \neq \emptyset$. By Corollary 2.2, $\mathcal{L} \square \mathcal{M} \backslash \mathcal{L}' \square \mathcal{M}' \neq \emptyset$. ■

As a first application of the Syntactic Lemma, we prove a weak converse of Theorem V.5.6.

**Theorem 2.4** *Let $\mathcal{L}$ be a cone and let $\mathcal{M}$ be a cone closed under union. If $\mathcal{L} \square \mathcal{M}$ is principal, then $\mathcal{L}$ or $\mathcal{M}$ is principal.*

Note that as for the Syntactic Lemma, the conclusion : $\mathcal{L}$ and $\mathcal{M}$ are principal is false (see Exercise 3.1).

Proof. Let $A$ be a generator of $\mathcal{L}\square\mathcal{M}: \mathcal{L}\square\mathcal{M}=A\Gamma$. Then $A=\sigma(L)$ with $L\in\mathcal{L}$ and $\sigma: X_L^*\to X_A^*$ a $\mathcal{M}$-substitution. By Proposition V.5.4, there is a language $M\in\mathcal{M}\,\underline{\mathrm{Elm}}=\mathcal{M}$ such that $A\leqslant L\uparrow M$. Since $L\uparrow M\in L\Gamma\square M\Gamma$ and $L\Gamma\square M\Gamma$ is a cone, it follows that $A\in L\Gamma\square M\Gamma$ and consequently $\mathcal{L}\square\mathcal{M}\subset L\Gamma\square M\Gamma$. In view of Corollary 2.3, $\mathcal{L}\subset L\Gamma$ or $\mathcal{M}\subset M\Gamma$. Since $L\Gamma\subset\mathcal{L}$ and $M\Gamma\subset\mathcal{M}$, this implies $\mathcal{L}=L\Gamma$ or $\mathcal{M}=M\Gamma$. ∎

**Corollary 2.5** *Let $\mathcal{L}$ be a cone closed under union. Then $\mathcal{L}$ is principal iff $\mathcal{L}\square\mathcal{L}$ is principal.* ∎

**Exercises**

**2.1** Show that in the Syntactic Lemma, the "or" in the conclusion cannot be replaced by "and". (Hint. Consider a language $A$ which can be written in two different way as syntactic substitution: $A=L\uparrow M=L'\uparrow M'$ with $L\neq L'$ and $M\neq M'$.)

**2.2** Let $A_i, B_i$ be languages with $X_{A_i}\cap X_{B_i}=\emptyset, (i=1,2)$. Show that $A_1\uparrow B_1\leqslant A_2\uparrow B_2$ implies $A_1\leqslant A_2$ or $B_1\leqslant B_2$.

**2.3** Let $\mathcal{L}$ and $\mathcal{M}$ be incomparable cones. Show that $\mathcal{L}\square\mathcal{M}$ and $\mathcal{M}\square\mathcal{L}$ are incomparable.

## VII.3 Substitution Closure

This section contains applications of the Syntactic Lemma both to cones closed under substitution, and to cones not closed under substitution. For principal cones closed under substitution, the existence of a largest full AFL properly contained in it will be proved. Cones not closed under substitution are shown to produce infinite ascending chains of cones and thus nonprincipal cones.

A cone closed under substitution is a full AFL (Proposition V.5.5). Thus cones closed under substitution are idempotent elements in the semigroup of all families of languages with respect to the product $\square$. We first show that such an idempotent cannot be a nontrivial product of two other cones.

**Proposition 3.1** *Let $\mathcal{L}$ and $\mathcal{M}$ be cones, and set $\mathcal{N}=\mathcal{L}\square\mathcal{M}$. If $\mathcal{N}$ is closed under substitution, then $\mathcal{N}=\mathcal{L}$ or $\mathcal{N}=\mathcal{M}$.*

Proof. Since $\mathcal{N}$ is closed under substitution, $\mathcal{N}\square\mathcal{N}\subset\mathcal{L}\square\mathcal{M}$. Hence $\mathcal{N}\subset\mathcal{L}$ or $\mathcal{N}\subset\mathcal{M}$ by Corollary 2.3. Since $\mathcal{L}\subset\mathcal{N}$ and $\mathcal{M}\subset\mathcal{N}$, it follows that $\mathcal{N}=\mathcal{L}$ or $\mathcal{N}=\mathcal{M}$. ∎

From Proposition 3.1, we obtain:

**Proposition 3.2** *Let $\mathcal{L}$ be a substitution closed full* AFL, *and let $\mathcal{M}$ be a family of languages. If $\mathcal{L}=\mathcal{M}\hat{\Gamma}$, then $\mathcal{L}=\mathcal{M}\Gamma$.*

Proof. Observe that $\mathscr{L}=\mathscr{M}\hat{\Gamma}=\mathscr{M}\Gamma\,\underline{\mathrm{Rat}}=\mathrm{Rat}\,\square\,\mathscr{M}\Gamma$. Thus by Proposition 3.1, $\mathscr{L}=\mathrm{Rat}$ or $\mathscr{L}=\mathscr{M}\Gamma$. In the first case $\mathscr{M}\subset\mathrm{Rat}$. Thus in both cases, $\mathscr{M}\Gamma=\mathscr{M}\hat{\Gamma}$. ∎

Proposition 3.2 is interesting when $\mathscr{L}$ is principal.

**Corollary 3.3** *Let $\mathscr{L}$ be a substitution closed full principal* AFL. *If $\mathscr{L}=A\hat{\Gamma}$, then $\mathscr{L}=A\Gamma$.* ∎

Thus the set of generators of the cone $\mathscr{L}$ is equal to the set of generators of the full AFL $\mathscr{L}$.

**Example 3.1** The cone Alg is closed under substitution. Thus any context-free language $L$ such that $L\hat{\Gamma}=\mathrm{Alg}$ also satisfies $\mathrm{Alg}=L\Gamma$.

For any cone $\mathscr{L}$, define the sets of generators and nongenerators of $\mathscr{L}$ as follows:

$$\mathrm{Gen}(\mathscr{L})=\{L\in\mathscr{L}\mid L\Gamma=\mathscr{L}\}\qquad \mathrm{Ng}(\mathscr{L})=\{L\in\mathscr{L}\mid L\Gamma\subsetneq\mathscr{L}\}=\mathscr{L}\setminus\mathrm{Gen}(\mathscr{L}).$$

Clearly $\mathrm{Gen}(\mathscr{L})\neq\emptyset$ iff $\mathscr{L}$ is principal, and $\mathrm{Ng}\ (\mathscr{L})\neq\{\emptyset\}$ iff $\mathscr{L}\neq\mathrm{Rat}$. If both conditions are satisfied, then $\mathrm{Gen}(\mathscr{L})$ and $\mathrm{Ng}(\mathscr{L})$ are families of languages. In view of the corollary above, if $\mathscr{L}$ is closed under substitution and principal, then $\mathrm{Gen}(\mathscr{L})=\{L\in\mathscr{L}\mid L\hat{\Gamma}=\mathscr{L}\}$.

**Proposition 3.4** *Let $\mathscr{L}\neq\mathrm{Rat}$ be a principal cone. Then $\mathrm{Ng}(\mathscr{L})$ is a cone, and any cone properly contained in $\mathscr{L}$ is contained in $\mathrm{Ng}(\mathscr{L})$.*

Proof. Let $L\in\mathrm{Ng}(\mathscr{L})$, and let $M$ be a language such that $L\geqslant M$. Then $M\Gamma\subset L\Gamma\subsetneq\mathscr{L}$, whence $M\in\mathrm{Ng}(\mathscr{L})$. Next let $\mathscr{M}$ be a cone contained in $\mathscr{L}$. If $\mathscr{M}$ is not contained in $\mathrm{Ng}(\mathscr{L})$, then there is a language $A$ in $\mathscr{M}\cap\mathrm{Gen}(\mathscr{L})$. Consequently $\mathscr{L}=A\Gamma\subset\mathscr{M}$, and $\mathscr{M}=\mathscr{L}$. ∎

If $\mathscr{L}$ is substitution closed, then a more precise result can be obtained.

**Theorem 3.5** *Let $\mathscr{L}\neq\mathrm{Rat}$ be a substitution closed full principal* AFL. *Then $\mathrm{Ng}(\mathscr{L})$ is a substitution closed full* AFL *and is the maximal full* AFL *properly contained in $\mathscr{L}$.*

Proof. In view of the previous proposition, $\mathrm{Ng}(\mathscr{L})$ is the maximal cone properly contained in $\mathscr{L}$. In view of Proposition V.5.5, it suffices to show that $\mathrm{Ng}(\mathscr{L})$ is substitution closed. Define $\mathscr{M}=\mathrm{Ng}(\mathscr{L})\,\square\,\mathrm{Ng}(\mathscr{L})$. Then $\mathscr{M}$ is a cone, and $\mathscr{M}\subset\mathscr{L}\,\square\,\mathscr{L}=\mathscr{L}$. Thus either $\mathscr{M}\subset\mathrm{Ng}(\mathscr{L})$ whence $\mathscr{M}=\mathrm{Ng}(\mathscr{L})$, or $\mathscr{M}=\mathscr{L}$. Assume the second case holds, i.e. $\mathscr{L}=\mathrm{Ng}(\mathscr{L})\,\square\,\mathrm{Ng}(\mathscr{L})$. Then Proposition 3.1 implies that $\mathscr{L}=\mathrm{Ng}(\mathscr{L})$, contrary to the definition. Thus $\mathscr{M}=\mathrm{Ng}(\mathscr{L})$ and $\mathrm{Ng}(\mathscr{L})$ is closed under substitution. ∎

**Example 3.1** (continued). Theorem 3.5 asserts the existence of a maximal full AFL Ng(Alg) properly contained in Alg. Thus $\mathrm{Lin}\subset\mathrm{Ng}(\mathrm{Alg})$, and since Ng(Alg) is closed under substitution, $\mathrm{Lin}\ \overline{\mathrm{Lin}^*}$ is contained in Ng(Alg). It is not known whether Ng(Alg) is principal or not (see also below).

We now apply the Syntactic Lemma to the construction of nonprincipal cones or AFLs. This will be done by showing the existence of infinite strictly ascending chains of cones. In the sequel of this section, we use the simplified notation $\mathscr{L}\mathscr{M}$ for $\mathscr{L} \square \mathscr{M}$. Thus in particular

$$\mathscr{L}\mathscr{L} = \mathscr{L} \square \mathscr{L}, \quad \mathscr{L}^n = \mathscr{L} \square \mathscr{L}^{n-1} = \mathscr{L}\bar{\mathscr{L}}^{n-1} \qquad (n \geq 2), \quad \mathscr{L}^1 = \mathscr{L}.$$

**Theorem 3.6** *Let $\mathscr{L}$ be a cone. If $\mathscr{L}$ is not closed under substitution, then $\mathscr{L}^k \subsetneq \mathscr{L}^{k+1}$ for any $k \geq 1$ and $\mathscr{L}\bar{\mathscr{L}}^*$ is nonprincipal.*

Since $\mathscr{L}\bar{\mathscr{L}}^*$ is closed under substitution (Corollary VI.3.3) the nonprincipality as a cone or as a full AFL are equivalent by Corollary 3.3.

Proof. By assumption, $\mathscr{L} \subsetneq \mathscr{L}^2$. Suppose that $\mathscr{L}^k = \mathscr{L}^{k+1}$ for some $k$, and let $n$ be the least integer such that $\mathscr{L}^n = \mathscr{L}^{n+1}$. Then $n \geq 2$. Next $\mathscr{L}^n = \mathscr{L}^{n+l}$ for $l \geq 1$. Thus $\mathscr{L}^n = \mathscr{L}^{2n}$. Hence $\mathscr{L}^n$ is substitution closed. Since $n \geq 2$, $\mathscr{L}^n = \mathscr{L}\mathscr{L}^{n-1}$. In view of Proposition 3.1, $\mathscr{L}^n = \mathscr{L}$ or $\mathscr{L}^n = \mathscr{L}^{n-1}$. The first equality is impossible since $\mathscr{L} \subsetneq \mathscr{L}^2 \subset \mathscr{L}^n$, and the second is ruled out by the minimality assumption on $n$. Thus $\mathscr{L}^k \subsetneq \mathscr{L}^{k+1}$ for any $k \geq 1$.
Next assume $\mathscr{M} = \mathscr{L}\bar{\mathscr{L}}^* = \bigcup_{k \geq 1} \mathscr{L}^k$ is principal, and let $A$ be a generator of $\mathscr{M}$. Then $\mathscr{M} = A\hat{\Gamma} = A\Gamma$. There exists an integer $k$ such that $A \in \mathscr{L}^k$. Consequently, $\mathscr{L}^{k+1} \subset \mathscr{M} = A\Gamma \subset \mathscr{L}^k\Gamma = \mathscr{L}^k$, in contradiction with the first part of the proof. Hence $\mathscr{M}$ is nonprincipal. ■

**Example 3.2** The cone Lin of linear languages is principal and is not closed under substitution. Consequently, there is an ascending chain

$$\mathrm{Lin} \subsetneq (\mathrm{Lin})^2 \subsetneq \cdots \subsetneq (\mathrm{Lin})^k \subsetneq \cdots$$

of principal cones, and the substitution closure

$$\mathrm{Lin}\,\overline{\mathrm{Lin}}^* = \bigcup_{k \geq 1} (\mathrm{Lin})^k$$

is not principal. $\mathrm{Lin}\,\overline{\mathrm{Lin}}^* = \mathrm{Qrt}$ is the family of quasi-rational or standard matching choice or nonexpansive or derivation bounded languages. The cone $(\mathrm{Lin})^k$ is denoted by $\mathrm{Qrt}(k)$ and is called the family of quasi-rational languages of rank $k$. Quasi-rational languages will be studied in more detail in Section 5.

**Example 3.3** In the next section, we shall consider the cone generated by the Dyck language $D_1'^*$. This is the cone of restricted one counter languages. We shall see later that $D_1'^*\Gamma$, also denoted Rocl, is not a full AFL. Thus we obtain an ascending chain $\mathrm{Rocl} \subsetneq \cdots \subsetneq (\mathrm{Rocl})^n \subsetneq \cdots$ of principal cones, and the family $\mathrm{Fcl} = \mathrm{Rocl}\,\overline{\mathrm{Rocl}}^*$ of finite counter languages is not principal and therefore is contained in Ng(Alg). We note $\mathrm{Fcl}(k)$ the cone $(\mathrm{Rocl})^k$. In Section VIII.7 we prove that Rocl and Lin are incomparable, and even that Rocl is incomparable with Qrt and symmetrically, that Fcl and Lin are incomparable. These results cannot be derived from the general methods described in this chapter (see also Exercise 3.3). The combination of Lin and Rocl gives another example of a substitution closed full AFL properly contained in Alg. We use the following proposition.

**Proposition 3.7** *Let* $\mathscr{L}_1$ *and* $\mathscr{L}_2$ *be two incomparable cones, and set* $\mathscr{L} = \mathscr{L}_1 \cup \mathscr{L}_2$. *Then* $\mathscr{L}\bar{\mathscr{L}}^*$ *is nonprincipal.*

Proof. It suffices to show that the cone $\mathscr{L}$ is not a full AFL, and to apply Theorem 3.6. We verify that $\mathscr{L}$ is not closed under union. Consider indeed languages $L_1 \in \mathscr{L}_1 \backslash \mathscr{L}_2$, and $L_2 \in \mathscr{L}_2 \backslash \mathscr{L}_1$. Since $\mathscr{L}_1$ and $\mathscr{L}_2$ are cones, we may assume $L_1 \subset X_1^+$, $L_2 \subset X_2^+$ for some disjoint alphabets $X_1, X_2$. Let $L = L_1 \cup L_2$. Assume $L \in \mathscr{L}_1$. Then $L \cap X_2^* = L_2 \in \mathscr{L}_1$. Thus $L \notin \mathscr{L}_1$ and similarly $L \notin \mathscr{L}_2$. Thus $\mathscr{L}$ is not closed under union. ■

**Example 3.4** The family $\mathrm{Gre} = (\mathrm{Rocl} \cup \mathrm{Lin})\overline{(\mathrm{Rocl} \cup \mathrm{Lin})}^*$ is the substitution closure of the restricted one counter languages and the linear languages. By Proposition VIII.7.1, the rational cones Rocl and Lin are incomparable. Thus in view of the previous proposition, Gre is a substitution closed nonprincipal full AFL properly contained in the principal AFL of context-free languages. Greibach [1970] conjectured that Gre = Ng(Alg). Boasson [1973b] has proved that the inclusion of Gre in Ng(Alg) is proper. A characterization of Ng(Alg) is still lacking.

## Exercises

**3.1** Give examples of cones $\mathscr{L}$ and $\mathscr{M}$ such that $\mathscr{L} \square \mathscr{M}$ is principal and either $\mathscr{L}$ or $\mathscr{M}$ is nonprincipal. (Hint. Take Qrt for $\mathscr{L}$ or $\mathscr{M}$.)

**3.2** Let $\mathscr{L}$ be a cone which is not a full AFL. Show that $\mathscr{L}^k \subsetneq \mathscr{L}^k \hat{\Gamma} \subsetneq \mathscr{L}^{k+1}$ for $k \geqslant 1$.

**3.3** Let $\mathscr{L}_1$ and $\mathscr{L}_2$ be two incomparable cones. Show that if none of $\mathscr{L}_1$ and $\mathscr{L}_2$ is substitution closed, then $\mathscr{L}_1^n$ and $\mathscr{L}_2^n$ are incomparable for any $n \geqslant 2$. (Thus $(\mathrm{Rocl})^n$ and $(\mathrm{Lin})^n$ are incomparable.)

**3.4** Let $\mathscr{L}$ be a substitution closed full principal AFL, and let $L$ be a generator of $\mathscr{L}$. Show that for any regular language $K$, $L \cap K$ or $L \cap (X_L^* \backslash K)$ is a generator of $\mathscr{L}$.

**3.5** (Greibach [1970]) Let $\mathscr{L}$ and $\mathscr{M}$ be two incomparable cones not closed under substitution. Assume further that $\mathscr{L} \not\subset \mathscr{M}\bar{\mathscr{M}}^*$ and $\mathscr{M} \not\subset \mathscr{L}\bar{\mathscr{L}}^*$. (This situation is realized for $\mathscr{L} = \mathrm{Lin}$, $\mathscr{M} = \mathrm{Rocl}$. See Section VIII.7.)

a) Show that $\mathscr{M}^p \subset \mathscr{M}^q \mathscr{L}^n$ (resp. $\mathscr{M}^p \subset \mathscr{L}^n \mathscr{M}^q$) iff $p \leqslant q$, $(p, q, n \geqslant 1)$.

b) Show that $\mathscr{L}^n \mathscr{M}^p \subset \mathscr{L}^m \mathscr{M}^q$ iff $n \leqslant m$ and $p \leqslant q$, $(p, q, n, m \geqslant 1)$.

For $k \geqslant 1$, $n_1, n_2, \ldots, n_k \geqslant 1$ define

$$\mathscr{L}(n_1, n_2, \ldots, n_k) = \begin{cases} \mathscr{L}^{n_1}\mathscr{M}^{n_2} \cdots \mathscr{L}^{n_{k-1}}\mathscr{M}^{n_k} & \text{if } k \text{ even,} \\ \mathscr{L}^{n_1}\mathscr{M}^{n_2} \cdots \mathscr{L}^{n_k} & \text{if } k \text{ odd.} \end{cases}$$

c) Show that $\mathscr{L}(n_1, \ldots, n_k, 1) \not\subset \mathscr{L}(m_1, \ldots, m_k)$, and $\mathscr{ML}(n_1, \ldots, n_k) \not\subset \mathscr{L}(m_1, \ldots, m_k)$ for all $n_1, \ldots, n_k, m_1, \ldots, m_k \geq 1$.

d) Show that $\mathscr{L}(2n_1, 2n_2, \ldots, 2n_k) \subset \mathscr{L}(2m_1, 2m_2, \ldots, 2m_k)$ iff $n_1 \leq m_1$, $n_2 \leq m_2$, $\ldots, n_k \leq m_k$. (This yields hierarchies of cones which are order-isomorphic to $\mathbb{N}^k$.)

## VII.4 One Counter Languages

We announced in Section 1 that the restricted Dyck language $D_1'^*$ is not a generator of the family of context-free languages. This will be proved in the next chapter. The present section is concerned with the study of the cone and the full AFL generated by $D_1'^*$. Indeed, these two families can be characterized by restrictions on the pushdown automata (pda) recognizing them. More precisely, $D_1'^*\Gamma$ is the family of languages recognized by pda's having just one pushdown symbol. For $D_1'^*\hat{\Gamma}$ the condition is slightly weaker.

If a pda has just one pushdown symbol, then only the length of the words on the pushdown store can be used for computations, and the pushdown store works as a counter. Therefore the languages in $D_1'^*\hat{\Gamma}$ and $D_1'^*\Gamma$ are called one counter and restricted one counter languages. Note that the restriction on pda's defining one counter languages is in some sense dual to the restriction on grammars defining linear languages.

We start with the definition of one counter pda's, introduced by Greibach [1969]. Since in this section we manipulate integers, the empty word will be denoted by $\varepsilon$ in order to avoid confusion with the integer 1.

**Definition** A one counter pushdown automaton $M = \langle X, Q, q_-, Q_+, Y, y_0, \mathfrak{R}\rangle$ is composed of an input alphabet $X$, a finite set $Q$ of states, an initial state $q_- \in Q$, a set $Q_+ \subset Q$ of final states, an alphabet $Y$ consisting of two pushdown symbols $y, y_0$, and a finite set of rules $\mathfrak{R} \subset Q \times (X \cup \varepsilon) \times Y \times Q \times Y^*$. A rule $(q, u, z, q', \gamma)$ is written $(q, u, z) \to (q', \gamma)$, and $\mathfrak{R}$ is subject to the following conditions:

$$(q, u, y) \to (q', \gamma) \in \mathfrak{R} \Rightarrow \gamma \in y^*;$$
$$(q, u, y_0) \to (q', \gamma) \in \mathfrak{R} \Rightarrow \gamma \in y^* y_0. \tag{4.1}$$

If $(q, u, z) \to (q', \gamma)$ is a rule, then $z$ is called the pushdown symbol and $u$ is the input symbol of the rule. A configuration is a triple $(q, w, \gamma) \in Q \times X^* \times Y^*$. A move is a couple of configurations denoted by

$$(q_1, w_1, \gamma_1) \models (q_2, w_2, \gamma_2)$$

such that there is a rule $(q_1, u, z) \to (q_2, \gamma)$ in $\mathfrak{R}$ with $w_1 = uw_2$, $\gamma_1 = z\gamma'$, $\gamma_2 = \gamma\gamma'$ for some $\gamma'$. Note that $\gamma_1 \neq \varepsilon$. A computation of length $k \geq 0$ is a sequence of $k+1$ configurations $(q_i, w_i, \gamma_i)$, $(0 \leq i \leq k)$ such that

$(q_{i-1}, w_{i-1}, \gamma_{i-1}) \models (q_i, w_i, \gamma_i)$ for $i = 1, \ldots, k$. A computation is usually written

$$(q_0, w_0, \gamma_0) \models (q_1, w_1, \gamma_1) \models \cdots \models (q_k, w_k, \gamma_k) \tag{4.2}$$

or shorter

$$(q_0, w_0, \gamma_0) \overset{k}{\models} (q_k, w_k, \gamma_k).$$

As usual, $\overset{+}{\models}$ and $\overset{*}{\models}$ denote the transitive and reflexive and transitive closure of $\models$. Note that if $\gamma_0 \in y^* y_0$ in (4.2), then $\gamma_1, \ldots, \gamma_k \in y^* y_0$ by (4.1), and if $\gamma_0 \in y^+$, then $\gamma_1, \ldots, \gamma_{k-1} \in y^+$ and $\gamma_k \in y^*$.

We associate to $\boldsymbol{M}$ two languages:

$$\|\boldsymbol{M}\| = \{w \in X^* \mid \exists q_+ \in Q_+, (q_-, w, y_0) \overset{+}{\models} (q_+, \varepsilon, y_0)\};$$

$$|\boldsymbol{M}| = \{w \in X^* \mid \exists q_+ \in Q_+, (q_-, w, y) \overset{+}{\models} (q_+, \varepsilon, \varepsilon)\}.$$

Clearly only rules with pushdown symbol equal to $y$ are involved in computations of words in $|\boldsymbol{M}|$. Thus if $\mathfrak{R}'$ is the set of rules with pushdown symbol $y$, then $|\boldsymbol{M}| = |\boldsymbol{M}'|$, where $\boldsymbol{M}' = \langle X, Q, q_-, Q_+, Y, y_0, \mathfrak{R}'\rangle$. Note that $\|\boldsymbol{M}\|$ is defined as the set of words which leave $y_0$ on the pushdown store. The usual definition is rather to consider the words which empty the pushdown store (see the books in the Bibliography). The definition above stems from the cited paper of Greibach. It is easy to see that both definitions are equivalent (Exercise 4.1). The present definition allows reinitialization.

A language $L$ is a one counter language (a restricted one counter language) if $L = \|\boldsymbol{M}\|$ ($L = |\boldsymbol{M}|$) for some one counter pda $\boldsymbol{M}$. The families of one counter languages and of restricted one counter languages are denoted by Ocl and Rocl. Clearly both families are closed under copy.

**Example 4.1** The Lukasiewicz language $Ł \subset \{a, b\}^*$ is a restricted one counter language. Let $\boldsymbol{M}$ be the one counter pda with unique state $q$, and with rules

$$(q, a, y) \to (q, y^2), \qquad (q, b, y) \to (q, \varepsilon).$$

Then it is easily seen that $w \in |\boldsymbol{M}|$ iff $|w|_a = |w|_b - 1$ and $|w'|_a \geq |w'|_b$ for any proper right factor $w'$ of $w$. Consequently $|\boldsymbol{M}| = Ł$ by Proposition II.4.1.

**Example 4.2** $D_1'^* \subset \{x, \bar{x}\}^*$ is a restricted one counter language. Let indeed $\boldsymbol{M}$ be the one counter pda with the two states $q_-, q$, with final state $q$ and with rules

$$(q_-, x, y) \to (q_-, y^2), \qquad (q_-, \bar{x}, y) \to (q_-, \varepsilon), \qquad (q_-, \varepsilon, y) \to (q, \varepsilon).$$

Then $|\boldsymbol{M}| = D_1'^*$ follows immediately from Corollary II.4.2.

**Example 4.3** $D_1^* \subset \{x, \bar{x}\}^*$ is a one counter language. Consider indeed the one counter pda $\boldsymbol{M}$ with states $q_-, q, \bar{q}$, final state $q_-$ and with following rules:

$$(q_-, x, y_0) \to (q, yy_0), \qquad (q, x, y) \to (q, y^2), \qquad (q, \bar{x}, y) \to (q, \varepsilon);$$
$$(q_-, \bar{x}, y_0) \to (\bar{q}, yy_0), \qquad (\bar{q}, \bar{x}, y) \to (\bar{q}, y^2), \qquad (\bar{q}, x, y) \to (q, \varepsilon);$$
$$(q, \varepsilon, y_0) \to (q_-, y_0), \qquad (\bar{q}, \varepsilon, y_0) \to (q_-, y_0).$$

Then the equality $\|\boldsymbol{M}\| = D_1^*$ follows from the relation $D_1^* = (D_1' \cup \bar{D}_1')^*$, where $\bar{D}_1'$ is obtained from $D_1'$ by exchanging $x$ and $\bar{x}$. This relation is a consequence of formula (1.2).

**Proposition 4.1** Rocl $\subset$ Ocl.

Proof. Let $\boldsymbol{M} = \langle X, Q, q_-, Q_+, Y, y_0, \mathfrak{R}\rangle$ be a one counter automaton and set $L = |\boldsymbol{M}|$. We may assume that all rules of $\boldsymbol{M}$ have pushdown symbol $y$. Let $\bar{q}$ be a new state, and define $\bar{\boldsymbol{M}} = \langle X, Q \cup \bar{q}, \bar{q}, Q_+, Y, y, \bar{\mathfrak{R}}\rangle$ by

$$\bar{\mathfrak{R}} = \mathfrak{R} \cup \{(\bar{q}, \varepsilon, y_0) \to (q_-, yy_0)\}.$$

For any computation in $\boldsymbol{M}$

$$(q_-, w, y) \models (q_1, w_1, y^{n_1}) \models \cdots \models (q_k, w_k, y^{n_k}) \models (q_+, \varepsilon, \varepsilon) \tag{4.3}$$

$$(\bar{q}, w, y_0) \models (q_-, w, yy_0) \models (q_1, w_1, y^{n_1} y_0)$$
$$\models \cdots \models (q_k, w_k, y^{n_k} y_0) \models (q_+, \varepsilon, y_0) \tag{4.4}$$

is a computation in $\bar{\boldsymbol{M}}$. Conversely, any computation $(\bar{q}, w, y_0) \overset{+}{\models} (q_+, \varepsilon, y_0)$ in $\bar{\boldsymbol{M}}$ factorizes into (4.4), and then (4.3) is a computation in $\boldsymbol{M}$. Thus $|\boldsymbol{M}| = \|\bar{\boldsymbol{M}}\|$. ■

**Proposition 4.2** *The family* Ocl *is closed under union, product and the plus operation.*

Proof. The closure under union is straightforward. Let $L, L' \subset X^*$ be two one counter languages, and let $L = \|\boldsymbol{M}\|$, $L' = \|\boldsymbol{M}'\|$ for

$$\boldsymbol{M} = \langle X, Q, q_-, Q_+, Y, y_0, \mathfrak{R}\rangle, \qquad \boldsymbol{M}' = \langle X, Q', q'_-, Q'_+, Y, y_0, \mathfrak{R}'\rangle.$$

Clearly, we may assume $Q \cap Q' = \emptyset$. Define a one counter pda

$$\boldsymbol{M}'' = \langle X, Q \cup Q', q_-, Q'_+, Y, y_0, \mathfrak{R}''\rangle$$

with $\mathfrak{R}'' = \mathfrak{R} \cup \mathfrak{R}' \cup \{(q_+, \varepsilon, y_0) \to (q'_+, y_0) \mid q_+ \in Q_+\}$.
Then obviously $\|\boldsymbol{M}''\| = LL'$. Next let

$$\boldsymbol{M}^+ = \langle X, Q, q_-, Q_+, Y, y_0, \mathfrak{R}^+\rangle$$

with $\quad \mathfrak{R}^+ = \mathfrak{R} \cup \{(q_+, \varepsilon, y_0) \to (q_-, y_0) \mid q_+ \in Q_+\}$.

Then $\|\boldsymbol{M}^+\| = L^+$. ■

Note that these proofs are simplified by the possibility to reinitialize in the definition of $\boldsymbol{M}$.

**Theorem 4.3** Ocl *is the rational closure of* Rocl : Ocl = Rocl Rat.

Proof. By Propositions 4.1 and 4.2, we have

$$\text{Rocl } \underline{\text{Rat}} \subset \text{Ocl } \underline{\text{Rat}} \subset \text{Ocl}.$$

To prove conversely Ocl ⊂ Rocl $\underline{\text{Rat}}$, let $\boldsymbol{M} = \langle X, Q, q_-, Q_+, Y, y_0, \mathfrak{R}\rangle$ be a one counter automaton. For $p, q \in Q$ and $k \geqslant 0$, a computation

$$(p, w, y_0) \models (q_1, w_1, \gamma_1) \models \cdots \models (q_k, w_k, \gamma_k) \models (q, w', y_0)$$

is called proper if $|\gamma_1|, \ldots, |\gamma_k| > 1$, thus if $\gamma_1, \ldots, \gamma_k \in y^+y_0$. Define languages $L_{p,q}$ by

$$w \in L_{p,q} \quad \Leftrightarrow \quad \text{there is a proper computation } (p, w, y_0) \overset{+}{\models} (q, \varepsilon, y_0).$$

Any computation $(q_-, w, y_0) \overset{+}{\models} (q_+, \varepsilon, y_0)$ with $q_+ \in Q_+$ factorizes in a sequence of proper computations. Thus there exist states $q_1, q_2, \ldots, q_n$, $(n \geqslant 0)$, and a factorization $w = u_0u_1u_2 \cdots u_n$ such that $u_0 \in L_{q_+,q_1}$, $u_1 \in L_{q_1,q_2}, \ldots, u_n \in L_{q_n,q_+}$. Consider the alphabet $Z = Q \times Q$, and let $K \subset Z^*$ be the local regular language

$$K = (AZ^* \cap Z^*B) \backslash Z^*CZ^*,$$

$$A = \{q_-\} \times Q, \qquad B = Q \times Q_+, \qquad C = \{(q_1, q_2)(q_1', q_2') \mid q_1' \neq q_2\}.$$

Define a substitution $\sigma : Z^* \to X^*$ by

$$\sigma((p, q)) = L_{p,q}.$$

Then $\sigma(K) = \|\boldsymbol{M}\|$, and to prove the theorem, it suffices to show that $L_{p,q} \in$ Rocl. Define a one counter pda

$$\boldsymbol{M}_{p,q} = \langle X, Q \cup \bar{p}, \bar{p}, \{q\}, Y, y_0, \mathfrak{R}_{p,q}\rangle$$

where $\bar{p}$ is a new state, by

$$\mathfrak{R}_{p,q} = \mathfrak{R} \cup \{(\bar{p}, u, y) \to (q', \gamma) \mid (p, u, y_0) \to (q', \gamma y_0) \in \mathfrak{R}\}.$$

Then it is easily seen that $\|\boldsymbol{M}_{p,q}\| = L_{p,q}$. ■

For the rest of this section, we only consider restricted one counter languages. We therefore introduce some simplified notations. First, since the symbol $y_0$ will no longer be relevant, a one counter pda will be denoted by $\boldsymbol{M} = \langle X, Q, q_-, Q_+, \mathfrak{R}\rangle$. Next, all rules in $\mathfrak{R}$ are assumed to have pushdown symbol $y$, and we write

$$(q, u) \to (q', m) \quad \text{instead of} \quad (q, u, y) \to (q', y^{1+m}) \qquad (m \geqslant -1)$$

Thus $m$ denotes the increase or decrease of the pushdown store. With these notations, the computation (4.3) takes the form:

$$(q_-, w, 1) \models (q, w_1, n_1) \models \cdots \models (q_k, w_k, n_k) \models (q_+, \varepsilon, 0).$$

**Theorem 4.4** *The family* Rocl *is a rational cone and* Ocl *is a full* AFL.

Proof. The second part of the statement follows from the first part by the previous theorem. Thus it suffices to show that Rocl is a cone. Let $\boldsymbol{M} = \langle X, Q, q_-, Q_+, \mathfrak{R}\rangle$ be a one counter automaton and $L = |\boldsymbol{M}|$.

a) Let $\varphi: X^* \to Y^*$ be an alphabetic morphism. Define a pda $\varphi\boldsymbol{M} = \langle Y, Q, q_-, Q_+, \varphi\mathfrak{R}\rangle$ by

$$\varphi\mathfrak{R} = \{(q, \varphi u) \to (q', m) \mid (q, u) \to (q', m) \in \mathfrak{R}\}.$$

Then clearly $|\varphi\boldsymbol{M}| = \varphi\,|\boldsymbol{M}|$.

b) Let $K \subset X^*$ be a rational language, and let $\mathbf{A} = \langle X, P, p_-, P_+\rangle$ be a finite automaton recognizing $K$. Define $\boldsymbol{M}_K = \langle X, Q \times P, (q_-, p_-), Q_+ \times P_+, \mathfrak{R}_K\rangle$ by

$$\mathfrak{R}_K = \{((q, p), u) \to ((q', p'), m) \mid (q, u) \to (q', m) \in \mathfrak{R} \text{ and } p \cdot u = p'\}.$$

Then one has $|\boldsymbol{M}_K| = L \cap K$.

c) Let $Z = X \cup z$, with $z \notin X$, and let $\pi: Z^* \to X^*$ be the projection. The first idea for the construction of a one counter pda $\boldsymbol{M}'$ with $|\boldsymbol{M}'| = \pi^{-1}(L)$ is to add to $\mathfrak{R}$ the rules $(q, z) \to (q, 0)$ for $q \in Q$. This method does not work. Indeed, consider for instance a word $w = w'z$ with $w' \in L$. Then $w$ would not be accepted by $\boldsymbol{M}'$ since the pushdown store is empty after the lecture of $w'$. Therefore the presence of letters $z$ has to be anticipated. This yields the following construction. For each $(q, u) \in Q \times (X \cup \varepsilon)$, let $q_u$ be a new state. Then define $Q' = \{q_u \mid (q, u) \in Q \times (X \cup \varepsilon)\}$ and $\pi^{-1}\boldsymbol{M} = \langle Z, Q \cup Q', q_-, Q_+, \mathfrak{R}'\rangle$, where $\mathfrak{R}'$ is given by

$$\mathfrak{R}' = \{(q_-, z) \to (q_-, 0)\} \cup \mathfrak{R}''$$

$$\mathfrak{R}'' = \{(q, u) \to (q_u, 0), (q_u, z) \to (q_u, 0), (q_u, \varepsilon) \to (q', m) \mid (q, u) \to (q', m) \in \mathfrak{R}\}.$$

A straightforward verification shows that $|\pi^{-1}\boldsymbol{M}| = \pi^{-1}L$.

By the previous constructions, Rocl is closed under alphabetic morphism, inverse projection and intersection with rational languages. Thus Rocl is a rational cone. ■

In order to prove that Rocl is a principal cone, a "normal form" of one counter pda's will appear to be useful. A one counter pda $\boldsymbol{M} = \langle X, Q, q_-, Q_+, \mathfrak{R}\rangle$ is called normalized iff for any rule $(q, u) \to (q', m) \in \mathfrak{R}$, either $m = 1$ or $m = -1$. Thus the length of the pushdown store always increases or decreases by 1.

**Lemma 4.5** *For any one counter pda* $\boldsymbol{M}$, *there is a normalized one counter* pda $\boldsymbol{M}'$ *such that* $|\boldsymbol{M}| = |\boldsymbol{M}'|$.

Proof. Let $\boldsymbol{M} = \langle X, Q, q_-, Q_+, \mathfrak{R}\rangle$, and define $\boldsymbol{M}' = \langle X, Q', q_-, Q_+, \mathfrak{R}'\rangle$ with $Q' \supset Q$ as follows. Let $(q, u) \to (p, m) \in \mathfrak{R}$ be a rule. Then (i) if $m = 1$ or $m = -1$, then $(q, u) \to (p, m) \in \mathfrak{R}'$; (ii) if $m = 0$, then let $\bar{q}$ be a new state and add to $\mathfrak{R}'$ the two rules $(q, u) \to (\bar{q}, 1)$, $(\bar{q}, \varepsilon) \to (p, -1)$; (iii) if $m \geqslant 2$, then let $\bar{q}_1, \bar{q}_2, \ldots, \bar{q}_{m-1}$ be new states and add to $\mathfrak{R}'$ the rules $(q, u) \to (\bar{q}_1, 1)$, $(\bar{q}_1, \varepsilon) \to (\bar{q}_2, 1), \ldots, (\bar{q}_{m-2}, \varepsilon) \to (\bar{q}_{m-1}, 1), (\bar{q}_{m-1}, \varepsilon) \to (p, 1)$. Such a "factorization" of the rules of $\mathfrak{R}$ yields a normalized pda $\boldsymbol{M}'$, and clearly $|\boldsymbol{M}'| = |\boldsymbol{M}|$. ■

**Theorem 4.6** *The rational cone* Rocl *is principal.* $D_1'^*, D_1', D_1, Ł$ *are generators of the cone* Rocl.

Proof. By Proposition 1.3, $D_1'^* \approx D_1' \approx D_1$. Next if $D_1'^*, Ł \subset \{a, b\}^*$ then we noted in Section II.4 that $D_1'^* b = Ł$. Thus $D_1'^* \approx Ł$, and the four languages are rationally equivalent. All these languages are restricted one counter languages, and since Rocl is a rational cone, we have $Ł\Gamma \subset$ Rocl. Thus it suffices to show the converse inclusion, or $Ł \geqslant L$ for any $L \in$ Rocl.

Let $L \subset X^*$, $L \in$ Rocl, and let $\boldsymbol{M} = \langle X, Q, q_-, Q_+, \mathfrak{R}\rangle$ be a normalized one counter automaton with $L = |\boldsymbol{M}|$. The proof is in several steps. Note that only minor changes will be made with respect to a proof showing that the languages recognized by general pda's are context-free.

a) First we prove that $L$ is the homomorphic image of a language recognized by a "deterministic" one counter pda. Number the rules of $\mathfrak{R}$ from 1 to $N = \mathrm{Card}(\mathfrak{R})$. Let $\pi_i$, $(1 \leqslant i \leqslant N)$ be the $i$-th rule. Then $\mathfrak{R} = \{\pi_1, \ldots, \pi_N\}$. Let $T = \{t_1, \ldots, t_N\}$ be an alphabet and define a one counter pda $\boldsymbol{M}' = \langle T, Q, q_-, Q_+, \mathfrak{R}'\rangle$ by

$$(q, t_i) \to (q', m) \in \mathfrak{R}' \quad \Leftrightarrow \quad \pi_i = (q, u) \to (q', m) \qquad 1 \leqslant i \leqslant N.$$

Then clearly $\boldsymbol{M}'$ is normalized, and $\boldsymbol{M}'$ is deterministic in the sense that there is no rule with left side of the form $(q, \varepsilon)$, and

$$\text{each } t \in T \text{ is the input symbol of exactly one rule in } \mathfrak{R}' \qquad (4.5)$$

Let $\alpha : T^* \to X^*$ be the morphism defined by $\alpha(t_i) = u$ iff $u$ is the input symbol of the $i$-th rule in $\mathfrak{R}$. Let $L' = |\boldsymbol{M}'|$. We prove that $L = \alpha(L')$. Consider indeed a computation in $\boldsymbol{M}'$:

$$(q_-, t_{i_1} \cdots t_{i_k}, 1) \models (q_1, t_{i_2} \cdots t_{i_k}, n_1) \models \cdots \models (q_{k-1}, t_{i_k}, n_k) \models (q_+, \varepsilon, 0).$$

Since $\pi_{i_1} = (q_-, \alpha(t_{i_1})) \to (q_1, n_1 - 1)$, $\pi_{i_r} = (q_{r-1}, \alpha(t_{i_r})) \to (q_r, n_{i_r} - n_{i_{r-1}})$, $(2 \leqslant r \leqslant k-1)$ and $\pi_{i_k} = (q_{k-1}, \alpha(t_{i_k})) \to (q_+, -n_k)$, it follows that $\alpha(t_{i_1} \cdots t_{i_k}) \in L$. Thus $\alpha(L') \subset L$. Conversely, let $w \in L$ and consider a computation in $\boldsymbol{M} : (q_-, w, 1) \overset{k}{\models} (q_+, \varepsilon, 0)$ of length $k$ and let $\pi_{i_1}, \ldots, \pi_{i_k}$ be the sequence of rules used in this computation. Then each $\pi_{i_r}$ is of the form $\pi_{i_r}=(q,\alpha(t_{i_r}))\rightarrow(q',m)$ for some $t_{i_r}\in T$, and $(q,t_{i_r})\rightarrow(q',m)$ is in $\mathfrak{R}'$. Therefore $\alpha(t_{i_1}\cdots t_{i_k})=w$ and $t_{i_1}\cdots t_{i_k}\in L'$. Thus $L\subset\alpha(L')$.

b) Next we show that $L'=|\boldsymbol{M}''|\cap K$ where $K\subset T^*$ is a regular language and $\boldsymbol{M}''$ has just one state. Define a one counter pda $\boldsymbol{M}''=\langle T,\{p\},p,\{p\},\mathfrak{R}''\rangle$ and a finite state automaton $\mathbf{A}=\langle T,Q,q_-,Q_+\rangle$ by

$$\left.\begin{array}{l}(p,t)\rightarrow(p,m)\in\mathfrak{R}''\\ q\cdot t=q'\end{array}\right\}\quad\Leftrightarrow\quad(q,t)\rightarrow(q',m)\in\mathfrak{R}'$$

Then it is easily seen by induction that

$$(q_-,w,1)\overset{k}{\models}(q_+,\varepsilon,0)\text{ in }\boldsymbol{M}'$$

$$\text{iff}\qquad (p,w,1)\overset{k}{\models}(p,\varepsilon,0)\text{ in }\boldsymbol{M}''\text{ and }q_-\cdot w=q_+.$$

Consequently $L'=L''\cap K$, with $L''=|\boldsymbol{M}''|$ and $K=|\mathbf{A}|$.

c) The set $\mathfrak{R}''$ of rules of $\boldsymbol{M}''$ defines a morphism $\upsilon$ from $T^*$ into the additive group $\mathbb{Z}$ by

$$\upsilon(t)=m\quad\Leftrightarrow\quad(p,t)\rightarrow(p,m)\in\mathfrak{R}''\qquad(t\in T).\tag{4.6}$$

Note that $\upsilon$ is well defined in view of (4.5).

Let $w=z_1\cdots z_k\in T^*$, $(z_i\in T)$. Assume $w\in L''$. Then there is a computation in $\boldsymbol{M}''$:

$$(p,z_1,\cdots z_k,n_0)\models(p,z_2\cdots z_k,n_1)\models\cdots\models(p,z_k,n_{k-1})\models(p,\varepsilon,n_k)\tag{4.7}$$

with $n_0=1$, $n_k=0$. By (4.6), $\nu(z_i)=n_i-n_{i-1}$ for $i=1,\ldots,k$, whence $\nu(z_1\cdots z_i)=n_i-1$ for $i=0,\ldots,k$. Since $n_i>0$ for $i=0,\ldots,k-1$, we have:

$$\upsilon(w)=-1\text{ and }\upsilon(w')\geqslant0\text{ for any proper left factor }w'\text{ of }w.\tag{4.8}$$

Assume conversely that (4.8) holds and define $n_i=1+\upsilon(z_1\cdots z_i)$ for $i=0,\ldots,k$. Then by (4.6), (4.7) is a computation for $w$ and therefore $w\in L''$. This proves that $w\in L''$ iff (4.8) holds. Finally let $Ł\subset\{a,b\}^*$ and let $\beta:T^*\rightarrow\{a,b\}^*$ be the morphism defined by

$$\beta(t)=\begin{cases}a & \text{if }\upsilon(t)=1\\ b & \text{if }\upsilon(t)=-1\end{cases}\qquad(t\in T)$$

(Recall that the integer $m$ in (4.6) is equal to 1 or $-1$ since the automaton $\boldsymbol{M}$ was supposed to be normalized.) Then Proposition II.4.1 and (4.8) show that $w\in L''$ iff $\beta(w)\in Ł$. Consequently $L''=\beta^{-1}(Ł)$. The three steps of the proof show that

$$L=\alpha(L')=\alpha(L''\cap K)=\alpha(\beta^{-1}(\L)\cap K).$$

Thus $L \leqslant \L$. ∎

We shall see in the next chapter that the principal cones Rocl and Lin are incomparable. A direct proof of this fact is long and tedious. A short proof uses results on iterative pairs.

The language $S_= = \{x^n y^n \mid n \geqslant 0\}$ is both a linear and a one counter language. Thus $S_=\Gamma \subset \mathrm{Rocl} \cap \mathrm{Lin}$. It is an open question whether this inclusion is strict. A related open problem is the following: Is the intersection of two principal cones a principal cone?

### Exercises

**4.1** (Greibach [1969]) Show that a language is context-free iff it is of the form $\|\boldsymbol{M}\| = \{w \in X^* \mid \exists q_+ \in Q_+ : (q_-, w, y_0) \overset{+}{\models} (q_+, \varepsilon, y_0)\}$, where $\boldsymbol{M}$ is a general pda (as defined for instance in Ginsburg [1966]).

**4.2** Let $X_n = \{x_0, x_1, \ldots, x_n\}$, $(n \geqslant 1)$. Define $L_n \subset X_n^*$ by

$$L_n = \bigcup_{k=1}^{n} x_k (L_n)^k \cup x_0.$$

Show that the languages $L_n$, $(n \geqslant 2)$ are generators of the cone Rocl.

**4.3** (Greibach [1969]) Let $\boldsymbol{M} = \langle X, Q, q_-, Q_+, Y, y_0, \mathfrak{R}\rangle$ be a pda. Define $P(\boldsymbol{M}) = \{h \in Y^* \mid \exists w \in X^*, \exists q \in Q \mid (q_-, w, y_0) \overset{+}{\models} (q, \varepsilon, h)\}$ to be the set of all words appearing on the push-down store during computations. Then $\boldsymbol{M}$ is an $n$-counter pda iff

$$P(\boldsymbol{M}) \subset y_0 y_1^* \cdots y_n^*, (y_1, \ldots, y_n \in Y).$$

Show that $L \in \mathrm{Fcl}(n)\hat{\Gamma}$ iff $L = \|\boldsymbol{M}\| = \{w \in X^* \mid \exists q_+ \in Q, (q_-, w, y_0) \overset{+}{\models} (q, \varepsilon, y_0)\}$ for some $n$-counter pda $\boldsymbol{M}$.

## VII.5 Quasi-Rational Languages

The family of quasi-rational languages is the closure under substitution of the family of linear languages. This remarkable family admits several quite different characterizations and therefore has also several denominations: quasi-rational languages are known as standard matching choice, or nonexpansive, or derivation bounded languages.

In this section, we prove the equivalence of some of these definitions. Papers on quasi-rational languages include Yntema [1967], Nivat [1967], Ginsburg and Spanier [1968], Salomaa [1969a], Gruska [1971a, b]. Results concerning subfamilies have been proved in Boasson, Crestin and Nivat [1973], Crestin [1978].

**Definition** The family Qrt($k$) of quasi-rational languages of rank $k \geq 1$ is defined inductively by:

$$\mathrm{Qrt}(1)=\mathrm{Lin}, \qquad \mathrm{Qrt}(k+1)=\mathrm{Lin} \square \mathrm{Qrt}(k) \qquad (k \geq 1).$$

The family of quasi-rational languages is

$$\mathrm{Qrt}=\bigcup_{k \geq 1} \mathrm{Qrt}(k).$$

In view of the associativity of substitution (Theorem V.5.1),

$$\mathrm{Qrt}(k)=\mathrm{Qrt}(k-1) \square \mathrm{Lin}=\mathrm{Qrt}(n) \square \mathrm{Qrt}(m)$$

for all $k \geq 2$, $n, m \geq 1$ with $k=n+m$. By Theorem VI.5.7, Qrt($k$) is a principal cone, hence closed under union. In view of Example 3.2 and Exercise 3.2,

$$\mathrm{Qrt}(k) \subsetneq \mathrm{Qrt}(k)\hat{\Gamma} \subsetneq \mathrm{Qrt}(k+1) \qquad k \geq 1. \tag{5.1}$$

**Example 5.1** Let $X=\{x, y, z\}$, let $S_{=}=\{x^n y^n : n \geq 0\}$, and consider the language

$$L=\{x^{n_1}y^{n_1}x^{n_2}y^{n_2} \cdots x^{n_p}y^{n_p}z^p : p \geq 0, n_1, \ldots, n_p \geq 0\}.$$

Then $L=\theta(S_{=})$, where $\theta:\{x, y\}^* \to X^*$ is given by $\theta(x)=S_{=}$, $\theta(y)=z$. Thus $L \in \mathrm{Qrt}(2)$. The iteration lemma for linear languages (Proposition V.6.6) easily shows that $L$ is not linear.

**Example 5.2** No generator of the cone Alg is quasi-rational since otherwise it would belong to Qrt($k$) for some $k \geq 1$, and (5.1) would be violated. It is much harder to prove that $D_1'^*$ (and consequently also $Ł$, $D_1'$, $D_1$) are not quasi-rational. This will be done in the next chapter (Theorem VIII.7.14).

We now give a first characterization of the quasi-rational languages. For this we need a definition.

**Definition** A context-free grammar $G=\langle V, X, P\rangle$ is nonexpansive if for every $\xi \in V$, and $w \in (V \cup X)^*$, $\xi \xrightarrow{*} w$ implies that $w$ contains at most one occurrence of $\xi$. A language $L$ is nonexpansive if there exists a nonexpansive context-free grammar generating $L$.

Let Nexp denote the family of nonexpansive languages.

**Theorem 5.1** Nexp = Qrt.

Proof. In order to verify the inclusion Qrt ⊂ Nexp, we show that Qrt($k$) ⊂ Nexp by induction on $k$. Clearly any linear grammar is nonexpansive. Thus Qrt(1) ⊂ Nexp. Next let $L \subset Y^*$, $L \in \mathrm{Qrt}(k+1)=\mathrm{Lin} \square \mathrm{Qrt}(k)$. Then there are a linear language $A \subset X^*$, and a Qrt($k$)-substitution $\theta: X^* \to Y^*$ such that $\theta(A)=L$. Let $G=\langle V, X, P\rangle$ be a linear grammar such that $A=L_G(\sigma)$ for a

$\sigma \in V$. By induction, there are nonexpansive grammars $G_x = \langle V_x, Y, P_x \rangle$ such that $\theta(x) = L_{G_x}(\sigma_x)$ with $\sigma_x \in V_x$ for each $x \in X$. Let $H = \langle W, Y, Q \rangle$ be the grammar generating $L$ constructed in Section II.2. We claim that $H$ is nonexpansive. Otherwise there is a $\xi \in W = V \cup \bigcup_{x \in X} V_x$ such that

$$\xi \xrightarrow[H]{*} a\xi b\xi c \tag{5.2}$$

for some words $a, b, c$. The variable $\xi$ is in none of the sets $V_x$, since otherwise (5.2) would hold in $G_x$, and $G_x$ would be expansive. Thus $\xi \in V$. But then (5.2) cannot hold since $G$ is linear. Thus $H$ is nonexpansive and $L \in \text{Nexp}$.
Conversely, let $L$ be a nonexpansive language. Let $G = \langle V, X, P \rangle$ be a nonexpansive grammar with $L = L_G(\sigma)$ for some $\sigma \in V$. We prove that $L$ is quasi-rational by induction on $n = \text{Card}(V)$. If $n = 1$, then $G$ is linear and $L$ is quasi-rational. Thus assume $n \geqslant 2$, and that the conclusion holds for all nonexpansive grammars with less than $n$ variables. Clearly $G$ is still nonexpansive if it reduced in $\sigma$ (in the sense of Section II.2). Thus we assume that $G$ is reduced in $\sigma$. Define $V' \subset V$ to be the set of variables $\xi$ such that $\xi \xrightarrow{*} a\sigma b$ for some $a, b \in (V \cup X)^*$. Since $\sigma \in V'$, $V'$ is nonempty. Define a grammar

$$G' = \langle V', X \cup (V \backslash V'), P' \rangle, \qquad P' = \{\xi \to \alpha \in P \mid \xi \in V'\}.$$

We claim that $G'$ is linear. Assume the contrary. Then there exists a production $\xi \to f\eta_1 g\eta_2 h$ in $P'$ with $\xi, \eta_1, \eta_2 \in V'$. Consequently $\eta_1 \xrightarrow{*} a_1\sigma b_1$, $\eta_2 \xrightarrow{*} a_2\sigma b_2$ for some words $a_1, b_1, a_2, b_2$, and since $G$ is reduced, $\sigma \xrightarrow{*} u\xi v$ for some words $u, v$. Combining these derivations, we obtain

$$\sigma \xrightarrow{*} u\xi v \xrightarrow{*} uf\eta_1 g\eta_2 hv \xrightarrow{*} ufa_1\sigma b_1 ga_2\sigma b_2 hv,$$

and $G$ is expansive, contrary to the assumption. Thus $G'$ is linear. If $V' = V$, then $G = G'$ is linear and $L$ is quasi-rational. Thus suppose $V' \neq V$ and consider the grammar

$$\bar{G} = \langle V \backslash V', X, P \backslash P' \rangle.$$

Note that $\bar{G}$ is indeed a grammar, i.e. $\eta \to \alpha \in P \backslash P'$ implies $\alpha \in (X \cup V \backslash V')^*$, since otherwise $\alpha = a\zeta b$ for some $\zeta \in V'$, hence $\alpha \to a'\sigma b'$ for some $a', b'$, and $\eta \in V'$. Since $\sigma \in V'$, $\bar{G}$ has less than $n$ variables, and obviously is nonexpansive. Thus by induction $L_\xi = L_{\bar{G}}(\xi)$ is quasi-rational for each $\xi \in V \backslash V'$. Next $L = \theta(L_{G'}(\sigma))$, where $\theta : (X \cup V \backslash V')^* \to X^*$ is the substitution defined by $\theta(x) = x$ for $x \in X$, and $\theta(\xi) = L_\xi$ for $\xi \in V \backslash V'$. Thus $L \in \text{Lin} \mathbin{\square} \text{Qrt} = \text{Qrt}$ and $L$ is quasi-rational. ■

We now consider another property of grammars which will appear to give a characterization of the quasi-rational languages of rank $k$. Let $G=\langle V, X, P\rangle$ be a context-free grammar. The **index** of a derivation

$$w_0 \to w_1 \to \cdots \to w_n$$

is the maximum number of occurrences of variables in $w_0, \ldots, w_n$, i.e. the greatest of the numbers $|w_0|_V, \ldots, |w_n|_V$. Given a word $w \in (V \cup X)^*$ such that $\xi \xrightarrow{*} w$, the **index of $w$ with respect to** $\xi \in V$ is the least of the indices of all derivations $\xi \xrightarrow{*} w$. Denote this number by $\text{ind}(\xi, w)$. Clearly $\text{ind}(\xi, w) \geqslant 1$. Next define

$$L_G^{(k)}(\xi) = \{w \in X^* \mid \text{ind}(\xi, w) \leqslant k\}.$$

Thus $L_G^{(k)}(\xi)$ is the set of all words $w \in X^*$ for which there exists a derivation $\xi \xrightarrow{*} w$ with index at most $k$. Obviously

$$L_G^{(k)}(\xi) \subset L_G^{(k+1)}(\xi), \qquad L_G(\xi) = \bigcup_{k \geqslant 1} L_G^{(k)}(\xi).$$

The grammar $G$ has **index $k$ in $\xi$** if $k$ is the least integer such that $L_G(\xi) = L_G^{(k)}(\xi)$, and $G$ has **infinite index in $\xi$** if $L_G(\xi) \neq L_G^{(k)}(\xi)$ for all $k \geqslant 1$. Let $\text{ind}(G, \xi)$ denote the index of $G$ in $\xi$. Thus $\text{ind}(G, \xi)$ is either a positive integer or infinite.

**Definition** Let $L$ be an algebraic language. Then the **index** of $L$ is defined as

$$\text{ind}(L) = \min \text{ind}(G, \xi)$$

where $(G, \xi)$ ranges over all algebraic grammars such that $L = L_G(\xi)$. Thus the index of $L$ is either a positive integer or infinite. Let $\text{Ind}(k)$ be the family of all algebraic languages of index $\leqslant k$, $(k \geqslant 1)$.

**Example 5.3** Any linear language has index 1, since any linear grammar has index 1 in all its variables.

**Example 5.4** Let $Ł \subset \{a, b\}^*$ be the Lukasiewicz language. $Ł$ is generated by the grammar with productions $\xi \to a\xi\xi + b$. We claim that this grammar has infinite index.

Define indeed a sequence of words $(f_m)_{m \geqslant 1}$ by $f_1 = b$, $f_{m+1} = af_m f_m$. We verify that $\text{ind}(\xi, f_m) = m$. Since $\xi \to b$, we have $\text{ind}(\xi, f_1) = 1$. Arguing by induction, assume $\text{ind}(\xi, f_m) = m$. Then there exists a derivation

$$\xi = w_0 \to w_1 \to \cdots \to w_r = f_m$$

of index $m$. Consequently the derivation

$$\xi \to a\xi\xi \to aw_1\xi \to \cdots \to aw_r\xi \to aw_r w_1 \to \cdots \to aw_r w_r = f_{m+1}$$

has index $m+1$, and $\text{ind}(\xi, f_{m+1}) \leqslant m+1$. Conversely, consider a derivation

$$\xi = w_0 \to w_1 \to \cdots \to w_p = f_{m+1}.$$

Then $w_1 = a\xi\xi$, and by Lemma II.1.1, each $w_i$ admits a factorization $w_i = au_iv_i$ such that

$$\xi = u_1 \xrightarrow{k_2} u_2 \to \cdots \to u_{p-1} \xrightarrow{k_p} u_p$$

$$\xi = v_1 \xrightarrow{l_2} v_2 \to \cdots \to v_{p-1} \xrightarrow{l_p} v_p$$

with $k_i + l_i = 1$ for $i = 2, \ldots, p$. Since $f_{m+1} = au_pv_p$, and $u_p, v_p \in Ł$, and since $Ł$ is prefix (Proposition II.4.1), it follows that $u_p = v_p = f_m$. Next either $k_p = 0$ or $l_p = 0$. Assume $k_p = 0$, then $l_p = 1$, and $|v_{p-1}|_\xi = 1$. Consequently $|v_i|_\xi \geqslant 1$ for each $i = 1, \ldots, p-1$. Next $u_{p-1} = u_p = f_m$, and by induction, there is an integer $i$, $(1 \leqslant i \leqslant p-1)$ such that $|u_i|_\xi \geqslant m$. Thus $|w_i|_\xi \geqslant m+1$, showing that $\text{ind}(\xi, f_{m+1}) \geqslant m+1$.

Note that we did not prove that the language $Ł$ has infinite index. We only proved that a particular grammar generating $Ł$ has infinite index. The proof that $Ł$ has infinite index is much more difficult. It is a consequence of Theorem VIII.7.14 in view of the following result.

**Theorem 5.2** *For any* $k \geqslant 1$, $\text{Ind}(k) = \text{Qrt}(k)$.

The proof of this theorem is in two parts. First we prove

**Lemma 5.3** *For any* $k \geqslant 1$, $\text{Qrt}(k) \subset \text{Ind}(k)$.

Proof. The proof is by induction on $k$. Any linear grammar has index 1. Hence $\text{Lin} \subset \text{Ind}(1)$. Since by induction

$$\text{Qrt}(k+1) = \text{Lin} \,\square\, \text{Qrt}(k) \subset \text{Lin} \,\square\, \text{Ind}(k),$$

it suffices to show the inclusion $\text{Lin} \,\square\, \text{Ind}(k) \subset \text{Ind}(k+1)$. For this, consider a language $L \subset Y^*$ in $\text{Lin} \,\square\, \text{Ind}(k)$. Then $L = \theta(A)$, where $A \subset X^*$ is linear, and $\theta: X^* \to Y^*$ is a $\text{Ind}(k)$-substitution. Let $G = \langle V, X, P \rangle$ be a linear grammar in canonical form (as defined in Section V.6), and let $\sigma \in V$ be such that $A = L_G(\sigma)$. Next for each $x \in X$, let $G_x = \langle V_x, Y, P_x \rangle$ be a grammar such that $\theta(x) = L_{G_x}(\sigma_x)$, and having index $\leqslant k$ in $\sigma_x$. Let $H = \langle W, Y, Q \rangle$ be the grammar such that $L = L_H(\sigma)$ deduced from $G$ and the $G_x$, as given in Section II.2.

Observe that for each production $\xi \to x\eta$ in $P$, the production $\xi \to \sigma_x\eta$ is in $Q$. Thus for all $w \in \theta(x)$, there is a derivation $\xi \xrightarrow{*} w\eta$ of index at most $k+1$ in $H$. The same conclusion holds for the other types of productions in $P$. Consequently, for each derivation $\sigma \xrightarrow{*} f$ in $G$, and for each $w \in \theta(f)$, there exists a derivation $\sigma \xrightarrow{*} w$ of index at most $k+1$ in $H$. Thus $L \in \text{Ind}(k+1)$. ■

The proof of the converse inclusion, namely $\mathrm{Ind}(k) \subset \mathrm{Qrt}(k)$, is more involved and requires some investigations of the index of derivations. For this, we first introduce an auxiliary notation: Given a grammar $G$ and words $u$, $v$, we write

$$u \mapsto v$$

as an abbreviation for

$$u = v \quad \text{or} \quad u \to v.$$

Thus $u \mapsto v$ iff $u \overset{n}{\to} v$ with $n = 0$ or $n = 1$.

We now prove a general lemma concerning decompositions of derivations, and their relation to the notion of index. Let $G = \langle V, X, P \rangle$ be a context-free grammar, and consider a derivation in $G$:

$$w_1 \to w_2 \to \cdots \to w_n \tag{5.3}$$

with $w_n \in X^*$. Assume further that

$$w_1 = u_0 \xi_1 u_1 \cdots u_{s-1} \xi_s u_s, \qquad \xi_1, \ldots, \xi_s \in V, \qquad u_0, \ldots, u_s \in X^*.$$

In view of Lemma II.1.1, each word $w_i$ admits a factorization

$$w_i = u_0 h_1^{(i)} u_1 h_2^{(i)} \cdots u_{s-1} h_s^{(i)} u_s$$

such that

$$\begin{aligned} \xi_1 &= h_1^{(1)} \mapsto h_1^{(2)} \mapsto \cdots \mapsto h_1^{(n)} \\ \xi_2 &= h_2^{(1)} \mapsto h_2^{(2)} \mapsto \cdots \mapsto h_2^{(n)} \\ &\vdots \\ \xi_s &= h_s^{(1)} \mapsto h_s^{(2)} \mapsto \cdots \mapsto h_s^{(n)} \end{aligned} \tag{5.4}$$

and further for each $i = 1, \ldots, n-1$, $h_t^{(i)} \to h_t^{(i+1)}$ for exactly one $t$. The derivations (5.4) are said to be **induced** by (5.3).

The following lemma is an extension of the argument used in Example 5.4.

**Lemma 5.4** *Assume that the derivation* (5.3) *has index* $\leqslant p$. *Then for* $t = 1, \ldots, s$, *at least* $t$ *of the induced derivations* (5.4) *have index* $\leqslant p - s + t$.

Proof. Assume the contrary. Then there is some $t$ such that less than $t$ induced derivations have index $\leqslant p - s + t$. Since the index of the other derivations is greater than $p - s + t$, this means that there exist $r = s - t + 1$ induced derivations which all have an index at least $p - s + t + 1 = p - r + 2$. For simplicity, we assume that these derivations are the first ones in (5.4). Next, since $h_1^{(1)}, \ldots, h_r^{(1)} \in V$, and $h_1^{(n)}, \ldots, h_r^{(n)} \in X^*$, there is a largest integer $m < n$ such that

$$|h_1^{(m)}|_V \geqslant 1, \qquad |h_2^{(m)}|_V \geqslant 1, \ldots, |h_r^{(m)}|_V \geqslant 1,$$

and there is some $j \in \{1, \ldots, r\}$ such that $h_j^{(m+1)} \in X^*$. Consider this $j$, and the corresponding induced derivation $\xi_j \xrightarrow{*} h_j^{(n)}$ in (5.4). The index of this derivation is $\geq p-r+2$. Thus there is some exponent $l$ with $1 \leq l \leq m$ and such that

$$|h_j^{(l)}|_V \geq p-r+2.$$

Since $|h_i^{(l)}|_V \geq 1$ for $i=1, \ldots, r$, $i \neq j$, we have

$$|w_l|_V \geq (p-r+2)+(r-1)=p+1.$$

Thus the index of the derivation (5.3) is $\geq p+1$, and this is a contradiction. ∎

**Proposition 5.5** (Ginsburg and Spanier [1968]) *Let $G=\langle V, X, P\rangle$ be a context-free grammar. Then $L_G^{(k)}(\xi)$ is a quasi-rational language of rank $k$ for all $\xi \in V$ and $k \geq 1$.*

The proposition claims that the languages $L_G^{(k)}(\xi)$ are context-free. Ginsburg and Spanier [1968] show that this holds even if $G$ is an arbitrary phrase-structure grammar.

Proof. Let $k \geq 1$ be fixed. We first construct a context-free grammar $H$ generating $L_G^{(k)}(\xi)$, which proves that these languages are context-free. The analysis of the grammar $H$ then shows that the languages are quasi-rational of rank $k$.

Define a grammar

$$H=\langle W, X, Q\rangle$$

as follows: $W=V \times \{1, \ldots, k\}$. (An element of $W$ is noted $[\xi, p]$ instead of $(\xi, p)$.) Next, for each production

$$\xi \to u_0\xi_1u_1\xi_2 \cdots u_{s-1}\xi_su_s \in P, \tag{5.5}$$

with $\xi_1, \ldots, \xi_s \in V$, $u_0, \ldots, u_s \in X^*$ and $s \leq k$, the productions

$$[\xi, p] \to u_0[\xi_1, p_1]u_1[\xi_2, p_2] \cdots u_{s-1}[\xi_s, p_s]u_s$$

are in $Q$ for all $p \geq s$, and for all permutations $(p_1, p_2, \ldots, p_s)$ of $(p-s+1, \ldots, p)$.

Note that the $p_j$'s are distincts, and that $\{p_1, \ldots, p_s\}=\{p-s+1, \ldots, p\}$. Note also that if $s>k$ in (5.5), then this production gives no contribution to $Q$. Finally, observe that if $s=0$, then $[\xi, p] \to u_0 \in Q$ for all $p=1, \ldots, k$. We claim:

$$L_H([\xi, p])=L_G^{(p)}(\xi) \quad \text{for} \quad \xi \in V, \quad 1 \leq p \leq k.$$

(i) We first prove the inclusion $L_G^{(p)}(\xi) \subset L_H([\xi, p])$ by induction on the length of derivations. Let $w \in L_G^{(p)}(\xi)$, and consider a derivation of index $\leq p$:

$$\xi \to w_1 \to w_2 \to \cdots \to w_n = w \quad \text{in } G.$$

If $n=1$, then $\xi \to w \in P$, and, in view of a previous remark, $[\xi, p] \to w \in Q$ for $p=1, \ldots, k$. Thus $w \in L_H([\xi, p])$.
Assume $n>1$, and set

$$w_1 = u_0\xi_1 u_1 \xi_2 \cdots \xi_s u_s, \qquad \xi_1, \ldots, \xi_s \in V, \qquad u_0, \ldots, u_s \in X^*.$$

Obviously $s \leqslant p$, since the derivation has index $\leqslant p$.
In view of Lemma 5.4, there is at least one induced derivation with index $\leqslant p-s+1$. Next, there is at least one other derivation which has index $\leqslant p-s+2$. Continuing in this way, one determines, for each $j=1, \ldots, s$, an integer $\pi(j) \in \{1, \ldots, s\}$, distinct from the previous ones, such that the induced derivation $\xi_{\pi(j)} \xrightarrow{*} h_{\pi(j)}^{(n)}$ has index at most $p-s+j$. Since the length of the induced derivations is smaller than $n$, it follows by induction that

$$h_{\pi(j)}^{(n)} \in L_H([\xi_{\pi(j)}, p-s+j]) \qquad j=1, \ldots, s.$$

Note that $\pi$ is a permutation of $\{1, \ldots, s\}$. Setting $p_i = p-s+\pi^{-1}(i)$ for $i=1, \ldots, s$ we get a permutation $(p_1, \ldots, p_s)$ of $(p-s+1, \ldots, p)$, and

$$h_i^{(n)} \in L_H([\xi_i, p_i]) \qquad i=1, \ldots, s.$$

Finally

$$[\xi, p] \to u_0[\xi_1, p_1]u_1 \cdots u_{s-1}[\xi_s, p_s]u_s \in Q,$$

and consequently $w \in L_H([\xi, p])$.
(ii) We now prove the inclusion $L_H([\xi, p]) \subset L_G^{(p)}(\xi)$, by induction on the length of derivations. Let $w \in L_H([\xi, p])$, and consider a derivation

$$[\xi, p] \to w_1 \to w_2 \to \cdots \to w_n = w \quad \text{in } H.$$

If $n=1$, then $\xi \to w \in P$, and $w \in L_G^{(1)}(\xi)$. Otherwise

$$w_1 = u_0[\xi_1, p_1]u_1[\xi_2, p_2] \cdots u_{s-1}[\xi_s, p_s]u_s,$$

with $s \leqslant p$, and with $(p_1, \ldots, p_s)$ a permutation of $(p-s+1, \ldots, p)$. There is no loss of generality in assuming $p_1 = p-s+1, \ldots, p_s = p$. By Lemma II.1.1,

$$w = u_0 h_1 u_1 h_2 \cdots u_{s-1} h_s u_s,$$

with

$$[\xi_j, p_j] \xrightarrow[H]{*} h_j \qquad j=1, \ldots, s.$$

By induction, $h_j \in L_G^{(p_j)}(\xi_j)$ for $j=1, \ldots, s$, and therefore there are derivations

$$\xi_j \xrightarrow[G]{*} h_j \tag{5.6}$$

with index $\leqslant p_j = p-s+j$ for $j=1, \ldots, s$. Consider now the derivation

$$\begin{aligned} \xi &\underset{G}{\rightarrow} u_0\xi_1u_1\xi_2 \cdots u_{s-1}\xi_s u_s \\ &\underset{G}{\overset{*}{\rightarrow}} u_0h_1u_1\xi_2 \cdots u_{s-1}\xi_s u_s \\ &\underset{G}{\overset{*}{\rightarrow}} u_0h_1u_1h_2 \cdots u_{s-1}\xi_s u_s \\ &\vdots \\ &\underset{G}{\overset{*}{\rightarrow}} u_0h_1u_1h_2 \cdots u_{s-1}h_s u_s = w, \end{aligned} \tag{5.7}$$

obtained by concatenating the derivations (5.6) in increasing order. The $j$-th derivation (5.6) contributes in (5.7) to words containing at most $p_j + (s-j) = p$ nonterminals. Thus (5.7) has index at most $p$. This shows that $w$ is in $L_G^{(p)}(\xi)$.

(iii) It remains to prove that $L_H([\xi, p]) \in \mathrm{Qrt}(p)$. Observe first that for each production $[\xi, p] \rightarrow \alpha \in Q$, and for each nonterminal $[\eta, q]$ appearing in $\alpha$, we have $q \leqslant p$. Thus the grammar

$$H_p = \langle W_p, X, Q_p \rangle$$

with

$$W_p = V \times \{1, \ldots, p\}, \qquad Q_p = \{[\xi, q] \rightarrow \alpha \in Q \mid q \leqslant p\}$$

is well-defined, and further

$$L_{H_p}([\xi, q]) = L_H([\xi, q]) \quad \text{for } q \leqslant p.$$

For $p=1$, the grammar $L_{H_1}$ is linear by construction. Thus $L_H([\xi, 1]) \in \mathrm{Lin}$. Assume $p>1$, and consider the grammar

$$\bar{H}_p = \langle W_p \backslash W_{p-1}, X \cup W_{p-1}, Q_p \backslash Q_{p-1} \rangle.$$

By the previous remark, $\bar{H}_p$ is well-defined and by the construction of $H$, $\bar{H}_p$ is linear. Further, it is easily seen that

$$L_H([\xi, p]) = \theta(L_{\bar{H}_p}([\xi, p])),$$

where $\theta : (W_{p-1} \cup X)^* \rightarrow X^*$ is the substitution defined by

$$\theta(x) = x \qquad (x \in X);$$

$$\theta([\xi, q]) = L_H([\xi, q]) \qquad q < p.$$

Arguing by induction, $\theta$ is a Qrt $(p-1)$-substitution, and $L_H([\xi, p]) \in \mathrm{Qrt}(p)$. ∎

Proof of Theorem 5.2 The inclusion $\mathrm{Qrt}(k) \subset \mathrm{Ind}(k)$ holds by Lemma 5.3. Conversely, let $L \in \mathrm{Ind}(k)$. Then $L = L_G(\xi) = L_G^{(k)}(\xi)$ for some grammar $G$. By Proposition 5.5, $L_G^{(k)}(\xi) \in \mathrm{Qrt}(k)$. Thus $L \in \mathrm{Qrt}(k)$, and $\mathrm{Qrt}(k) = \mathrm{Ind}(k)$. ∎

There is another characterization of quasi-rational languages using the bracket operation defined in Section V.6, due to Nivat [1967] and Yntema [1967]. See the following exercises.

## Exercises

**5.1** Show that for a rational relation $R$, and for a quasi-rational language $L$ of rank $k$, the language $[R, L]$ is quasi-rational of the rank $k$. (The bracket $[R, L]$ was defined in Section V.6.)

**5.2** (Yntema [1967], Nivat [1967]) The set of quasi-rational relations of rank $k$ over $X$ and $Y$ is the least family of subsets of $X^* \times Y^*$ containing the relations $L \times L'$, $(L \subset X^*, L' \in Y^*)$ with $L, L' \in \mathrm{Qrt}(k)$ and closed under union, product and the star operation.
a) Show that for any quasi-rational relation $R$ of rank $k$, and for any quasi-rational language $L$ of rank $k$, $[R, L]$ is in $\mathrm{Qrt}(k+1)$.
b) Show that conversely any language $A \in \mathrm{Qrt}(k+1)$ can be written as $A = [R, 1]$, where $R$ is a quasi-rational relation of rank $k$. Give such a relation of rank 1 for the language of Example 5.1.

# VIII Iterative Pairs

In Chapter V it was shown that languages like $S_< = \{x^n y^m \mid 0 \leq n < m\}$ and $S_> = \{x^n y^m \mid 0 \leq m < n\}$ are rationally incomparable. In the previous chapter, the Syntactic Lemma gave infinite chains of rational cones which appear to be of increasing "complexity". Thus rational transductions classify languages according to some measure of complexity. In the present chapter, we give a description of this complexity measure in so far as context-free languages are concerned.

Since any algebraic language is a rational image of $D_2'^*$, it is obtained by adjoining additional restrictions to that language. The resulting language is then subjected to two types of constraints, some of "context-free" nature, the other of "regular" nature. The algebraic constraints are expressed by counting restrictions or by restriction on parenthesization. The rational constraints concern repetition and transcription. Rational transductions should ignore rational constraints, and detect modification of algebraic constraints.

The concepts introduced by Boasson [1976] and Beauquier [1978a] to describe the algebraic constraints are the notions of iterative pair and of system of iterative pairs. The central theorem of this chapter (Theorem 6.1) can be rephrased informally as follows: if $A$ and $B$ are algebraic languages and if $A \geq B$, then $A$ has systems of iterative pairs of greater complexity than those of $B$.

The systems of iterative pairs in a language $L$ are defined by a combinatorial property of $L$. But if $L$ is context-free, then these systems are closely related to some nesting of the self-embedding variables of any algebraic grammar generating $L$. Thus systems of iterative pairs give, through a property of the words of a language, a description of a "structural" feature common to all grammars for this language. In view of this relationship, the types of systems of iterative pairs can really be considered as a good measure of complexity. By the way, these facts constitute a justification a posteriori of the use of rational transductions as a tool for comparison of languages.

The first three sections of this chapter are concerned with iterative pairs and the Transfer Theorem for these pairs. As an illustration, we easily obtain most of the results of Section V.7, and we show a new infinite chain of rational cones. Sections 4–7 contain the generalization of the results to systems of iterative pairs and applications. The material for this chapter is from Boasson [1976] and Beauquier [1978a]. The second author gives, in Beauquier

[1977] another generalization, not included here, of iterative pairs, which he calls "multipairs".

## VIII.1 Types of Iterative Pairs

This section contains the definition and examples of iterative pairs. A classification of iterative pairs is given and some simple lemmas are proved.

**Definition** Let $X$ be an alphabet, and let $L \subset X^*$ be a language. A sequence

$$\pi = (a, u, b, v, c)$$

of words of $X^*$ is an iterative pair in $L$ if

(i) $au^n bv^n c \in L$ for $n \geqslant 0$;
(ii) $|uv| > 0$.

The words $u$ and $v$ are the iterating elements of $\pi$.

In view of (i) the word $w = aubvc$ belongs to $L$. Then $\pi$ is called an interative pair of $w$ in $L$, and $w$ is said to admit the iterative pair $\pi$ in $L$. We are interested in the language $L \cap au^*bv^*c$ or equivalently in the set of integers $k, l$ such that $au^k bv^l c \in L$. Therefore we define the set of exponents of $\pi$ in $L$ by

$$\mathrm{Exp}(\pi, L) = \{(k, l) \in \mathbb{N}^2 : au^k bv^l c \in L\}.$$

This set describes the local constraint of $L$ on the factorization $\pi$ of $w$. By definition, $\mathrm{Exp}(\pi, L)$ contains the diagonal $\Delta = \{(n, n) : n \in \mathbb{N}\}$. If $\pi$ is an iterative pair in $L$, and if $L \subset L'$, then clearly $\pi$ is also an iterative pair in $L'$ and $\mathrm{Exp}(\pi, L) \subset \mathrm{Exp}(\pi, L')$. Note that

$$\mathrm{Exp}(\pi, L) = \mathrm{Exp}(\pi, L \cap au^*bv^*c).$$

**Example 1.1** Let $L \subset X^*$ be context-free. The iteration lemmas assert that any long enough word $w$ in $L$ admits an iterative pair in $L$.

**Example 1.2** Let $X = \{x, y\}$ and $S_{\leqslant} = \{x^n y^m \mid 0 \leqslant n \leqslant m\}$. Then $\pi = (1, x, 1, y, 1)$ is an iterative pair of $xy$ in $S_{\leqslant}$, and $\mathrm{Exp}(\pi, S_{\leqslant}) = \{(n, m) \mid 0 \leqslant n \leqslant m\}$.

**Example 1.3** Let $L = \{x^n y^m \mid 0 \leqslant n \leqslant m \leqslant n+3\}$. The word $xy$ admits the iterative pair $\pi = (1, x, 1, y, 1)$ in $L$. The word $x^4y^4$ admits the iterative pair $\pi' = (1, x^4, 1, x^4, 1)$ in $L$, and $\mathrm{Exp}(\pi', L) = \Delta$.

**Example 1.4** Consider the language

$$L = \{x^k y^p \bar{y}^q z^n \bar{z}^m \bar{x}^k \mid 0 \leqslant k, 0 \leqslant p \leqslant q, 0 \leqslant m \leqslant n\}.$$

The word $w = xy\bar{y}z\bar{z}\bar{x}$ admits at least the following iterative pairs in $L$:

$$\pi_1 = (1, x, y\bar{y}z\bar{z}, \bar{x}, 1); \quad \pi_2 = (x, y, 1, \bar{y}, z\bar{z}\bar{x}); \quad \pi_3(xy\bar{y}, z, 1, \bar{z}, \bar{x});$$
$$\pi_4 = (xy, \bar{y}, 1, 1, z\bar{z}\bar{x}); \quad \pi_5 = (xy, 1, 1, \bar{y}, z\bar{z}\bar{x}); \quad \ldots$$

**Example 1.5** The language $\{x^n y^n z^n \mid n \geq 0\}$ has no iterative pairs.

We now introduce a classification of iterative pairs according to the nature of the set of exponents. Let

$$\pi = (a, u, b, v, c)$$

be an iterative pair in $L$, and define for $k, l \geq 0$:

$$G(l) = \{k : au^k bv^l c \in L\}; \qquad D(k) = \{l : au^k bv^l c \in L\}.$$

**Definition** The iterative pair $\pi$ in $L$ is called:
left strict if $G(l)$ is finite for any $l \geq 0$;
right strict if $D(k)$ is finite for any $k \geq 0$;
strict if it is both left strict and right strict;
very strict if $\mathrm{Exp}(\pi, L) = \Delta$;
nondegenerated if it is left strict or right strict.

Thus the following implications hold:

$$\text{very strict} \Rightarrow \text{strict} \begin{matrix} \nearrow \text{left strict} \searrow \\ \searrow \text{right strict} \nearrow \end{matrix} \text{nondegenerated.}$$

This terminology does not fit exactly with the terminology of Boasson and Beauquier. For Boasson [1976], a very strict pair satisfies a somewhat weaker condition, and Beauquier [1978a] calls strict the pairs we call very strict.

Note that if $\pi = (a, u, b, v, c)$ is nondegenerated in $L$, then both $u \neq 1$ and $v \neq 1$. Indeed, assume for instance that $u = 1$. Then clearly $au^* bv^l c \in L$ for any $l \geq 0$ and thus $\pi$ is not left strict. Further $aubv^l c = au^l bv^l c \in L$ for $l \geq 0$ and consequently $\pi$ is not right strict.

We now define the type $\chi(\pi, L)$ of a nondegenerated iterative pair $\pi$ in $L$ as a word of length 2 over a fixed alphabet $\Xi = \{e, \bar{e}, s, \bar{s}, l, \bar{l}, r, \bar{r}\}$:

$$\chi(\pi, L) = \begin{cases} e\bar{e} & \text{if } \pi \text{ is very strict ("}e\text{" for equal);} \\ s\bar{s} & \text{if } \pi \text{ is strict, but not very strict;} \\ l\bar{l} & \text{if } \pi \text{ is left strict, but not right strict;} \\ r\bar{r} & \text{if } \pi \text{ is right strict, but not left strict.} \end{cases}$$

Types are ordered by $e\bar{e} > s\bar{s} > l\bar{l}$, $s\bar{s} > r\bar{r}$.

This rather lengthy definition will be useful in later sections where we consider systems of iterative pairs.

**Example 1.2** (continued). The pair $\pi$ is left strict, but not right strict in $S_{\leqslant}$. Thus $\chi(\pi, S_{\leqslant}) = l\bar{l}$.

**Example 1.3** (continued). The pair $\pi$ is strict, but not very strict. The pair $\pi'$ is very strict.

**Example 1.4** (continued). $\chi(\pi_1) = e\bar{e}$, $\chi(\pi_2) = l\bar{l}$, $\chi(\pi_3) = r\bar{r}$.
The pairs $\pi_4$ and $\pi_5$ are degenerated. The iterative pairs $\pi_1, \pi_2, \pi_3$ can be considered to form a "system". The type of this "system" will be defined as $el\bar{l}r\bar{r}\bar{e}$. This is the reason for our notation for types.

**Example 1.6** Any iterative pair in a regular language is degenerated.

**Example 1.7** The equation $S_2 = xS_2\bar{x} \cup yS_2\bar{y} \cup 1$ defines the symmetric language $S_2$ over the alphabet $Z = X \cup \bar{X}$, with $X = \{x, y\}$, $\bar{X} = \{\bar{x}, \bar{y}\}$. We claim that any iterative pair in $S_2$ is very strict. Let indeed

$$\pi = (a, u, b, v, c)$$

be an iterative pair in $S_2$. Since $au^nbv^nc \in S_2$ for any $n \geqslant 0$, and $S_2 \subset X^*\bar{X}^*$, it follows that $u, v \in X^* \cup \bar{X}^*$. Next $f \in S_2$ implies $|f|_X = |f|_{\bar{X}}$. Thus if $u \in \bar{X}^*$ then $u = v = 1$ and $\pi$ is not an iterative pair. Consequently $u \in X^*$, and similarly $v \in \bar{X}^*$. Therefore $|u| = |v| > 0$ and

$$\text{Exp}(\pi, S_2) = \Delta.$$

**Example 1.8** Any iterative pair in the restricted Dyck language $D_1'^* \subset \{x, \bar{x}\}^*$ is either degenerated or very strict. Define indeed $\|f\| = |f|_x - |f|_{\bar{x}}$ for $f \in \{x, \bar{x}\}^*$ and consider an iterative pair

$$\pi = (a, u, b, v, c)$$

in $D_1'^*$. In view of the characterization of $D_1'^*$ given by Corollary II.4.2, we have $0 = \|aubvc\| = \|abc\| + \|u\| + \|v\| = \|u\| + \|v\|$, whence $\|u\| = -\|v\|$. Next $\|u\| \geqslant 0$, since otherwise $\|au^n\| = \|a\| + n\|u\| < 0$ for sufficiently large $n$, contrary to the fact that $au^nbv^nc \in D_1'^*$. If $\|u\| = 0 = \|v\|$, then $0 \leqslant \|au'\| = \|au^nu'\|$ for any left factor $u'$ of $u$, and $0 \leqslant \|aubv'\| = \|au^nbv^mv'\|$ for any left factor $v'$ of $v$. Thus $au^*bv^*c \subset D_1'^*$ and $\pi$ is degenerated. Otherwise $\|u\| > 0$ and $\|v\| < 0$ and $\|au^nbv^mc\| = (n-m)\|u\|$. Thus $au^nbv^mc \in D_1'^*$ iff $n = m$. Consequently $\pi$ is very strict.

We conclude this section by some simple lemmas concerning the transfer of iterative pairs.

**Lemma 1.1** *Let $\pi$ be an iterative pair in $L \cap K$, where $L$ is a language and $K$ is a local regular language. Then $\pi$ is an iterative pair in $L$, and* $\text{Exp}(\pi, L) = \text{Exp}(\pi, L \cap K)$.

Proof. Let $\pi=(a, u, b, v, c)$. Since $L\cap K\subset L$, $\mathrm{Exp}(\pi, L\cap K)\subset \mathrm{Exp}(\pi, L)$. Next $au^nbv^nc\in K$ for all $n\geq 0$. Since $K$ is local, this implies $au^*bv^*c\subset K$ (see Exercise I.4.1). Thus $L\cap K\supset L\cap au^*bv^*c$ and

$$\mathrm{Exp}(\pi, L)=\mathrm{Exp}(\pi, L\cap au^*bv^*c)\subset \mathrm{Exp}(\pi, L\cap K). \qquad \blacksquare$$

Let $\varphi: X^*\to Y^*$ be a morphism. For a sequence

$$\pi=(a, u, b, v, c)$$

of words of $X^*$, define

$$\varphi\pi=(\varphi a, \varphi u, \varphi b, \varphi v, \varphi c).$$

**Lemma 1.2** *Let $M\subset Y^*$, $L=\varphi^{-1}(M)$. If $\pi$ is a nondegenerated iterative pair in $L$, then $\varphi\pi$ is an iterative pair in $M$ and* $\mathrm{Exp}(\pi, L)=\mathrm{Exp}(\varphi\pi, M)$.

Proof. Let $\pi=(a, u, b, v, c)$. If $au^kbv^lc\in L$, then $\varphi(au^kbv^lc)\in M$. Thus $\mathrm{Exp}(\pi, L)\subset \mathrm{Exp}(\varphi\pi, M)$. Conversely, if $\varphi(au^kbv^lc)\in M$, then $au^kbv^lc\in \varphi^{-1}(M)=L$. Consequently $\mathrm{Exp}(\pi, L)=\mathrm{Exp}(\varphi\pi, M)$. It remains to show that $|\varphi u\varphi v|>0$. Suppose the contrary. Then $\mathrm{Exp}(\varphi\pi, M)=\mathbb{N}^2$ and $\pi$ is degenerated, contrary to the assumption. $\blacksquare$

There is a lack of a lemma of the following kind: If $L\subset X^*$ and $M=\varphi(L)$, and if $\pi'$ is an iterative pair in $M$, then there is an iterative pair $\pi$ in $L$ such that $\varphi\pi=\pi'$. Such a lemma does not exist.

**Example 1.7** Let $L=\{x^ny^nz^n : n\geq 0\}$ and let $\varphi:\{x, y, z\}^*\to\{x, y\}^*$ be the projection. Then $M=\varphi(L)=\{x^ny^n : n\geq 0\}$ has iterative pairs but $L$ has no iterative pairs.

The next section will be concerned with a strengthened notion of iterative pairs for the case of algebraic languages.

## Exercises

**1.1** Let $\pi=(a, u, b, v, c)$ be an iterative pair in $L$. Show that $\pi^N=(a, u^N, b, v^N, c)$ is an iterative pair in $L$ for any $N\geq 1$. Show that $\chi(\pi^N, L)\geq\chi(\pi, L)$, and give an example where this inequality is strict.

**1.2** Let $\pi=(a, u, b, v, c)$ be a nondegenerated iterative pair in $L\subset X^*$. Show that the mapping $(k, l)\mapsto au^kbv^lc$ from $\mathbb{N}^2$ into $X^*$ is injective.(Hint. Use Exercise I.2.6.)

**1.3** (Beauquier [1977]) An iterative pair $\pi=(a, u, b, v, c)$ is bounded in $L$ iff there is an integer $K$ such that $G(l)$ and $D(k)$ have at most $K$ elements for any $l, k\geq 0$. Show that if $L$ is context-free and $\pi$ is bounded in $L$, then there exists an iterative pair $\pi'$ which is very strict in $L$. (Hint. Use the fact, proved for instance in Ginsburg [1966], that $\mathrm{Exp}(\pi, L)$ is a rational subset of $\mathbb{N}^2$ for context-free $L$.)

**1.4** Let $\pi=(a, u, b, v, c)$ be an iterative pair in $L$. Show that if $L\cap au^*bv^*c$ is a regular language, then $\pi$ is degenerated (the converse is false).

## VIII.2 Grammatical Pairs

The iteration lemma for context-free languages shows the existence of iterative pairs for any infinite context-free language $L$. The iterative pairs obtained in that way depend on the grammar $G=\langle V, X, P\rangle$ generating $L$. More precisely, if $L=L_G(\xi_0)$, then the iteration lemma gives an iterative pair $\pi=(a, u, b, v, c)$ together with a variable $\xi \in V$ such that

$$\xi_0 \xrightarrow{*} a\xi c, \qquad \xi \xrightarrow{*} u\xi v, \qquad \xi \xrightarrow{*} b.$$

It follows that two different grammars for a same language can give two distinct iterative pairs of the same word. This raises the question how an iterative pair $\pi$ of a word $w$ in $L$ is linked to the iterative pairs of $w$ which are supplied by the grammars generating $L$. A rather precise answer to this question will be given in this section. It shows how a "combinatorial" property of the language $L$, described by the iterative pair, is related to a common "generative" property of all grammars generating $L$. We first give a definition.

**Definition** Let $L \subset X^*$ be an algebraic language, let $G=\langle V, X, \boldsymbol{P}\rangle$ be an algebraic grammar such that $L=L_G(\xi_0)$ for $\xi_0 \in V$. An iterative pair $\pi=(a, u, b, v, c)$ in $L$ is grammatical with respect to $G$ if there exists a variable $\xi \in V$ such that

$$\xi_0 \xrightarrow{*} a\xi c, \qquad \xi \xrightarrow{*} u\xi v, \qquad \xi \xrightarrow{*} b.$$

The variable $\xi$ is called the pivot of $\pi$ in $G$.

**Example 2.1** Let $X=\{x, y, z\}$, and let

$$L=\{x^{3n}zy^{3n} \mid n \geqslant 0\}.$$

The word $w=x^6zy^6$ admits the three iterative pairs in $L$:

$$\pi_1=(1, x^6, z, y^6, 1), \qquad \pi_2=(x^2, x^3, xzy, y^3, y^2),$$
$$\pi_3=(x, x^3, x^2zy, y^3, y^2).$$

Next consider the grammar $G$ with productions

$$\xi_0 \to x^3\xi y^3+\xi; \qquad \xi \to x^6\xi y^6+z.$$

Then $\pi_1$ is grammatical with respect to $G$, and neither $\pi_2$ nor $\pi_3$ are grammatical in $G$.

The example shows however that $\pi_2$ and $\pi_3$ can be transformed into grammatical pairs first by taking a power of the iterating elements and then by "shifting" these new iterating elements. We shall see that this procedure is valid in the general case. First we prove some lemmas concerning transfer of grammatical pairs.

**Lemma 2.1** *Let $L \subset X^*$ be generated by an algebraic grammar $G$, and let $K \subset X^*$ be a rational language. If $\pi$ is an iterative pair in $L \cap K$ grammatical with respect to $G_K$, then $\pi$ is grammatical with respect to $G$.*

Recall that $G_K$ is the grammar generating $L \cap K$ defined in Section II.2.

Proof. Let $\pi = (a, u, b, v, c)$. Assume that $G$ and $G_K$ are as in Section II.2. Then there exist derivations in $G_K$:

$$\hat{\sigma} \xrightarrow{*} a(q, \xi, q')c; \qquad (q, \xi, q') \xrightarrow{*} u(q, \xi, q')v; \qquad (q, \xi, q') \xrightarrow{*} b.$$

The first derivation factorizes into

$$\hat{\sigma} \to (q_-, \sigma, q_+) \xrightarrow{*} a(q, \xi, q')c$$

for some $q_+$. Consequently there exist, in the grammar $G$, derivations

$$\sigma \xrightarrow{*} a\xi c; \qquad \xi \xrightarrow{*} u\xi v; \qquad \xi \xrightarrow{*} b,$$

showing that $\pi$ is grammatical with respect to $G$. ■

**Lemma 2.2** *Let $\varphi : X^* \to Y^*$ be an alphabetic morphism, let $M \subset Y^*$ be generated by some algebraic grammar $G$, and let $L = \varphi^{-1}(M)$. If $\pi$ is a nondegenerated iterative pair in $L$ grammatical with respect to $\varphi^{-1}(G)$, then $\varphi\pi$ is grammatical with respect to $G$.*

Proof. Let $\pi = (a, u, b, v, c)$. In view of Lemma 1.2, $\varphi\pi$ is an iterative pair in $M$. With the notations of Section II.2, there are in the grammar $\varphi^{-1}G$ derivations

$$\sigma \xrightarrow{*} a\eta c, \qquad \eta \xrightarrow{*} u\eta v, \qquad \eta \xrightarrow{*} b.$$

In order to show that there are derivations

$$\sigma \xrightarrow{*} \varphi(a)\eta\varphi(c), \qquad \eta \xrightarrow{*} \varphi(u)\eta\varphi(v), \qquad \eta \xrightarrow{*} \varphi(b)$$

in $G$, it suffices to show that the nonterminal $\eta$ is distinct from $\omega$. Now if $\eta = \omega$, then $\varphi u = \varphi v = 1$, and $\varphi\pi$ is not an iterative pair. This achieves the proof. ■

The following lemma is important and will be used in the next section.

**Lemma 2.3** *Let $\psi : X^* \to Y^*$ be a morphism, let $L \subset X^*$ be generated by an algebraic grammar $G$, and let $M = \psi(L)$. If $\pi$ is an iterative pair in $M$ grammatical with respect to $\psi G$, then there exists an iterative pair $\pi'$ in $L$, grammatical with respect to $G$, such that $\psi\pi' = \pi$. Further* $\mathrm{Exp}(\pi', L) \subset \mathrm{Exp}(\pi, M)$.

This lemma shows that iterative pairs can be transferred through a morphism, provided they are grammatical with respect to some specific grammar.

Proof. Let $\pi=(a, u, b, v, c)$. By assumption there are in the grammar $\psi G$ derivations

$$\sigma \xrightarrow{*} a\xi c, \qquad \xi \xrightarrow{*} u\xi v, \qquad \xi \xrightarrow{*} b.$$

Consequently there are words $a', u', b', v', c'$ in $X^*$ such that $\psi a'=a$, $\psi u'=u$, $\psi b'=b$, $\psi v'=v$, $\psi c'=c$, and

$$\sigma \xrightarrow{*} a'\xi c', \qquad \xi \xrightarrow{*} u'\xi v', \qquad \xi \to b'$$

in the grammar $G$. Thus $\pi'=(a', u', b', v', c')$ is an iterative pair in $L$ grammatical with respect to $G$. The inclusion $\text{Exp}(\pi', L) \subset \text{Exp}(\pi, M)$ is obvious. ■

We now rely nondegenerated pairs to grammatical pairs. First we need a definition.

**Definition** Let $\pi=(a, u, b, v, c)$ and $\bar{\pi}=(\bar{a}, \bar{u}, \bar{b}, \bar{v}, \bar{c})$ be two sequences of words. Then $\bar{\pi}$ is deduced from $\pi$ if the two following conditions hold:

(i) there is an integer $N \geq 1$ such that

$$\bar{a}\bar{u}\bar{b}\bar{v}\bar{c}=au^Nbv^Nc;$$

(ii) there are integers $p, q \geq 0$, and factorizations $u=u_1u_2$, $v=v_1v_2$ such that

$$\bar{a} \in au^*u_1, \quad \bar{u}=(u_2u_1)^p, \quad \bar{b} \in u_2u^*bv^*v_1, \quad \bar{v}=(v_2v_1)^q, \quad \bar{c} \in v_2v^*c.$$

Condition (ii) implies that

$$\bar{a}\bar{u}^{k+1}\bar{b}\bar{v}^{l+1}\bar{c}=au^{N+pk}bv^{N+ql}c \quad \text{for } k+1, l+1 \geq 0.$$

Assume next that $\pi$ and $\bar{\pi}$ are iterative pairs in some language $L$. Then

$$(k, l) \in \text{Exp}(\bar{\pi}, L) \quad \Leftrightarrow \quad (N+p(k-1), N+q(l-1)) \in \text{Exp}(\pi, L). \qquad (2.1)$$

This implies immediately

**Lemma 2.4** *Let $\pi$ and $\bar{\pi}$ be iterative pairs in L, and assume that $\bar{\pi}$ is deduced from $\pi$. If $\pi$ is nondegenerated, then $\bar{\pi}$ is nondegenerated and $\chi(\bar{\pi}, L) \geq \chi(\pi, L)$.* ■

**Example 2.1** (continued). The pair $\bar{\pi}=(x^3, x^6, x^3zy^3, y^6, y^3)$ is deduced from $\pi_2$, and is deduced from $\pi_3$. It is grammatical with respect to $G$.

**Theorem 2.5** *Let $L \subset X^*$ be an algebraic language, and let $\pi$ be a nondegenerated iterative pair in L. For any algebraic grammar generating L there exists an iterative pair $\bar{\pi}$ deduced from $\pi$ and grammatical with respect to G.*

Proof. We first prove the theorem for iterative pairs of a special kind and derive then the general case from this special case.

Assume $X = Z \cup \{t, x, y\}$, with $t, x, y \notin Z$, and suppose that $\pi$ has the form

$$\pi = (at, xu, tbt, yv, tc)$$

with $a, u, b, v, c \in Z^*$. Define

$$L' = L \cap K, \qquad K = at(xu)^* tbt(yv)^* tc.$$

Recall that $\mathrm{Exp}(\pi, L) = \mathrm{Exp}(\pi, L')$. Next let $H = G_K$ be the grammar generating $L'$. By Lemma 2.1, it suffices to prove the existence of an iterative pair grammatical with respect to $H$. Let $N$ be the integer associated to $H$ by Ogden's Iteration Lemma II.2.5, and define

$$g = at(xu)^N tbt(yv)^N tc.$$

Since $\pi$ is nondegenerated, it is left strict or right strict. Suppose first $\pi$ left strict and mark the $N$ occurrences of the letter $x$ in $g$. In view of Lemma II.2.5, $g$ admits in $L'$ an iterative pair

$$\bar{\pi} = (\bar{a}, \bar{u}, \bar{b}, \bar{v}, \bar{c})$$

which is grammatical with respect to $H$, and $\bar{a}$, $\bar{u}$, $\bar{b}$ or $\bar{b}$, $\bar{v}$, $\bar{c}$ contain occurrences of $x$. We have to prove that $\bar{\pi}$ is deduced from $\pi$. First note that neither $\bar{u}$ nor $\bar{v}$ contain a letter $t$ since otherwise $\bar{a}\bar{u}^2\bar{b}\bar{v}^2\bar{c} \notin K$. Next neither $\bar{u}$ nor $\bar{v}$ can be nonempty factors of $a$ or $b$ or $c$. We prove that $\bar{v}$ contains no occurrence of $x$. Assume indeed $|\bar{v}|_x > 0$. Then $\bar{v}$ is a segment of $(xu)^N$. This implies that $\bar{a}\bar{u}^k\bar{b}\bar{v}^k \in at(xu)^*$ for $k \geqslant 0$. Hence the number of $x$'s in the words $\bar{a}\bar{u}^k\bar{b}\bar{v}^k\bar{c}$ increases with $k$, and the number of $y$'s remains constant equal to $N$. This is in contradiction with the fact that $\pi$ is left strict in $L$, hence in $L'$. Thus $|\bar{v}|_x = 0$.

Since $\bar{v}$ contains no occurences of the letter $x$, $\bar{u}$ contains occurrences of $x$, and $\bar{u}$ is a segment of $(xu)^N$ and similarly $\bar{v}$ is a segment of $(yv)^N$. Thus there are integers $p, q \geqslant 0$ such that

$$\bar{u} = u_2(xu)^p xu_1, \qquad \bar{v} = v_2(yv)^q yv_1$$

for words $u_1, u_2, v_1, v_2 \in Z^*$. Since $\bar{a}\bar{u}^2\bar{b}\bar{v}^2\bar{c} \in K$, this implies that $u = u_1u_2$, $v = v_1v_2$. It follows immediately that $\bar{\pi}$ is deduced from $\pi$. This proves the special case for left strict pairs. The case of right strict pairs is proved in the same manner by marking the occurrences of $y$ in $g$.

Consider now the general case of a nondegenerated iterative pair $\pi = (a, u, b, v, c)$ in $L \subset X^*$. Let $t, x, y$ be new letters, set $Y = X \cup \{t, x, y\}$, and let $\varphi : Y^* \to X^*$ be the projection. Let $L' = \varphi^{-1}(L)$ and $K = at(xu)^* tbt(yv)^* tc$,

$$\pi' = (at, xu, tbt, yv, tc).$$

Then $\varphi\pi' = \pi$, and obviously $\mathrm{Exp}(\pi', L') = \mathrm{Exp}(\pi, L)$. By the first part of the proof, there exists a pair $\hat{\pi}$ deduced from $\pi'$ which is grammatical with respect to the grammar $\varphi^{-1}(G)$. Then $\bar{\pi} = \varphi\hat{\pi}$ is deduced from $\pi = \varphi\pi'$, and is grammatical with respect to the grammar $G$ in view of Lemma 2.2. ∎

## VIII.3. Transfer of Iterative Pairs

We are now ready for the proof of the main theorem concerning transfer of iterative pairs through rational transductions.

**Theorem 3.1** (Transfer Theorem) *Let A and B be algebraic languages such that $A \geqslant B$. For any nondegenerated iterative pair $\pi$ in B, there is a nondegenerated iterative pair $\pi'$ in A such that $\chi(\pi', A) \geqslant \chi(\pi, B)$.*

As already noted, this theorem is false for nonalgebraic languages. The result gives a necessary condition for rational domination of algebraic languages. But the Transfer Theorem can also be regarded as a result on rational transductions: Rational transductions only decrease and never increase the types of iteration pairs occurring in algebraic languages. Thus rational transductions decrease the "complexity" of context-free languages in so far as the type of the iterative pairs of a language is considered as a measure of its complexity.

Let $A \subset X^*$, $B \subset Y^*$, and assume $A \geqslant B$. Then by Nivat's Theorem, there is an alphabet $Z$ such that

$$B = \psi(\varphi^{-1}(A) \cap K)$$

where $\psi: Z^* \to Y^*$ is a morphism, $\varphi: Z^* \to X^*$ can be chosen to be an alphabetic morphism and $K \subset Z^*$ is a regular language which we may suppose to be local. We first prove

**Proposition 3.2** *Let G be an algebraic grammar generating A. For any nondegenerated iterative pair $\pi = (a, u, b, v, c)$ in B grammatical with respect to the grammar $\psi[\varphi^{-1}(G)_K]$, there is an iterative pair $\pi' = (a', u', b', v', c')$ in A grammatical with respect to G such that* $\mathrm{Exp}(\pi', A) \subset \mathrm{Exp}(\pi, B)$. *Moreover*

$$au^k bv^l c \in \psi[\varphi^{-1}(a'u'^k b'v'^l c') \cap K] \qquad \text{for } k, l \geqslant 0.$$

Proof. Set $C = \varphi^{-1}(A) \cap K$. In view of Lemma 2.3, there exists an iterative pair $\pi_1 = (a_1, u_1, b_1, v_1, c_1)$ in $C$, grammatical with respect to $(\varphi^{-1}G)_K$, such that $\psi\pi_1 = \pi$. Further $\mathrm{Exp}(\pi_1, C) \subset \mathrm{Exp}(\pi, B)$. Since $K$ is local $\mathrm{Exp}(\pi_1, C) = \mathrm{Exp}(\pi_1, \varphi^{-1}(A))$ by Lemma 1.1. Next $\pi_1$ is grammatical with respect to $\varphi^{-1}G$ by Lemma 2.1. Define $\pi' = \varphi\pi_1$. Since $\pi_1$ is nondegenerated, $\pi'$ is an iterative pair in $A$, and $\mathrm{Exp}(\pi_1, \varphi^{-1}(A)) = \mathrm{Exp}(\pi', A)$ by Lemma 1.2. Finally $\pi'$ is grammatical with respect to $G$ in view of Lemma 2.2. Further $\mathrm{Exp}(\pi', A) \subset \mathrm{Exp}(\pi, B)$. Now consider a word $w = a'u'^k b'v'^l c'$. Then $w_1 = a_1 u_1^k b_1 v_1^l c_1 \in \varphi^{-1}(w)$ and since $K \supset a_1 u_1^* b_1 v_1^* c_1$, $w_1 \in \varphi^{-1}(w) \cap K$. Consequently $\psi w_1 = au^k bv^l c \in \psi(\varphi^{-1}(w) \cap K)$. ■

Proof of Theorem 3.1. Let $\varphi, \psi, K, Z$ be as above, and let $G$ be a grammar generating $A$. Let $\pi$ be a nondegenerated iterative pair in $B$. In view of Theorem 2.5, there exists an iterative pair $\bar{\pi}$ deduced from $\pi$, and grammatical with respect to the grammar $\psi[\varphi^{-1}(G)_K]$. Further $\chi(\bar{\pi}, B) \geqslant \chi(\pi, B)$ by Lemma 2.4. In view of the Proposition 3.2 above, there is an iterative pair $\pi'$ in $A$ such that $\mathrm{Exp}(\pi', A) \subset \mathrm{Exp}(\bar{\pi}, B)$. Consequently $\chi(\pi', A) \geqslant \chi(\bar{\pi}, B) \geqslant \chi(\pi, B)$. ■

The Transfer Theorem can be strengthened in the following way.

**Corollary 3.3.** *Let $A$ and $B$ be algebraic languages. If $A \geqslant B$, then for any nondegenerated iterative pair $\pi$ in $B$, there are a nondegenerated iterative pair $\pi'$ in $A$, and integers $N, p, q \geqslant 1$ such that*

$$(k, l) \in \mathrm{Exp}(\pi', A) \Rightarrow (N + p(k-1), N + q(l-1)) \in \mathrm{Exp}(\pi, B).$$

Proof. With the notations of the proof of Theorem 3.1, the grammatical iterative pair $\bar{\pi}$ deduced from $\pi$ satisfies formula (2.1). The corollary follows then immediately from Proposition 3.2. ■

As an illustration, we show that some of the results of Section V.7 are simple applications of the Transfer Theorem.

**Example 3.1** The languages $S_>$ and $S_<$ are incomparable. Indeed, the language $S_>$ has right strict iterative pairs and $S_<$ has no right strict iterative pair. Thus $S_<$ does not dominate $S_>$. The symmetric result is obtained by considering left strict pairs in $S_<$.

**Example 3.2** $S_{\neq}$ dominates none of the languages $S_=, S_>, S_<$. Indeed, the languages $S_=, S_>, S_<$ all have nondegenerated iterative pairs. On the contrary, all iterative pairs in $S_{\neq}$ are degenerated.
Note that the Transfer Theorem cannot be used to prove that $S_>$ and $S_<$ do not dominate $S_{\neq}$ (Proposition V.7.2), since $S_{\neq}$ only has degenerated pairs.

**Example 3.3** Any generator of the cone Alg or of Lin or Rocl has at least one very strict iterative pair. Indeed, any generator of one of these cones must dominate $S_=$.

There is an interesting interpretation of this example: It has been proved (Alt [1977], Alt and Mehlhorn [1976]) that if $L$ is a context-free language which admits a nondegenerated iterative pair, then the membership problem for $L$ (i.e. the problem to decide whether $u \in L$ or not) requires $\log n$ space ($n = |u|$) infinitely many often on a nondeterministic Turing machine. Thus the languages of Example 3.3 all have space complexity at least $\log n$. On the other hand, languages like the Dyck languages can be analyzed in space $\log n$ (see Hotz and Messerschmidt [1974]).

As an illustration of the more precise version of the Transfer Theorem formulated in Corollary 3.3, we now give another infinite chain of rational cones which all lie in Rocl ∩ Lin. Moreover we prove the existence of infinitely many pairwise incomparable rational cones.

Let $X=\{x, y\}$. For rational numbers $0<r\leqslant s$ define a language $M(r, s)\subset x^*y^*$ by

$$M(r, s)=1\cup\left\{x^ny^m \mid n, m\in\mathbb{N}\backslash 0, r\leqslant\frac{m}{n}\leqslant s\right\}.$$

Thus $M(1, 1)=S_=$. In Exercise V.7.2, it is stated that $M(r, s)\approx M(qr, qs)$ for any rational number $q>0$, whence $M(r, s)\approx M(r/s, 1)$. Thus only languages of the form $M(q, 1)$, $(0<q\leqslant 1)$, $q$ rational have to be considered.

**Proposition 3.4** (Berstel [1973]) *Let $q, q'$ be rational numbers with $0<q, q'\leqslant 1$. If $M(q', 1)\geqslant M(q, 1)$, then $q'\geqslant q$.*

The converse property, namely $q'\geqslant q\Rightarrow M(q', 1)\geqslant M(q, 1)$ is stated in Exercise V.7.2. Thus

**Corollary 3.5** *For rational numbers $0<q, q'\leqslant 1$, $M(q', 1)\geqslant M(q, 1)$ iff $q'\geqslant q$.* ∎

In particular, $q'>q$ implies $M(q', 1)\Gamma\supsetneq M(q, 1)\Gamma$. Hence the rational cones $M(q, 1)\Gamma$ $(0<q\leqslant 1)$ form an infinite chain of principal cones which is order-isomorphic to the interval $0<q\leqslant 1$ of rational numbers. In particular, for any two rational numbers $1\geqslant q'>q>0$, there exist infinitely many cones $\mathscr{L}$ such that $M(q', 1)\Gamma\supsetneq\mathscr{L}\supsetneq M(q, 1)\Gamma$.

Proof. Assume $M(q', 1)\geqslant M(q, 1)$. The word $xy$ admits in $M(q, 1)$ the strict iterative pair $\pi=(1, x, 1, y, 1)$, and

$$\operatorname{Exp}(\pi, M(q, 1))=\{(0, 0)\}\cup\left\{(n, m)\in(\mathbb{N}\backslash 0)^2 \mid q\leqslant\frac{m}{n}\leqslant 1\right\}.$$

By Corollary 3.3, there exist an iterative pair $\pi'$ in $M(q', 1)$, and integers $N, a, b\geqslant 1$ such that

$$\begin{aligned}&(n, m)\in\operatorname{Exp}(\pi', M(q', 1))\Rightarrow\\ &\qquad (N+(n-1)a, N+(m-1)b)\in\operatorname{Exp}(\pi, M(q, 1)).\end{aligned}\tag{3.1}$$

Since $\pi'$ is strict, it has the form

$$\pi'=(x^{n_1}, x^{a'}, x^{n_2}y^{m_1}, y^{b'}, y^{m_2})\qquad n_1, n_2, m_1, m_2\geqslant 0,\qquad a', b'\geqslant 1.$$

Set $c=N-a$, $d=N-b$, $c'=n_1+n_2$, $d'=m_1+m_2$. Then

$$\operatorname{Exp}(\pi', M(q', 1))\supset\left\{(n, m)\in(\mathbb{N}\backslash 0)^2 \mid q'\leqslant\frac{d'+mb'}{c'+na'}\leqslant 1\right\}\neq\varnothing,$$

and by (3.1)

$$q'\leqslant\frac{d'+mb'}{c'+na'}\leqslant 1\Rightarrow q\leqslant\frac{d+mb}{c+na}\leqslant 1\qquad (n, m>0).\tag{3.2}$$

In order to exploit this relation, define functions $\gamma_r, \gamma'_r$ for rational $r$ by

$$\gamma_r(n)=nr\frac{a}{b}+\frac{rc-d}{b}; \qquad \gamma'_r(n)=nr\frac{a'}{b'}+\frac{rc'-d'}{b'}. \tag{3.3}$$

A straightforward computation shows that

$$q\leqslant\frac{d+mb}{c+na}\leqslant 1 \quad \text{iff} \quad \gamma_q(n)\leqslant m\leqslant\gamma_1(n)$$

and similarly for $\gamma'_{q'}$. Thus (3.2) is equivalent to

$$\gamma'_{q'}(n)\leqslant m\leqslant\gamma'_1(n)\Rightarrow\gamma_q(n)\leqslant m\leqslant\gamma_1(n) \qquad (n>0). \tag{3.4}$$

The values of the functions $\gamma, \gamma'$ are rational, not necessarily integral numbers. Hence (3.4) implies

$$\gamma'_{q'}(n)\geqslant\gamma_q(n)-1; \qquad \gamma'_1(n)\leqslant\gamma_1(n)+1 \qquad (n>0).$$

In view of (3.3), it follows that $q'\frac{a'}{b'}\geqslant q\frac{a}{b}$ and $\frac{a'}{b'}\leqslant\frac{a}{b}$, whence $q'\geqslant q$. ■

For rational numbers $0<r<s<t<1$, define

$$K(r, s, t)=M(r, s)\cup M(t, 1).$$

**Lemma 3.6** *If $K(r', s', t')\geqslant K(r, s, t)$, and $t=t'$, $\frac{r}{s}=\frac{r'}{s'}$, then $r=r'$, $s=s'$.*

Proof. By the same argument as in the preceding proof, the inequality $K(r', s', t)\geqslant K(r, s, t)$ implies for $n, m>0$:

$$\begin{aligned} &\text{if} && \gamma'_{r'}(n)\leqslant m\leqslant\gamma'_{s'}(n) \quad\text{or}\quad \gamma'_{t'}(n)\leqslant m\leqslant\gamma'_1(n)\\ &\text{then} && \gamma_r(n)\leqslant m\leqslant\gamma_s(n) \quad\text{or}\quad \gamma_t(n)\leqslant m\leqslant\gamma_1(n)\end{aligned} \tag{3.5}$$

As above, this implies the inequalities

$$r'\frac{a'}{b'}\geqslant r\frac{a}{b}, \qquad \frac{a'}{b'}\leqslant\frac{a}{b} \tag{3.6}$$

whence $r\leqslant r'$, $s\leqslant s'$. In order to derive the equality, we claim that there exists an integer $n_0$ such that

$$\gamma'_{r'}(n)\leqslant m\leqslant\gamma'_{s'}(n)\Rightarrow\gamma_r(n)\leqslant m\leqslant\gamma_s(n) \qquad n\geqslant n_0 \tag{3.7}$$

$$\gamma'_{t'}(n)\leqslant m\leqslant\gamma'_1(n)\Rightarrow\gamma_t(n)\leqslant m\leqslant\gamma_1(n) \qquad n\geqslant n_0. \tag{3.8}$$

Note first that for rational numbers $z, z'$, $\gamma_z(n)\leqslant\gamma'_{z'}(n)$ implies

$$n\left(z\frac{a}{b}-z'\frac{a'}{b'}\right)\leqslant\frac{z'c'-d'}{b'}-\frac{zc-d}{b},$$

thus if $n$ is sufficiently large,

$$\gamma_z(n) \leqslant \gamma'_{z'}(n) \Rightarrow z\frac{a}{b} \leqslant z'\frac{a'}{b'} \quad \text{and} \quad \gamma_z(n) \geqslant \gamma'_{z'}(n) \Rightarrow z\frac{a}{b} \geqslant z'\frac{a'}{b'}. \tag{3.9}$$

Let $n_1$ be such that (3.9) holds for $z \in \{r, s, t, 1\}$, $z' \in \{r', s', t', 1\}$. Next since $r' < s'$, $\gamma'_{s'}(n) - \gamma'_{r'}(n) \geqslant 1$ for sufficiently large $n$, and consequently there is an integer $m$ such that $\gamma'_{r'}(n) \leqslant m \leqslant \gamma'_{s'}(n)$. Let $n_2$ be such that $\gamma'_{s'}(n) - \gamma'_{r'}(n) \geqslant 1$ and $\gamma'_1(n) - \gamma'_t(n) \geqslant 1$ for all $n \geqslant n_2$, and set $n_0 = \max(n_1, n_2)$.
Assume now that (3.7) is false. Then there are integers $n \geqslant n_0$, $m$ such that, by (3.5),

$$m \leqslant \gamma'_{s'}(n) \quad \text{and} \quad \gamma_t(n) \leqslant m,$$

whence $\gamma_t(n) \leqslant \gamma'_{s'}(n)$ and by (3.9), $t\frac{a}{b} \leqslant s'\frac{a'}{b'}$. Since $t = t'$, (3.6) implies $t' \leqslant s'$ which is impossible. Thus (3.7) holds. Consequently $\gamma'_{s'}(n) \leqslant \gamma_s(n) + 1$ for all $n \geqslant n_0$, and therefore $s'\frac{a'}{b'} \leqslant s\frac{a}{b}$. In view of (3.6) and since $\frac{r}{s} = \frac{r'}{s'}$,

$$r\frac{a}{b} = r'\frac{a'}{b'}, \qquad s\frac{a}{b} = s'\frac{a'}{b'} \tag{3.10}$$

Next assume (3.8) is false. Then as above there is an integer $n \geqslant n_0$ such that $\gamma'_{t'}(n) \leqslant \gamma_s(n)$, and by (3.9) it follows that $t'\frac{a'}{b'} \leqslant s\frac{a}{b}$. By (3.10), this again implies $t' \leqslant s'$, which is impossible. This proves (3.8). (3.8) implies $\gamma'_{t'}(n) \geqslant \gamma_t(n) - 1$ for $n \geqslant n_0$, whence $t'\frac{a'}{b'} \geqslant t\frac{a}{b} = t'\frac{a}{b}$. Thus $\frac{a'}{b'} \geqslant \frac{a}{b}$. By (3.6), we obtain $\frac{a'}{b'} = \frac{a}{b}$, and by (3.10), it follows that $r = r'$, $s = s'$. ■

**Proposition 3.7** (Berstel [1973]) *Let* $0 < r < s < t < 1$. *For any two rational numbers* $q \neq q'$ *with* $0 < q, q' < 1$, *the rational cones generated by* $K(qr, qs, t)$ *and* $K(q'r, q's, t)$ *are incomparable.*

Proof. Assume for instance that $K(q'r, q's, t) \geqslant K(qr, qs, t)$. Then by Lemma 3.6, $q'r = qr$, whence $q = q'$. ■

Proposition 3.7 shows the existence of an infinity of pairwise incomparable rational cones all contained in Rocl ∩ Lin, since all languages $K(r, s, t)$ are dominated by $S_=$. It can be shown (Exercise 3.1) that the full AFLs generated by $K(qr, qs, t)$ and $K(q'r, q's, t)$ are also incomparable.

**Exercise**

**3.1** (Berstel [1974]) Let $L, M \subset x^*y^*$ be two context-free languages. Show that $L\Gamma \subset M\Gamma$ iff $L\hat{\Gamma} \subset M\hat{\Gamma}$. Prove that the full AFLs $K(qr, qs, t)\hat{\Gamma}$ and $K(q'r, q's, t)\hat{\Gamma}$ are incomparable for $q \neq q'$.

## VIII.4 Systems of Iterative Pairs

The Transfer Theorem for iterative pairs proved in the preceding section gives a first criterion for rational domination, but the notion of iterative pairs is not powerful enough to allow comparison of languages whose words admit simultaneously several iterative pairs. Consider for instance the languages

$$A = \{x^n y^n z^p t^p : n, p \geq 0\} \quad \text{and} \quad B = \{x^n y^p z^p t^n : n, p \geq 0\}.$$

All iterative pairs in these two languages are very strict, so the Transfer Theorem gives no information. We shall see that $A$ and $B$ are incomparable. This will follow from the consideration of "systems" of iterative pairs. In the present example, systems for $A$ are formed of two consecutive pairs, and systems for $B$ are formed of nested pairs. Thus the types of systems are incomparable, and in view of a Transfer Theorem to be proved, the languages are rationally incomparable.

The present section contains the definition and examples of systems of iterative pairs. Section 5 is concerned with grammatical systems, and Section 6 contains the proof of the Transfer Theorem for systems.

We shall be concerned in the sequel with sequences of words of the form

$$\theta = (a_0, u_1, a_1, \ldots, a_{m-1}, u_m, a_m) \tag{4.1}$$

and with sequences derived from (4.1) by concatenating some of its terms. A typical sequence related to (4.1) is

$$\theta' = (a_0 u_1 a_1 \cdots a_{i_1-1}, u_{i_1}, a_{i_1} u_{i_1+1} \cdots a_{i_2-1}, u_{i_2}, \ldots, u_{i_r}, a_{i_r} \cdots u_m a_m) \tag{4.2}$$

with $1 \leq i_1 < \cdots < i_r \leq m$. For sake of simplicity, we introduce the notation

$$\theta' = \langle u_{i_1}, u_{i_2}, \ldots, u_{i_r} \rangle \tag{4.3}$$

for (4.2). Thus $\langle u_1, u_2, \ldots, u_m \rangle$ is equivalent to (4.1), and $\pi = \langle u, v \rangle$ is an equivalent notation for $\pi = (a, u, b, v, c)$.

Next if $L$ is a language we define

$$\operatorname{Exp}(\theta, L) = \{(k_1, \ldots, k_m) \in \mathbb{N}^m | a_0 u_1^{k_1} a_2 \cdots a_{m-1} u_m^{k_m} a_m \in L\}.$$

Finally we introduce the following notation: For $P \subset \mathbb{N}^2$, $Q \subset \mathbb{N}^q$,

$$P * Q = \{(p, k_1, \ldots, k_q, p') \mid (p, p') \in P, (k_1, \ldots, k_q) \in Q\}.$$

Thus, with $\Delta=\{(n, n): n \geqslant 0\}$ we have for instance

$$\Delta * \Delta=\{(n, m, m, n): n, m \geqslant 0\}$$

while $\Delta \times \Delta=\{(n, n, m, m): n, m \geqslant 0\}$.

**Definition** Let $X$ be an alphabet, and let $L \subset X^*$ be a language. A sequence of words of $X^*$

$$\theta=(a_0, u_1, a_1, \ldots, a_{2n-1}, u_{2n}, a_{2n})$$

is a system of iterative pairs of length $n$ in $L$ if one of the following conditions is satisfied:

(i) $n=1$ and $\theta$ is a nondegenerated iterative pair in $L$;
(ii) $n>1$ and
(ii.1) $\pi=\langle u_1, u_{2n}\rangle$ is a nondegenerated iterative pair in $L$,
(ii.2) $\theta'=\langle u_2, \ldots, u_{2n-1}\rangle$ is a system of iterative pairs in $L$,
(ii.3) $\operatorname{Exp}(\theta, L) \supset \operatorname{Diag}(\theta, L)$, where $\operatorname{Diag}(\theta, L)=\Delta * \operatorname{Diag}(\theta', L)$.
(iii) $n>1$ and there is some integer $p$, $(1 \leqslant p<n)$ such that
(iii.1) $\theta_1=\langle u_1, \ldots, u_{2p}\rangle$ and $\theta_2=\langle u_{2p+1}, \ldots, u_{2n}\rangle$ are systems of iterative pairs in $L$,
(iii.2) $\operatorname{Exp}(\theta, L) \supset \operatorname{Diag}(\theta, L)$, where $\operatorname{Diag}(\theta, L)=\operatorname{Diag}(\theta_1, L) \times \operatorname{Diag}(\theta_2, L)$.

$\theta$ is called a system of iterative pairs of the word $w=a_0 u_1 \cdots u_{2n} a_{2n}$ in $L$, and $w$ is said to admit the system $\theta$ in $L$.

Notation. If $\theta$ is defined by condition (ii), then we write $\theta=\pi * \theta'$; if $\theta$ is defined by (iii), then we write $\theta=\theta_1 \times \theta_2$.

With this notation, any system of iterative pairs $\theta$ of length $n$ of a word $w$ has an expression as a combination of $n$ nondegenerated iterative pairs $\pi_1, \ldots, \pi_n$ of $w$, which are combined by the operations $*$ and $\times$. The pairs $\pi_1, \ldots, \pi_n$ are called the underlying pairs of $\theta$. Note that the operation $\times$ is associative.

**Definition** The type $\chi(\theta, L)$ of a system $\theta$ of iterative pairs in $L$ is

$$\chi(\theta, L)=\begin{cases}\chi(\pi, L) & \text{if } \theta=\pi \text{ is an iterative pair;} \\ z\chi(\theta', L)\bar{z} & \text{if } \theta=\pi * \theta' \quad \text{and} \quad \chi(\pi, L)=z\bar{z}; \\ \chi(\theta_1, L)\chi(\theta_2, L) & \text{if } \theta=\theta_1 \times \theta_2.\end{cases}$$

The type is clearly a word of the restricted Dyck language $D_4'^*$ over the alphabet $\Xi$. Types of the same length are ordered by the following conventions:

$$e>s>r, s>l; \qquad \bar{e}>\bar{s}>\bar{r}, \bar{s}>\bar{l};$$

if $h, h' \in \Xi^*$ and $|h| = |h'|$, then $h \geqslant h'$ iff

$$h_1 \geqslant h'_1, \qquad h_2 \geqslant h'_2$$

for all factorizations $h = h_1 h_2$, $h' = h'_1 h'_2$ such that $0 < |h_1| = |h'_1| < |h|$. This definition agrees with the order on types of iterative pairs given in Section 1.

**Definition** A system $\theta$ of iterative pairs in $L$ is independant if the following conditions (ii.4) and (iii.3) hold instead of (ii.3) and (iii.2) for all steps of the definition of $\theta$:

(ii.4) $\operatorname{Exp}(\theta, L) = \operatorname{Exp}(\pi, L) * \operatorname{Exp}(\theta', L)$;

(iii.3) $\operatorname{Exp}(\theta, L) = \operatorname{Exp}(\theta_1, L) \times \operatorname{Exp}(\theta_2, L)$.

These conditions express the property that simultaneous iteration of some of the underlying iterative pairs of $\theta$ does not create new elements in $\operatorname{Exp}(\theta, L)$.

Note that there may exist systems $\theta_1$, $\theta_2$ in $L$ such that $\operatorname{Exp}(\theta_1, L) = \operatorname{Exp}(\theta_2, L)$ and $\chi(\theta_1, L) \neq \chi(\theta_2, L)$ (see Example 4.3 below). In other terms, a system of iterative pairs $\theta$ is completely specified only when the expression in its underlying iterative pairs is supplied. This expression is described by the type $\chi(\theta, L)$. Nevertheless, if $\theta$ is a system of iterative pairs in $L$, and if $\theta' = (a'_0, u'_1, \dots, u'_{2n}, a'_{2n})$ is a sequence of words such that $\operatorname{Exp}(\theta, L) = \operatorname{Exp}(\theta', L')$, then an expression in iterative pairs in $L'$ can be given for $\theta'$ which is the same as the expression of $\theta$ in $L$, i.e. such that $\chi(\theta, L) = \chi(\theta', L)$.
Independent systems of iterative pairs have the useful property that their expression in iterative pairs is determined by the set of exponents. Thus the situation described above cannot occur for independent systems. More precisely we claim: If $\theta = (a_0, u_1, \dots, u_{2n}, a_{2n})$ and $\bar{\theta} = (\bar{a}_0, \bar{u}_1, \dots, \bar{u}_{2n}, \bar{a}_{2n})$ are systems of iterative pairs in $L$ with $\operatorname{Exp}(\theta, L) = \operatorname{Exp}(\bar{\theta}, L)$, and if $\theta$ is independent in $L$, then $\chi(\theta, L) = \chi(\bar{\theta}, L)$ (then $\bar{\theta}$ is also independent). Assume indeed the contrary. Then there is an underlying pair $\bar{\pi} = \langle \bar{u}_i, \bar{u}_j \rangle$ of $\bar{\theta}$ such that $\langle u_i, u_j \rangle$ is not an underlying pair of $\theta$. Consequently there is an integer $k \neq j$ such that either $\pi = \langle u_k, u_i \rangle$ or $\pi = \langle u_i, u_k \rangle$ is an underlying pair of $\theta$, according to $k < i$ or $i < k$. By conditions (ii.3) and (iii.2) applied to $\bar{\theta}$, we have

$$(q, \dots, q, p, q, \dots, q, p, q, \dots, q) \in \operatorname{Exp}(\bar{\theta}, L)$$

for all $q, p \geqslant 0$, where the $p$'s occur at the entries $i, j$. Since $\theta$ is independent, conditions (ii.4) and (iii.3) imply that either $(q, p) \in \operatorname{Exp}(\pi, L)$ or $(p, q) \in \operatorname{Exp}(\pi, L)$, according to $k < i$ or $i < k$. Thus in both cases, $\operatorname{Exp}(\pi, L) = \mathbb{N}^2$, and $\theta$ has an underlying pair which is degenerated, contrary to the definition.

**Example 4.1** Let $A = \{x^n y^n z^p t^p : n, p \geqslant 0\}$. Then

$$\theta = (1, x, 1, y, 1, z, 1, t, 1)$$

is a system of iterative pairs of $w = xyzt$ in $A$. The underlying iterative pairs are $\pi_1 = \langle x, y\rangle$ and $\pi_2 = \langle z, t\rangle$; thus $\theta = \pi_1 \times \pi_2$. Further

$$\mathrm{Exp}(\theta, A) = \mathrm{Diag}(\theta, A) = \Delta \times \Delta.$$

Thus $\theta$ is independent, and $\chi(\theta, A) = e\bar{e}e\bar{e}$.

**Example 4.2** Let $B = \{x^n y^p z^p t^n : n, p \geq 0\}$. Then

$$\theta' = (1, x, 1, y, 1, z, 1, t, 1)$$

is a system of iterative pairs in $B$, whose underlying iterative pairs are $\pi_3 = \langle x, t\rangle$ and $\pi_4 = \langle y, z\rangle$. Thus $\theta' = \pi_3 * \pi_4$. Since

$$\mathrm{Exp}(\theta', B) = \mathrm{Diag}(\theta', B) = \Delta * \Delta,$$

$\theta'$ is independent in $B$, and $\chi(\theta', B) = ee\bar{e}\bar{e}$. The types $\chi(\theta, A)$ and $\chi(\theta', B)$ are incomparable.

**Example 4.3** Let $C = A \cup B$, with $A$, $B$ as above. Then the word $xyzt$ admits the two systems $\theta = \pi_1 \times \pi_2$ and $\theta' = \pi_3 * \pi_4$ in $C$. But now

$$\mathrm{Exp}(\theta, C) = \mathrm{Exp}(\theta', C) = \Delta \times \Delta \cup \Delta * \Delta.$$

Hence neither $\theta$ nor $\theta'$ is independent in $C$.

**Example 4.4** Let $L = \{xyz^p t^p \mid p \geq 0\} \cup \{x^n y^n zt \mid n \geq 0\}$. The word $w = xyzt$ admits in $L$ the two very strict iterative pairs

$$\pi_1 = (xy, z, 1, t, 1) \quad \text{and} \quad \pi_2 = (1, x, 1, y, zt).$$

But $\quad \theta = \pi_1 \times \pi_2 = (1, x, 1, y, 1, z, 1, t, 1)$

is not a system of iterative pairs in $L$, since

$$\mathrm{Exp}(\theta, L) = \Delta \times \{(1, 1)\} \cup \{(1, 1)\} \times \Delta \not\supseteq \mathrm{Diag}(\theta, L) = \Delta \times \Delta.$$

**Example 4.5** Let $L = \{x^k y^p \bar{y}^q z^n \bar{z}^m \bar{x}^k \mid 0 \leq k, 0 \leq p \leq q, 0 \leq m \leq n\}$ be the language of Example 1.4. Then

$$\theta = (1, x, 1, y, 1, \bar{y}, 1, z, 1, \bar{z}, 1, \bar{x}, 1) = \pi_1 * (\pi_2 \times \pi_3)$$

with $\pi_1 = \langle x, \bar{x}\rangle$, $\pi_2 = \langle y, \bar{y}\rangle$, $\pi_3 = \langle z, \bar{z}\rangle$ is a system of iterative pairs in $L$. $\mathrm{Diag}(\theta, L) = \{k, p, p, n, n, k) \in \mathbb{N}^6 \mid k, p, n \geq 0\}$; further $\theta$ is easily seen to be independent in $L$. Finally

$$\chi(\theta, L) = ell\bar{l}r\bar{r}\bar{e}.$$

**Example 4.6** Let $S_2$ be the symmetric language over $Z = X \cup \bar{X}$, with $X = \{x, y\}$, $\bar{X} = \{\bar{x}, \bar{y}\}$ defined by

$$S_2 = xS_2\bar{x} \cup yS_2\bar{y} \cup 1.$$

We claim that any system of iterative pairs $\theta$ in $S_2$ is of the form

$$\theta = \pi_1 * \pi_2 * \cdots * \pi_n, \tag{4.4}$$

where $\pi_1, \ldots, \pi_n$ are iterative pairs. Indeed $\theta$ cannot be of the form $\theta = \theta_1 \times \theta_2$, since otherwise at least one of the underlying iterative pairs of $\theta$ would have its iterating elements both in $X^*$ or in $\bar{X}^*$, contrary to Example 1.7. Next all iterative pairs in (4.4) are very strict, thus $\theta$ has type $e^n\bar{e}^n$. $\theta$ may be dependent, but there exist independent systems of type $e^n\bar{e}^n$ in $S_2$ for any $n \geq 1$. Thus the systems

$$\theta = \langle xy, xy^2, \ldots, xy^n, \bar{y}^n\bar{x}, \ldots, \bar{y}^2\bar{x}, \bar{y}\bar{x}\rangle$$

or $$\theta = \langle x, y, x, y, \ldots, x, y, \bar{y}, \bar{x}, \ldots, \bar{y}, \bar{x}, \bar{y}, \bar{x}\rangle$$

are independent.

**Example 4.7** Let $D_1'^*$ be the restricted Dyck language over $X = \{x, \bar{x}\}$. We claim that any independent system of iterative pairs in $D_1'^*$ is of length 1, i.e. is a (nondegenerated) iterative pair in $D_1'^*$. Consider indeed a system of iterative pairs of length 2 in $D_1'^*$:

$$\theta = (a_0, u_1, a_1, u_2, a_2, u_3, a_3, u_4, a_4).$$

Two possibilities arise:
(i) $\theta = \pi_1 \times \pi_2$, with $\pi_1 = \langle u_1, u_2\rangle$, $\pi_2 = \langle u_3, u_4\rangle$.
By Example 1.8, these pairs are very strict. Set $k = \|u_1\| = -\|u_2\|$, $l = \|u_3\| = -\|u_4\|$. Then $k, l > 0$. Next

$$f = a_0 u_1^{1+l} a_1 u_2 a_2 u_3 a_3 u_4^{1+k} a_4 \in D_1'^*.$$

Indeed $\|f\| = l\,\|u_1\| + k\,\|u_4\| = 0$. Further if $f'$ is a left factor of $a_0 u_1^{1+l} a_1 u_2 a_2 u_3 a_3$, then $\|f'\| > 0$. Finally, if $f' = a_0 u_1^{1+l} a_1 u_2 a_2 u_3 a_3 u_4^m u'$ with $0 \leq m \leq k$ and $u'$ a left factor of $u_4$, then $\|f'\| = l\,\|u_1\| + m\|u_4\| + \|a_0 u_1 a_1 u_2 a_2 u_3 a_3 u'\| \geq l\|u_1\| + m\|u_4\| \geq 0$. Consequently

$$(1+l, 1, 1, 1+k) \in \mathrm{Exp}(\theta, D_1'^*)$$

and $\mathrm{Exp}(\theta, D_1'^*) \supsetneq \Delta \times \Delta$.
(ii) $\theta = \pi_1 * \pi_2$, with $\pi_1 = \langle u_1, u_4\rangle$, $\pi_2 = \langle u_2, u_3\rangle$.
Setting $k = \|u_1\| = -\|u_4\|$, $l = \|u_2\| = -\|u_3\|$, an identical argument shows that $f = a_0 u_1^{1+l} a_1 u_2 a_2 u_3^{1+k} a_3 u_4 a_4$ is in $D_1'^*$, which proves that $\Delta * \Delta$ is strictly contained in $\mathrm{Exp}(\theta, D_1'^*)$.
Note that there exist in $D_1'^*$ systems of iterative pairs of any length and of any type (in $\{e, \bar{e}\}^*$) (Exercise 4.1). The above proof shows that these systems are not independent.

We conclude this section by the analog of Lemmas 1.1 and 1.2.

**Lemma 4.1** *Let $\theta$ be system of iterative pairs in $L \cap K$ where $L$ is a language and $K$ is a local regular language. Then $\theta$ is a system of iterative pairs of the same type in $L$, and* $\operatorname{Exp}(\theta, L) = \operatorname{Exp}(\theta, L \cap K)$.

Proof. Let $\theta = (a_0, u_1, \ldots, u_{2n}, a_{2n})$. By conditions (ii.3) and (iii.2) of the definition, $a_0 u_1^k \cdots u_{2n}^k a_{2n} \in K$ for $k \geq 0$. Since $K$ is local, $a_0 u_1^* \cdots u_{2n}^* a_{2n} \subset K$. Thus

$$\operatorname{Exp}(\theta, L) = \operatorname{Exp}(\theta, L \cap a_0 u_1^* \cdots u_{2n}^* a_{2n}) \subset \operatorname{Exp}(\theta, L \cap K).$$

The converse inclusion is obvious. ■

Let $\varphi : X^* \to Y^*$ be a morphism, and let

$$\theta = (a_0, u_1, \ldots, u_{2n}, a_{2n}) \tag{4.5}$$

be a sequence of words of $X^*$. Define

$$\varphi(\theta) = (\varphi a_0, \varphi u_1, \ldots, \varphi u_{2n}, \varphi a_{2n}).$$

**Lemma 4.2** *Let $M \subset Y^*$, $L = \varphi^{-1}(M)$. If $\theta$ is a system of iterative pairs in $L$, then $\varphi(\theta)$ is a system of the same type in $M$ and* $\operatorname{Exp}(\theta, L) = \operatorname{Exp}(\varphi(\theta), M)$.

Proof. It suffices to prove the last equality. Let $\theta$ be given by (4.5). Clearly $\operatorname{Exp}(\theta, L) \subset \operatorname{Exp}(\varphi(\theta), M)$. Next if $\varphi(a_0 u_1^{k_1} \cdots u_{2n}^{k_{2n}} a_{2n}) \in M$, then $a_0 u_1^{k_1} \cdots u_{2n}^{k_{2n}} a_{2n} \in \varphi^{-1}(M) = L$. Thus $\operatorname{Exp}(\varphi(\theta), M) \subset \operatorname{Exp}(\theta, L)$. ■

### Exercises

**4.1** Show that in $D_1'^*$ there exist systems of iterative pairs of any length and of any type contained in $\{e, \bar{e}\}^*$.

**4.2** (Beauquier [1977]) Show that the underlying iterative pairs of any system in $D_n'^*(n \geq 1)$ are all very strict. Show that in $D_n'^*(n \geq 2)$ there exist independent systems of iterative pairs of any length and of any type contained in $\{e, \bar{e}\}^*$.

## VIII.5 Grammatical Systems

This section contains the extension of the notion of grammatical pairs to systems of iterative pairs. Such a system is formed of grammatical iterative pairs with the additional requirement that the pivots of the iterative pairs are simultaneously derivable from the nonterminal generating the language.

**Definition** Let $L \subset X^*$ be an algebraic language, and let $G = \langle V, X, P \rangle$ be an algebraic grammar such that $L = L_G(\sigma)$ for $\sigma \in V$. Let

$$\theta = (a_0, u_1, \ldots, u_{2n}, a_{2n})$$

be a system of iterative pairs in $L$. Then $\theta$ is grammatical with respect to $G$ and has pivot $w \in (V \cup X)^*$ if

(i) $n=1$, and $\theta$ is a grammatical iterative pair with pivot $w$,

(ii) $\theta = \pi * \theta'$, where $\pi$ is a grammatical pair with pivot $w$, $\theta'$ is a grammatical system with pivot $w'$, and

$$w \overset{*}{\underset{G}{\rightarrow}} a_1 w' a_{2n-1};$$

(iii) $\theta = \theta_1 \times \theta_2$, where $\theta_1$ is a grammatical system of length $k$ and pivot $w_1$, $\theta_2$ is a grammatical system with pivot $w_2$, and

$$w = w_1 a_{2k} w_2 \quad \text{and} \quad \sigma \overset{*}{\underset{G}{\rightarrow}} a_0 w a_{2n}.$$

It is easily seen by induction that the pivot $w$ is a word of $V(X^*V)^*$.

**Example 5.1** Let

$$\theta = (a_0, u_1, \ldots, u_{2n}, a_{2n}) = \pi_1 * \pi_2 * \cdots * \pi_n,$$

and assume $\theta$ is grammatical with respect to $G$. Then there are variables $\xi_1, \xi_2, \ldots, \xi_n \in V$ such that

$$\sigma \overset{*}{\rightarrow} a_0 \xi_1 a_{2n}; \qquad \xi_1 \overset{*}{\rightarrow} u_1 \xi_1 u_{2n};$$

$$\xi_1 \overset{*}{\rightarrow} a_1 \xi_2 a_{2n-1}; \qquad \xi_2 \overset{*}{\rightarrow} u_2 \xi_2 u_{2n-1};$$

$$\vdots \qquad\qquad \vdots$$

$$\xi_{n-1} \overset{*}{\rightarrow} a_{n-1} \xi_n a_{n+1}; \quad \xi_n \overset{*}{\rightarrow} u_n \xi_n u_{n+1}; \quad \xi_n \overset{*}{\rightarrow} a_n.$$

The pivot of $\theta$ is $\xi_1$.

**Example 5.2** Assume that $\theta$ is of the form

$$\theta = \pi_1 \times \pi_2 \times \cdots \times \pi_n.$$

If $\theta$ is grammatical with respect to $G$, then there are variables $\xi_1, \ldots, \xi_n \in V$ such that

$$\sigma \overset{*}{\rightarrow} a_0 \xi_1 a_2 \xi_2 a_4 \cdots a_{2n-2} \xi_n a_{2n};$$

$$\xi_i \overset{*}{\rightarrow} u_{2i-1} \xi_i u_{2i}; \quad \xi_i \overset{*}{\rightarrow} a_{2i-1} \qquad (i=1, \ldots, n).$$

The pivot of $\theta$ is $\xi_1 a_2 \xi_2 a_4 \cdots \xi_n$.

**Example 5.3** Let $D_1'^* \subset \{x, \bar{x}\}^*$. Then

$$\theta = (1, x, 1, \bar{x}, 1, x, 1, \bar{x}, 1) = \pi_1 \times \pi_2,$$

with $\pi_1 = (1, x, 1, \bar{x}, x\bar{x})$, $\pi_2 = (x\bar{x}, x, 1, \bar{x}, 1)$ is a (dependent) system of iterative pairs in $D_1'^*$. $\theta$ is grammatical with respect to the grammar with rules $\xi \to \xi\xi + x\xi\bar{x} + 1$. Indeed $\xi$ is pivot of both $\pi_1$ and $\pi_2$, and $\xi\xi$ is the pivot of $\theta$.

We now prove the analogues of Lemmas 2.1.–2.3.

**Lemma 5.1** *Let $L \subset X^*$ be generated by an algebraic grammar $G$, and let $K \subset X^*$ be a rational language. If $\theta$ is a system of iterative pairs in $L \cap K$ grammatical with respect to $G_K$, then $\theta$ is grammatical with respect to $G$.*

Proof. By induction on the length $n$ of $\theta = (a_0, u_1, \ldots, u_{2n}, a_{2n})$. For $n = 1$, the result is Lemma 2.1. We assume that $G$ and $G_K$ are as in Section II.2. Assume $\theta = \pi * \theta'$. Then in $G_K$:

$$\hat{\sigma} \to (q_-, \sigma, q_+) \xrightarrow{*} a_0(q, \xi, q')a_{2n};$$

$$(q, \xi, q') \xrightarrow{*} u_1(q, \xi, q')u_{2n}; \qquad (q, \xi, q') \xrightarrow{*} a_1 w' a_{2n-1},$$

where $w'$ is the pivot of $\theta'$ in $G_K$. Now $w'$ is of the form $w' = (q_1, \xi_1, q_1')v_1 \cdots v_{r-1}(q_r, \xi_r, q_r')$ with $v_1, \ldots, v_{r-1} \in X^*$. Assuming by induction that $w'' = \xi_1 v_1 \cdots v_{r-1}\xi_r$ is the pivot of $\theta'$ in $G$, we have, in $G$,

$$\sigma \xrightarrow{*} a_0 \xi a_{2n}, \qquad \xi \xrightarrow{*} u_1 \xi u_{2n}, \qquad \xi \xrightarrow{*} a_1 w'' a_{2n-1}.$$

Thus $\theta$ is grammatical in $G$ with pivot $\xi$. The case $\theta = \theta_1 \times \theta_2$ is proved in the same way. ■

**Lemma 5.2** *Let $\varphi: X^* \to Y^*$ be an alphabetic morphism, let $M \subset Y^*$ be generated by an algebraic grammar $G$, and set $L = \varphi^{-1}(M)$. If $\theta$ is a system of iterative pairs in $L$ grammatical with respect to $\varphi^{-1}(G)$, then $\varphi(\theta)$ is grammatical with respect to $G$.*

Proof. By induction on the length $n$ of $\theta = (a_o, u_1, \ldots, u_{2n}, a_{2n})$. For $n = 1$, the result is Lemma 2.2. Let $G$ and $\varphi^{-1}(G)$ be as in Section II.2, and suppose first $\theta = \theta_1 \times \theta_2$. Let $k$ be the length of $\theta_1$, and let $w, w_1, w_2$ be the pivots of $\theta, \theta_1, \theta_2$ in $\varphi^{-1}(G)$ respectively. Suppose by induction that no variable in $w_1$ and $w_2$ is equal to the variable $\omega$ of $\varphi^{-1}(G)$. Then the same is true for $w = w_1 a_{2k} w_2$. Suppose next by induction that $\varphi(w_1)$ and $\varphi(w_2)$ are the pivots in $G$ of $\varphi(\theta_1)$ and $\varphi(\theta_2)$. Then in $G$

$$\sigma \xrightarrow{*} \varphi(a_0)\varphi(w)\varphi(a_{2n})$$

and $\varphi(w)$ is the pivot of $\varphi(\theta)$ in $G$. Thus the lemma is proved in this case.

Suppose next $\theta=\pi*\theta'$. Then in $\varphi^{-1}G$:

$$\sigma \xrightarrow{*} a_0\eta a_{2n}; \qquad \eta \xrightarrow{*} u_1\eta u_{2n}; \qquad \eta \xrightarrow{*} a_1 w' a_{2n-1},$$

where $w'$ is the pivot in $\varphi^{-1}G$ of $\theta'$. Since $\pi$ is nondegenerated, $\eta$ is not the variable $\omega$. Assuming by induction that $\varphi w'$ is the pivot of $\varphi(\theta')$ in $G$, it follows that

$$\sigma \xrightarrow{*} \varphi(a_0)\eta\varphi(a_{2n}); \qquad \eta \xrightarrow{*} \varphi(u_1)\eta\varphi(u_{2n}); \qquad \eta \xrightarrow{*} \varphi(a_1)\varphi(w')\varphi(a_{2n-1})$$

in $G$, thus $\varphi\eta=\eta$ is the pivot of $\varphi(\theta)$ in $G$. ■

**Lemma 5.3** *Let $\psi: X^* \to Y^*$ be a morphism, let $L \subset X^*$ be generated by an algebraic grammar $G$, and let $M=\psi(L)$. If $\theta$ is a system of iterative pairs in $M$ grammatical with respect to $\psi G$, then there exists a system of iterative pairs $\hat{\theta}$ in $L$, grammatical with respect to $G$, such that $\psi(\hat{\theta})=\theta$. Further* $\mathrm{Exp}(\hat{\theta}, L) \subset \mathrm{Exp}(\theta, M)$ *and* $\chi(\hat{\theta}, L) \geqslant \chi(\theta, M)$.

Proof. The inclusion $\mathrm{Exp}(\hat{\theta}, L) \subset \mathrm{Exp}(\theta, M)$ is immediate. The inequality $\chi(\hat{\theta}, L) \geqslant \chi(\theta, M)$ is a consequence of the fact that for any iterative pair $\hat{\pi}$ underlying to $\hat{\theta}$, $\pi=\psi(\hat{\pi})$ is underlying to $\theta$ and therefore $\chi(\hat{\pi}, L) \geqslant \chi(\pi, M)$ in view of Lemma 2.3. For simplicity, set $m=2n$. Assume first $\theta=\pi*\theta'$, and let $\xi$ and $w'$ be the pivots of $\pi$ and $\theta'$ in $\psi G$. Then in $\psi G$:

$$\sigma \xrightarrow{*} a_0\xi a_m, \qquad \xi \xrightarrow{*} u_1\xi u_m, \qquad \xi \xrightarrow{*} a_1 w' a_{m-1}.$$

Thus there exist $\hat{a}_0, \hat{a}_m, \hat{u}_1, \hat{u}_m, \hat{a}_1, \hat{a}_{m-1}, \hat{w}'$ with $\psi\hat{a}_0=a_0$, $\psi\hat{a}_m=a_m$, $\psi\hat{u}_1=u_1$, $\psi\hat{u}_m=u_m$, $\psi\hat{a}_1=a_1$, $\psi\hat{a}_{m-1}=a_{m-1}$, $\psi\hat{w}'=w'$ such that in $G$:

$$\sigma \xrightarrow{*} \hat{a}_0\xi\hat{a}_m; \qquad \xi \xrightarrow{*} \hat{u}_1\xi\hat{u}_m; \qquad \xi \xrightarrow{*} \hat{a}_1\hat{w}'\hat{a}_{m-1}.$$

Next by induction there exists a grammatical system $\bar{\theta}'$ in $G$ with $\psi\bar{\theta}'=\theta'$ and with pivot $\bar{w}'$ verifying $\psi(\bar{w}')=w'$. $w'$ has the form

$$w'=\xi_1 b_1 \cdots b_{r-1}\xi_r,$$

where $\xi_1, \ldots, \xi_r$ are variables and $b_1, \ldots, b_{r-1}$ are among the $a_j$'s. Consequently

$$\bar{w}'=\xi_1\bar{b}_1 \cdots \bar{b}_{r-1}\xi_r \qquad \hat{w}'=\xi_1\hat{b}_1 \cdots \hat{b}_{r-1}\xi_r$$

with $\psi\bar{b}_i=\psi\hat{b}_i=b_i$ for $i=1, \ldots, r-1$. Since $\bar{w}'$ is the pivot of $\bar{\theta}'$, it follows that $\hat{w}'$ is the pivot of the system $\hat{\theta}'$ obtained by replacing $\bar{b}_i$ by $\hat{b}_i$ in $\bar{\theta}'$. This proves the lemma when $\theta=\pi*\theta'$. The case where $\theta=\theta_1\times\theta_2$ is similar. ■

Note that if $\theta$ is independent, it does not follow that $\hat{\theta}$ is independent (see Example 6.4).

The main result of this section is the generalization of Theorem 2.5 to systems of iterative pairs. We therefore define:

**Definition** Let

$$\theta=(a_0, u_1, a_1, \ldots, u_m, a_m) \quad \text{and} \quad \bar\theta=(\bar a_0, \bar u_1, \bar a_1, \ldots, \bar u_m, \bar a_m) \tag{5.1}$$

be sequences of words. Then $\bar\theta$ is deduced from $\theta$ if

(i) there is an integer $N \geq 1$ such that

$$\bar a_0 \bar u_1 \bar a_1 \bar u_2 \cdots \bar u_m \bar a_m = a_0 u_1^N a_1 u_2^N \cdots u_m^N a_m;$$

(ii) there exist factorizations

$$u_1 = u_1' u_1'', \ldots, u_m = u_m' u_m'',$$

words $c_1, \ldots, c_m, d_1, \ldots, d_m$ and integers $p_1, \ldots, p_m$, $(1 \leq p_i \leq N)$ with

$$u_i^N = c_i \bar u_i d_i; \qquad \bar u_i = (u_i'' u_i')^{p_i}; \qquad i = 1, \ldots, m \tag{5.2}$$

$$\bar a_0 = a_0 c_1; \qquad \bar a_i = d_i a_i c_{i+1} \quad (1 \leq i < m); \qquad \bar a_m = d_m a_m. \tag{5.3}$$

**Example 5.4** Let $\theta=(a_0, u_1, a_1, u_2, a_2, u_3, a_3, u_4, a_4,) = \pi_1 \times \pi_2$ be a system of iterative pairs in some language $L$, with $\pi_1 = \langle u_1, u_2\rangle$, $\pi_2 = \langle u_3, u_4\rangle$. Then $\bar\theta = (a_0, u_1^2, a_1, u_2^2, a_2, u_3^2, a_3, u_4^2, a_4)$ is deduced from $\theta$. Set

$$\bar\pi_1 = (a_0, u_1^2, a_1, u_2^2, a_2 u_3^2 a_3 u_4^2 a_4),$$
$$\bar\pi_2 = (a_0 u_1^2 a_1 u_2^2 a_2, u_3^2, a_3, u_4^2, a_4).$$

Then $\bar\pi_1$ and $\bar\pi_2$ are iterative pairs in $L$, but note that they are not deduced from $\pi_1$ and $\pi_2$. Next if both $\bar\pi_1$ and $\bar\pi_2$ are nondegenerated (this is not necessarily true, see the following example), then $\bar\theta = \bar\pi_1 \times \bar\pi_2$.

**Example 5.5** Let $L = \{x^n y^n z^p t^p \mid n, p \geq 0\} \cup x^* y^* z^2 t^2$. Then

$$\theta = (1, x, 1, y, 1, z, 1, t, 1) = \pi_1 \times \pi_2$$

is a system of iterative pairs, and $\pi_1 = \langle x, y\rangle$, $\pi_2 = \langle z, t\rangle$ are very strict iterative pairs. Next

$$\bar\theta = (1, x^2, 1, y^2, 1, z^2, 1, t^2, 1)$$

is deduced from $\theta$, but $\bar\pi = (1, x^2, 1, y^2, z^2 t^2)$ is a degenerated iterative pair in $L$.

Since we will deal now with several systems of iterative pairs simultaneously, we precise the notation (4.3) and write

$$\theta' = \langle u_{i_1}, \ldots, u_{i_r}\rangle_\theta$$

if $\theta'$ is a subsequence of a sequence $\theta$.

**Proposition 5.4** *Let $\theta$ be a system of iterative pairs in L and let $\bar{\theta}$ be deduced from $\theta$. If $\theta$ is independent, then $\bar{\theta}$ is an independent system of iterative pairs in L, and $\chi(\bar{\theta}, L) \geqslant \chi(\theta, L)$.*

Proof. Let $n$ be the length of $\theta$, and let $\theta, \bar{\theta}$ be as in (5.1) with $m = 2n$. First note that

$$(k_1, \ldots, k_m) \in \operatorname{Exp}(\bar{\theta}, L) \iff (N + p_1(k_1 - 1), \ldots, N + p_m(k_m - 1)) \in \operatorname{Exp}(\theta, L). \quad (5.4)$$

Then the proposition is true for $n = 1$. Arguing by induction, assume $\theta = \pi * \theta'$ and define

$$\bar{\pi} = \langle \bar{u}_1, \bar{u}_m \rangle_{\bar{\theta}}, \qquad \bar{\theta}' = \langle \bar{u}_2, \ldots, \bar{u}_{m-1} \rangle_{\bar{\theta}}.$$

Note that $\bar{\pi}$ and $\bar{\theta}'$ are not deduced from $\pi$ and $\theta$. But from the independence of $\theta$, it follows that

$$(k_1, k_m) \in \operatorname{Exp}(\bar{\pi}, L) \iff (N + p_1(k_1 - 1), N + p_m(k_m - 1)) \in \operatorname{Exp}(\pi, L)$$

$$(k_2, \ldots, k_{m-1}) \in \operatorname{Exp}(\bar{\theta}', L) \iff (N + p_2(k_2 - 1), \ldots, N + p_{m-1}(k_{m-1} - 1)) \in \operatorname{Exp}(\theta', L).$$

Thus, by (5.4),

$$\operatorname{Exp}(\bar{\theta}, L) = \operatorname{Exp}(\bar{\pi}, L) * \operatorname{Exp}(\bar{\theta}', L). \quad (5.5)$$

Consider next

$$\pi_1 = (a_0 c_1, \bar{u}_1, d_1 a_1 u_2 \cdots u_{m-1} a_{m-1} c_m, \bar{u}_m, d_m a_m)$$
$$\theta_1 = (a_0 u_1 c_2, \bar{u}_2, \bar{a}_2, \ldots, \bar{u}_{m-1}, d_{m-1} u_m a_m).$$

Then $\pi_1$ and $\theta_1$ are deduced from $\pi$ and $\theta'$. By induction $\theta_1$ is an independent system in $L$ and by Lemma 2.4 and by induction, $\chi(\pi_1, L) \geqslant \chi(\pi, L)$ and $\chi(\theta_1, L) \geqslant \chi(\theta', L)$. The independence of $\theta$ implies that

$$\operatorname{Exp}(\pi_1, L) = \operatorname{Exp}(\bar{\pi}, L), \qquad \operatorname{Exp}(\theta_1, L) = \operatorname{Exp}(\bar{\theta}', L).$$

Thus $\bar{\theta}$ is an independent system by (5.5) and $\chi(\bar{\theta}, L) \geqslant \chi(\theta, L)$. The proof is quite similar in the case where $\theta = \theta_1 \times \theta_2$. ∎

Note that (5.4) is a precise description of $\operatorname{Exp}(\bar{\theta}, L)$ in terms of $\operatorname{Exp}(\theta, L)$. We give a more compact formulation of (5.4) by introducing first two vectors of $\mathbb{N}^m$

$$\vec{p} = (p_1, \ldots, p_m), \qquad \vec{q} = (N - p_1, \ldots, N - p_m).$$

Then we agree to define the product of two vectors $\vec{a} = (a_1, \ldots, a_m)$ and $\vec{b} = (b_1, \ldots, b_m)$ of $\mathbb{N}^m$ by

$$\vec{a}\vec{b} = (a_1 b_1, \ldots, a_m b_m).$$

Then (5.4) is equivalent to

$$\vec{q}+\vec{p}\,\mathrm{Exp}(\bar{\theta}, L)=\mathrm{Exp}(\theta, L)\cap(\vec{q}+\vec{p}\mathbb{N}^{2n}). \tag{5.6}$$

We are now ready for the analogue of Theorem 2.5.

**Theorem 5.5** *Let $L\subset X^*$ be an algebraic language, and let $\theta$ be an independent system of iterative pairs in L. For any algebraic grammar G generating L, there exists a system $\bar{\theta}$ deduced from $\theta$ which is grammatical with respect to G.*

Proof. As for Theorem 2.5, the proof is in two parts. We first prove the theorem in a special case, and then reduce the general case to this special case. Let

$$\theta=(a_0, u_1, \ldots, u_{2n}, a_{2n})$$

and set $m=2n$. In the first part, we assume that $X$ is the disjoint union of an alphabet $Z$ and of letters $t, x_1, \ldots, x_m$ such that

$$a_0\in Z^*t, \qquad a_i\in tZ^*t \quad (i=1,\ldots,m-1), \qquad a_m\in tZ^*,$$
$$u_i\in x_iZ^* \quad (i=1,\ldots,m).$$

Let $\pi_1,\ldots,\pi_n$ be the iterative pairs underlying to $\theta$. For each $\pi_j$, there are integers $l_j, r_j$ with $1\leqslant l_j<r_j\leqslant m$ such that $\pi_j=\langle u_{l_j}, u_{r_j}\rangle$. Renumber the $\pi_j$'s in such a way that $l_1<l_2<\cdots<l_n$. Then there is a sequence $\theta_n,\ldots,\theta_1$ of independent systems of iterative pairs such that

$$\theta_n=\pi_n, \qquad \theta_1=\theta$$

and for $i=1,\ldots,n-1$, one of the three following possibilities hold:

$$\theta_i=\pi_i * \theta_{i+1}; \qquad \theta_i=\pi_i\times\theta_{i+1};$$
$$\theta_i=(\pi_i * \theta^{(1)})\times\theta^{(2)}, \qquad \theta_{i+1}=\theta^{(1)}\times\theta^{(2)}$$

where $\theta^{(1)}$ and $\theta^{(2)}$ are systems of iterative pairs. (Note that the two first cases could be considered as a special case of the third one, arising when one of the two systems $\theta^{(1)}$ or $\theta^{(2)}$ is the "empty system" of length 0.)
Let $K=a_0u_1^*a_1u_2^*\cdots u_m^*a_m$ and let $H=G_K=\langle V, X, P\rangle$ be the grammar generating $L\cap K$ obtained from $G$. Let $\sigma\in V$ be such that $L\cap K=L_H(\sigma)$. In view of Lemma 5.1, it suffices to prove the result for the grammar $H$. Let $N$ be the integer associated to $H$ by Ogden's Iteration Lemma II.2.5, and define

$$g=a_0v_1a_1v_2\cdots v_ma_m \quad \text{with } v_i=u_i^N \qquad (1\leqslant i\leqslant m).$$

We proceed by induction on $i=n, n-1,\ldots,1$. Consider first $i=n$. Then $\pi_n=\langle u_k, u_{k+1}\rangle$ for some $k$. By the proof of Theorem 2.5, there exists a variable

$\xi_n$ such that

$$\sigma \xrightarrow{*} g_n = a_0 v_1 \cdots a_{k-1} c_k \xi_n d_{k+1} v_{k+2} \cdots v_m,$$

$$\xi_n \xrightarrow{*} \bar{u}_k \xi_n \bar{u}_{k+1}, \qquad \xi_n \xrightarrow{*} d_k a_k c_{k+1},$$

where $c_k$, $d_k$, $c_{k+1}$, $d_{k+1}$, $\bar{u}_k$, $\bar{u}_{k+1}$ are as in (5.2), (5.3). Thus the word $g$ admits a grammatical pair $\bar{\theta}_n$ with pivot $\xi_n$. Suppose now $n > i$, and assume

$$\theta_i = (\pi_i * \theta^{(1)}) \times \theta^{(2)}, \qquad \theta_{i+1} = \theta^{(1)} \times \theta^{(2)}.$$

(The two other cases are only simplified versions of the present one.) Let $\pi_i = \langle u_l, u_r \rangle$. The order on the $\pi_j$'s implies that

$$\theta^{(1)} = \langle u_{l+1}, \ldots, u_{r-1} \rangle, \qquad \theta^{(2)} = \langle u_{q_1}, \ldots, u_{q_h} \rangle$$

for some $r < q_1 < q_2 < \cdots < q_h$. We now make the following induction hypotheses:

a) There exists a grammatical system of iterative pairs $\bar{\theta}_{i+1} = \bar{\theta}^{(1)} \times \bar{\theta}^{(2)}$ where

$$\bar{\theta}^{(1)} = (a_0 v_1 \cdots v_l a_l c_{l+1}, \bar{u}_{l+1}, \bar{a}_{l+1}, \ldots, \bar{u}_{r-1}, d_{r-1} a_{r-1} v_r \cdots v_m a_m)$$

$$\bar{\theta}^{(2)} = (a_0 v_1 \cdots a_{q_1} c_{q_1}, \bar{u}_{q_1}, d_{q_1} \cdots c_{q_2}, \bar{u}_{q_2}, \ldots, \bar{u}_{q_h}, d_{q_h} a_{q_h} \cdots v_m a_m)$$

such that if $w^{(1)}$ and $w^{(2)}$ are the pivots of $\theta^{(1)}$ and $\theta^{(2)}$, then

$$\sigma \xrightarrow{*} g_{i+1} = a_0 v_1 \cdots v_l a_l c_{l+1} w^{(1)} d_{r-1} a_{r-1} v_r \cdots c_{q_1} w^{(2)} d_{q_h} a_{q_h} \cdots v_m a_m.$$

This implies the existence of derivations

$$c_{l+1} w^{(1)} d_{r-1} \xrightarrow{*} v_{l+1} a_{l+1} \cdots v_{r-1};$$

$$c_{q_1} w^{(2)} d_{q_h} \xrightarrow{*} v_{q_1} a_{q_1} v_{q_1+1} \cdots a_{q_h-1} v_{q_h}. \tag{5.7}$$

b) Any variable occurring in $w^{(1)}$ and $w^{(2)}$ derives in a word having at least one $a_j$ as factor.

These induction hypotheses are clearly satisfied by $\bar{\theta}_n$. Next we proceed as in the proof of Theorem 2.5 and distinguish two cases according to whether $\pi_i$ is left strict or right strict. Suppose that $\pi_i$ is right strict and mark the positions of the letter $x_r$ in the word $g_{i+1}$ (and not in $g$!). Then $g_{i+1}$ admits a grammatical iterative pair

$$\bar{\pi}_i = (\bar{a}, \bar{u}, \bar{b}, \bar{v}, \bar{c}).$$

Consequently $g_{i+1} = \bar{a}\bar{u}\bar{b}\bar{v}\bar{c}$ and there are, in $H$, the derivations

$$\sigma \xrightarrow{*} \bar{a}\xi_i\bar{c}, \qquad \xi_i \xrightarrow{*} \bar{u}\xi_i\bar{v}, \qquad \xi_i \xrightarrow{*} \bar{b},$$

whence

$$\sigma \xrightarrow{*} \bar{a}\xi_i\bar{c} \to \bar{a}\bar{u}^k\bar{b}\bar{v}^k\bar{c} \qquad (k \geq 0). \tag{5.8}$$

Further $\bar{u}$ or $\bar{v}$ contains occurrences of the letter $x_r$. First we note that neither $\bar{u}$ nor $\bar{v}$ contains the letter $t$ since otherwise, by (5.8), the number of letters $t$ in the words of $L \cap K$ would not be bounded. Next in view of hypothesis b), neither $\bar{u}$ nor $\bar{v}$ contains an occurrence of a variable. By the form of $K$, neither $\bar{u}$ nor $\bar{v}$, if they are nonempty, occur as factors in one of the $a_i$'s. Thus $\bar{u}$ and $\bar{v}$ are factors either of some $v_j$'s or of $c_{l+1}, d_{r-1}, c_{q_1}, d_{q_h}$. Assume now that $\bar{u}$ contains an occurrence of the letter $x_r$. Then $\bar{u}$ is a factor of $v_r$, and $\bar{v}$ is a factor of $v_r$, or of some $v_j (j > r)$ or of $c_{q_1}$ or $d_{q_h}$. If $\bar{v}$ is not a factor of $v_r$, there is some factorization

$$v_r = c'\bar{u}d',$$

and in view of (5.7)

$$\bar{a}\bar{u}^k\bar{b}\bar{v}^k\bar{c} \xrightarrow{*} a_0v_1a_1v_2 \cdots a_{r-1}c'\bar{u}^kd'a_rf^{(k)} \in L \cap K$$

where $f^{(k)}$ is some word in $u_{r+1}^*a_{r+1} \cdots u_m^*a_m$, $(k \geq 0)$. Thus $\bar{u}$ is of the form $\bar{u} = (u''u')^p$ for some factorization $u_r = u'u''$, and the independence of $\theta$ implies that

$$(N, N + p(k-1)) \in \mathrm{Exp}(\pi_i, L \cap K) \qquad \text{for } k \geq 0.$$

This is impossible since $\pi_i$ is right strict. Similarly if $\bar{v}$ is also a factor of $v_r$, then there is a factorization

$$v_r = c''\bar{u}e\bar{v}d''$$

and $$\bar{a}\bar{u}^k\bar{b}\bar{v}^k\bar{c} \xrightarrow{*} a_0v_1 \ldots a_{r-1}c''\bar{u}^ke\bar{v}^kd''a_rv_{r+1} \cdots v_ma_m \in L \cap K$$

and the independence of $\theta$ leads to a contradiction. Thus $\bar{u}$ contains no occurrence of the letter $x_r$, therefore $\bar{v}$ is a factor of $v_r$, and by the same argument as above one proves that $\bar{u}$ is a nonempty factor of $v_l$. This gives factorizations

$$v_l = u_l^N = c_l\bar{u}d_l, \qquad v_r = u_r^N = c_r\bar{v}d_r, \qquad u_l = u_l'u_l'', \qquad u_r = u_r'u_r''$$

such that $\bar{u} = (u_l''u_l')^{p_l}$, $\bar{v} = (u_r''u_r')^{p_r}$ for some $p_l, p_r \geq 1$.

The same conclusion holds if $\pi_i$ is left strict by marking the positions of $x_l$ instead of $x_r$.

Thus there are derivations

$$\sigma \xrightarrow{*} g_i = a_0v_1 \cdots a_{l-1}c_l\xi_id_ra_r \cdots c_{q_1}w^{(2)}d_{q_h} \cdots v_ma_m,$$

$$\xi_i \xrightarrow{*} \bar{u}\xi_i\bar{v}, \qquad \xi_i \xrightarrow{*} d_la_lc_{l+1}w^{(1)}d_{r-1}a_{r-1}c_r.$$

This shows that the induction hypotheses a) and b) are satisfied by the system of iterative pairs

$$\bar{\theta}_i = (a_0 \cdots a_{l-1}c_l, \bar{u}_l, \bar{a}_l, \ldots, \bar{u}_r, d_ra_r \cdots v_ma_m) \times \bar{\theta}^{(2)}.$$

Thus the proof is complete in the case where $\theta$ has the special form described at the beginning.

Finally, and as in the proof of Theorem 2.5, consider the general case of an independent system of iterative pairs

$$\theta = (a_0, u_1, \ldots, u_{2n}, a_{2n}).$$

Let $t, x_1, \ldots x_m (m = 2n)$ be new letters, set $Y = X \cup \{t, x_1, \ldots, x_m\}$ and let $\varphi : Y^* \to X^*$ be the projection. Set $M = \varphi^{-1}(L)$ and

$$K = a_0 t(x_1 u_1)^* t a_1 t(x_2 u_2)^* \cdots (x_m u_n)^* t a_m,$$
$$\theta' = (a_0 t, x_1 u_1; t a_1 t, \ldots, t a_m).$$

Then $\varphi(\theta') = \theta$ and $\mathrm{Exp}(\theta', M) = \mathrm{Exp}(\theta, L)$. By the first part of the proof, there is a system of iterative pairs $\hat{\theta}$ deduced from $\theta$ which is grammatical with respect to the grammar $\varphi^{-1}(G)$. Then $\bar{\theta} = \varphi(\hat{\theta})$ is deduced from $\theta = \varphi(\theta')$ and $\bar{\theta}$ is grammatical with respect to the grammar $G$ in view of Lemma 5.2. ■

## VIII.6 Transfer of Systems

The following theorem was proved by Boasson [1976] for systems of length 2, and by Beauquier [1978a] in the general case.

**Theorem 6.1** (Transfer Theorem for Systems of Iterative Pairs) *Let $A$ and $B$ be algebraic languages such that $A \geq B$. For any independent system $\theta$ of iterative pairs in $B$, there exists a system $\theta'$ of iterative pairs in $A$ of the same length such that $\chi(\theta', A) \geq \chi(\theta, B)$.*

Note that $\theta'$ is not claimed to be independent, and in fact the theorem is false with this additional condition. We shall give some "substantial" counterexample below (Example 6.4). This restriction makes the theorem difficult to apply in its present form since dependent systems of great length or "complexity" are in general easily constructed in languages which have only simple independent systems. This was seen for instance for $D_1'^*$ in Example 4.7. We therefore state below several variations or special cases of the Transfer Theorem.

Let $A \subset X^*$, $B \subset Y^*$ and assume $A \geq B$. Then by Nivat's Theorem, there is an alphabet $Z$ such that

$$B = \psi(\varphi^{-1}(A) \cap K)$$

where $\psi : Z^* \to Y^*$ is a morphism, $\varphi : Z^* \to X^*$ is an alphabetic morphism and $K \subset Z^*$ is a local regular language. With these notations, we first prove

**Proposition 6.2** *Let A and B be algebraic languages such that $A \geqslant B$, and let G be an algebraic grammar generating A. For any system of iterative pairs*

$$\theta = (a_0, u_1, \ldots, u_{2n}, a_{2n})$$

*in B grammatical with respect to the grammar $\psi[\varphi^{-1}(G)_K]$, there exists a system of iterative pairs*

$$\theta' = (a'_0, u'_1, \ldots, u'_{2n}, a'_{2n})$$

*in A grammatical with respect to G such that $\chi(\theta', A) \geqslant \chi(\theta, B)$ and* $\mathrm{Exp}(\theta', A) \subset \mathrm{Exp}(\theta, B)$. *Further*

$$a_0 u_1^{k_1} \cdots u_{2n}^{k_{2n}} a_{2n} \in \psi[\varphi^{-1}(a'_0 u_1'^{k_1} \cdots u_{2n}'^{k_{2n}} a'_{2n}) \cap K], \text{ for } k_1, \ldots, k_{2n} \geqslant 0.$$

Proof. Set $C = \varphi^{-1}(A) \cap K$. Then $B = \psi(C)$. In view of Lemma 5.3 there exists a system of iterative pairs $\bar{\theta} = (\bar{a}_0, \bar{u}_1, \ldots, \bar{u}_{2n}, \bar{a}_{2n})$ in $C$, grammatical with respect to $(\varphi^{-1}G)_K$, such that $\psi(\bar{\theta}) = \theta$. Further $\mathrm{Exp}(\bar{\theta}, C) \subset \mathrm{Exp}(\theta, B)$. Since $K$ is local, $\mathrm{Exp}(\bar{\theta}, C) = \mathrm{Exp}(\bar{\theta}, \varphi^{-1}(A))$ by Lemma 4.1, and $\bar{\theta}$ is grammatical with respect to $\varphi^{-1}(G)$ by Lemma 5.1. Define $\theta' = \varphi(\theta)$. Then $\theta'$ is a system of iterative pairs in $A$ and $\mathrm{Exp}(\theta', A) = \mathrm{Exp}(\bar{\theta}, \varphi^{-1}(A))$ by Lemma 4.2, and $\theta'$ is grammatical with respect to $G$ by Lemma 5.2. Thus $\mathrm{Exp}(\theta', A) \subset \mathrm{Exp}(\theta, B)$. Next if $w = a'_0 u_1'^{k_1} \cdots u_{2n}'^{k_{2n}} a'_{2n}$, let $\bar{w} = \bar{a}_0 \bar{u}_1^{k_1} \cdots \bar{u}_{2n}^{k_{2n}} \bar{a}_{2n}$. Then $\bar{w} \in \varphi^{-1}(w)$, and $\bar{w} \in K$ since $K \supset \bar{a}_0 \bar{u}_1^* \cdots \bar{u}_{2n}^* \bar{a}_{2n}$. Consequently $\psi\bar{w} = a_0 u_1^{k_1} \cdots u_{2n}^{k_{2n}} a_{2n} \in \psi[\varphi^{-1}(w) \cap K]$. ■

Proof of Theorem 6.1. Let $\varphi, \psi, K, Z$ be as above, and let $G$ be a grammar generating $A$. In view of Theorem 5.5 there exists an (independent) system of iterative pairs $\bar{\theta}$ deduced from $\theta$ which is grammatical with respect to the grammar $\psi[\varphi^{-1}(G)_K]$. Further $\chi(\bar{\theta}, B) \geqslant \chi(\theta, B)$ by Lemma 5.4. In view of the proposition above, there is a system of iterative pairs $\theta'$ in $A$ such that $\mathrm{Exp}(\theta', A) \subset \mathrm{Exp}(\bar{\theta}, B)$. Consequently $\chi(\theta', A) \geqslant \chi(\bar{\theta}, B) \geqslant \chi(\theta, B)$. ■

We now prove the following more precise version of Theorem 6.1.

**Proposition 6.3** *Let A and B be algebraic languages such that $A \geqslant B$. For any independent system of iterative pairs $\theta$ in B of length n, there exists a system of iterative pairs $\theta'$ in A of the same length, and two vectors $\vec{q}, \vec{p} \in \mathbb{N}^{2n}$ such that*

$$\vec{q} + \vec{p}\, \mathrm{Exp}(\theta', A) \subset \mathrm{Exp}(\theta, B) \cap (\vec{q} + \vec{p}\mathbb{N}^{2n}). \tag{6.1}$$

Proof. With the notations of the proof of Theorem 6.1, we have by (5.6)

$$\vec{q} + \vec{p}\, \mathrm{Exp}(\bar{\theta}, B) = \mathrm{Exp}(\theta, B) \cap (\vec{q} + \vec{p}\mathbb{N}^{2n}).$$

Since $\mathrm{Exp}(\theta', A) \subset \mathrm{Exp}(\bar{\theta}, A)$, the conclusion follows. ■

There is another formulation of Proposition 6.2 which can be used for proving nondomination.

**Corollary 6.4** *Let A and B be algebraic languages such that $A \geqslant B$, and let $\alpha$ be a type of a system of iterative pairs (i.e. a word in $D_4'^*$ over $\Xi$). If for any algebraic grammar $G'$ generating B, there is a system $\theta'$ of iterative pairs in B grammatical with respect to $G'$ and of type $\chi(\theta', B) \geqslant \alpha$, then for any algebraic grammar G generating A, there exists a system $\theta$ of iterative pairs in A grammatical with respect to G such that $\chi(\theta, A) \geqslant \alpha$.*

Proof. Assume that there is a grammar $G$ for $A$, such that any system $\theta$ of iterative pairs in $A$ grammatical with respect to $G$ satisfies $\chi(\theta, A) \not\geqslant \alpha$. Then with the notations of Proposition 6.2, let $G' = \psi[\varphi^{-1}(G)_K]$. There is a system $\theta'$ in $B$ grammatical with respect to $G'$ such that $\chi(\theta', B) \geqslant \alpha$. By Proposition 6.2, there exists a system $\theta$ in $A$, grammatical with respect to $G$, such that $\chi(\theta, A) \geqslant \chi(\theta', B) \geqslant \alpha$. This yields the contradiction. ∎

There is a more striking statement of the previous corollary. Define the grammatical type $\gamma(A)$ of an algebraic language to be the following subset of $D_4'^*$ over $\Xi$: $\alpha \in \gamma(A)$ iff for any algebraic grammar $G$ generating $A$, there exists a system $\theta$ of iterative pairs in $A$ grammatical with respect to $G$ such that $\chi(\theta, A) \geqslant \alpha$. Then Corollary 6.4 can be restated as:

**Corollary 6.5** *Let A and B be algebraic languages. If $A \geqslant B$ then $\gamma(A) \supset \gamma(B)$.* ∎

This corollary expressed the influence of rational domination on the "structural" description of the languages. Indeed, the "complexity" of a context-free language $A$ can be defined as the features which are common to all grammars generating $A$. The types of grammatical systems shared by all grammars for $A$ reflect in some sense these common features since they described restrictions in the use of nonterminals which are common to all grammars. Thus the corollary says that if $A$ dominates $B$, the grammars for $B$ are not as rich in structure as the grammars for $A$.

However, the meaning of results like Corollary 6.5 concerning grammatical systems of iterative pairs must not be overestimated. Indeed, the converse of the corollary is false, since for instance $\gamma(S_=) = \gamma(S_2)$, where $S_2$ is the symmetric language and $S_= = \{x^n y^n \mid n \geqslant 0\}$. This is a consequence of the fact (Exercise 6.1) that whenever a context-free language $L$ has a grammatical iterative pair of type $e\bar{e}$ with respect to some grammar $G$, then it has also grammatical systems of iterative pairs of type $e^n\bar{e}^n$ with respect to $G$ for any $n \geqslant 2$. (Of course, these systems are not independent in general.) Thus the sets $\gamma(L)$ are not a satisfactory measure of complexity, since a language like $S_2$ is intuitively regarded as, and proved by rational domination to be more "complex" than $S_=$. Consequently, the consideration of grammatical systems of iterative pair is not sufficient, even if they are very useful in some cases (see for instance Theorem 7.14).

Results like Proposition 6.2 or Corollary 6.5 may be delicate to apply since they require some knowledge about all grammars generating a given language. Thus the description of special cases of the Transfer Theorem or of Proposition 6.2 may provide helpful results for comparison of languages. Such a special case is realized when all the underlying iterative pairs of a given system are very strict, and when further this system is independent. Therefore we define:

**Definition** Let $\theta$ be a system of iterative pairs in $L$. Then $\theta$ is called very strict if all underlying iterative pairs of $\theta$ are very strict.

Thus $\theta$ is very strict in $L$ iff $\chi(\theta, L) \in \{e, \bar{e}\}^*$. We first give a characterization of independence for very strict systems.

**Lemma 6.6** *Let $\theta$ be a very strict system of iterative pairs in L. Then $\theta$ is independent iff* $\mathrm{Exp}(\theta, L) = \mathrm{Diag}(\theta, L)$.

Proof. We first prove that the condition is necessary by induction on the length $n$ of $\theta$. For $n = 1$ there is nothing to prove. Assume $\theta = \pi * \theta'$. Then $\mathrm{Exp}(\pi, L) = \Delta$ since $\theta$ is very strict. Next $\mathrm{Exp}(\theta, L) = \Delta * \mathrm{Exp}(\theta', L)$ since $\theta$ is independent. By induction, $\mathrm{Exp}(\theta', L) = \mathrm{Diag}(\theta', L)$ and therefore $\mathrm{Exp}(\theta, L) = \Delta * \mathrm{Diag}(\theta', L) = \mathrm{Diag}(\theta, L)$. Next assume $\theta = \theta_1 \times \theta_2$. Then $\mathrm{Exp}(\theta, L) = \mathrm{Exp}(\theta_1, L) \times \mathrm{Exp}(\theta_2, L)$ since $\theta$ is independent. By induction, $\mathrm{Exp}(\theta_i, L) = \mathrm{Diag}(\theta_i, L)$ $i = 1.2$. Thus $\mathrm{Exp}(\theta, L) = \mathrm{Diag}(\theta_1, L) \times \mathrm{Diag}(\theta_2, L) = \mathrm{Diag}(\theta, L)$.
Conversely, assume $\mathrm{Exp}(\theta, L) = \mathrm{Diag}(\theta, L)$. If $n = 1$ there is nothing to prove. If $\theta = \pi * \theta'$, then $\mathrm{Exp}(\pi, L) = \Delta$ since $\theta$ is very strict. Next $\mathrm{Exp}(\theta, L) = \Delta * \mathrm{Diag}(\theta', L)$ and $\mathrm{Exp}(\theta, L) \supset \{(1, 1)\} * \mathrm{Exp}(\theta', L)$. Thus $\{(1, 1)\} * \mathrm{Exp}(\theta', L) \subset \{(1, 1)\} * \mathrm{Diag}(\theta', L)$, whence $\mathrm{Diag}(\theta', L) = \mathrm{Exp}(\theta', L)$. Thus $\theta'$ is independent by induction, and since $\mathrm{Exp}(\theta, L) = \Delta * \mathrm{Exp}(\theta', L)$, $\theta$ is independent. If $\theta = \theta_1 \times \theta_2$, then $\mathrm{Exp}(\theta, L) = \mathrm{Diag}(\theta_1, L) \times \mathrm{Diag}(\theta_2, L)$; Let $k$ be the length of $\theta_1$ and let $\vec{1}_k = (1, 1, \ldots, 1) \in \mathbb{N}^{2k}$. Then $\vec{1}_k \times \mathrm{Exp}(\theta_2, L) \subset \mathrm{Exp}(\theta, L)$, thus $\mathrm{Exp}(\theta_2, L) = \mathrm{Diag}(\theta_2, L)$. Similarly $\mathrm{Exp}(\theta_1, L) = \mathrm{Diag}(\theta_1, L)$. Consequently $\theta_1$ and $\theta_2$ are independent by induction, and $\mathrm{Exp}(\theta, L) = \mathrm{Exp}(\theta_1, L) \times \mathrm{Exp}(\theta_2, L)$. Thus $\theta$ is independent. ∎

**Corollary 6.7** *Let A and B be algebraic languages such that $A \geqslant B$. For any independent very strict system $\theta$ of iterative pairs in B, there exists an independent very strict system $\bar{\theta}$ of iterative pairs of the same type in A.*

Proof. The only new assertion is the independence of $\bar{\theta}$. Let $n$ be the length of $\theta$ and set $m = 2n$. Then in view of (6.1) and of the previous lemma, there is a very strict system $\bar{\theta}$ in $A$ of same type as $\theta$ such that

$$\vec{q} + \vec{p}\ \mathrm{Exp}(\bar{\theta}, A) \subset \mathrm{Diag}(\theta, B), \tag{6.2}$$

where as in Section 5:

$$\vec{p} = (p_1, \ldots, p_m), \qquad \vec{q} = (N - p_1, \ldots, N - p_m).$$

We prove by induction on $n$ that (6.2) implies:

$$\mathrm{Exp}(\bar{\theta}, A) = \mathrm{Diag}(\bar{\theta}, A). \tag{6.3}$$

The corollary follows then by the previous lemma. If $n=1$, then (6.3) holds by definition. Thus assume $n>1$, and suppose first $\theta=\theta_1\times\theta_2$, for some independent very strict systems $\theta_1$ and $\theta_2$. Let $l$ be the length of $\theta_1$. Since $\theta$ and $\bar{\theta}$ have the same type, $\bar{\theta}=\bar{\theta}_1\times\bar{\theta}_2$, where $\bar{\theta}_1$ and $\bar{\theta}_2$ are of the same type as $\theta_1$ and $\theta_2$ respectively. Next $(k_1,\ldots,k_{2l})\in\mathrm{Exp}(\bar{\theta}_1, A)$ implies that

$$\vec{q}+\vec{p}(k_1,\ldots,k_{2l},1,\ldots,1)\in\mathrm{Diag}(\theta, B)=\mathrm{Diag}(\theta_1, B)\times\mathrm{Diag}(\theta_2, B).$$

Thus (6.2) is verified for $\bar{\theta}_1$ and $\theta_1$, and similarly for $\bar{\theta}_2$ and $\theta_2$. By induction, $\mathrm{Exp}(\bar{\theta}_i, A)=\mathrm{Diag}(\bar{\theta}_i, A)$, $(i=1,2)$. Now let $(k_1,\ldots,k_m)\in\mathrm{Exp}(\bar{\theta}, A)$. Then by (6.2)

$$(N+(k_1-1)p_1,\ldots,N+(k_{2l}-1)p_{2l})\in\mathrm{Diag}(\theta_1, B)$$

$$(N+(k_{2l+1}-1)p_{2l+1},\ldots,N+(k_m-1)p_m)\in\mathrm{Diag}(\theta_2, B).$$

Consequently $(k_1,\ldots,k_m)\in\mathrm{Diag}(\bar{\theta}_1, A)\times\mathrm{Diag}(\bar{\theta}_2, A)$, whence

$$\mathrm{Exp}(\bar{\theta}, A)\subset\mathrm{Diag}(\bar{\theta}_1, A)\times\mathrm{Diag}(\bar{\theta}_2, A).$$

Suppose now $\theta=\pi*\theta'$. Then $\bar{\theta}=\bar{\pi}*\bar{\theta}'$. Since $\bar{\theta}$ is very strict, $\mathrm{Exp}(\bar{\pi}, A)=\Delta$. By the same argument as above, it is seen that $\mathrm{Exp}(\bar{\theta}', A)=\mathrm{Diag}(\bar{\theta}', A)$. It follows in the same way that

$$\mathrm{Exp}(\bar{\theta}, A)\subset\Delta*\mathrm{Diag}(\bar{\theta}, A).$$ ■

We now give some examples of the use of the Transfer Theorem and of its variations. Further applications are given in the next section.

**Example 6.1** The language $A=\{x^ny^nz^pt^p \mid n,p\geq 0\}(\approx(S_=)^2)$ strictly dominates $S_==\{x^ny^n \mid n\geq 0\}$. Obviously $A$ dominates $S_=$. To prove that the domination is strict, we observe that there exists in $A$ an independent very strict system $\theta$ of type $e\bar{e}e\bar{e}$, namely $\theta=(1,x,1,y,1,z,1,t,1)$. Clearly no word of $S_=$ admits an independent system of length 2. Thus by Corollary 6.7, $S_=$ does not dominate $A$. (Note that $S_=$ has very strict systems of type $ee\bar{e}\bar{e}$, for instance $(1,x,1,x,1,y,1,y,1)$, but these systems clearly are dependent.)

**Example 6.2** The languages $A=\{x^ny^nz^pt^p \mid n,p\geq 0\}$ and $B=\{x^ny^pt^pz^n \mid n,p\geq 0\}$ are rationally incomparable. Indeed, the independent very strict systems of length 2 have all type $e\bar{e}e\bar{e}$ in $A$, and type $ee\bar{e}\bar{e}$ in $B$. These types are distinct, so by Corollary 6.7, the languages $A$ and $B$ are incomparable.

**Example 6.3** Consider the languages

$$L=\{x^ny^mz^pt^q \mid n\geq m\geq 0, 0\leq p\leq q\}\approx S_{\geq}\cdot S_{\leq},$$

$$M=\{x^ny^mz^pt^q \mid m\geq n\geq 0, 0\leq q\leq p\}\approx S_{\leq}\cdot S_{\geq}.$$

We claim that $L$ and $M$ are incomparable. Assume for instance that $M$ dominates $L$. Clearly $xyzt$ admits, in $L$ an independent system of iterative pairs of type $r\bar{r}l\bar{l}$. Then the Transfer Theorem asserts the existence of a system of iterative pairs of length 2 in $M$, of type $\geqslant r\bar{r}l\bar{l}$. This system is necessarily of the form $\pi_1 \times \pi_2$, where $\pi_1$ and $\pi_2$ are (nondegenerated) iterative pairs. Clearly any such system has type $l\bar{l}r\bar{r}$. This yields the contradiction.

**Example 6.4** (Latteux [1977]) The restricted Dyck language $D_1'^* \subset \{x, \bar{x}\}^*$ strictly dominates the language $L$ of the preceding example. The fact that $L$ does not dominate $D_1'^*$ follows from the Transfer Theorem for iterative pairs, since $D_1'^*$ has very strict pairs and obviously $L$ has no pair of this type. The remarkable feature of this example is that $D_1'^*$ dominates $L$, whereas $L$ has independent systems of length 2, and $D_1'^*$ has none such system (Example 4.7). To prove $D_1'^* \geqslant L$, we first note that

$$D_1'^* \geqslant D_1'^* \cap x^*\bar{x}^*x^*\bar{x}^* \approx N$$

with $N = \{x^n y^m z^p t^q \mid n \geqslant m \geqslant 0, 0 \leqslant p \leqslant q, n+p = m+q\}$.
Next we verify that $L = x^*Nt^*$ which gives the desired conclusion. Obviously $x^*Nt^* \subset L$. Conversely, let $w = x^n y^m z^p t^q \in L$. Then $n \geqslant m$, $p \leqslant q$. If $k = (n-m-(q-p) \geqslant 0$, let $r = n-k$. Then $r = m+q-p \geqslant 0$, and $r-m = q-p \geqslant 0$. Thus $w = x^k w'$ with $w' = x^r y^m z^p t^q \in N$, whence $w \in x^*N$. Symmetrically, $k \leqslant 0$ implies $w \in Nt^*$.

### Exercise

**6.1** Suppose that a context-free language $L$ has an iterative pair of type $e\bar{e}$. Show that for any context-free grammar $G$ generating $L$, there is in $L$ a system of iterative pairs of type $e^n\bar{e}^n$ grammatical with respect to $G$, for any $n \geqslant 2$.

## VIII.7 Applications

In this section, we apply the Transfer Theorem and its corollaries to continue the classifications of context-free languages. Several infinite chains of cones are constructed. The main results are: The cone of restricted one counter languages and the cone of quasi-rational languages are incomparable; the cone of finite counter languages and the cone of linear languages are incomparable. We first prove a weaker result used later.

**Proposition 7.1** *The cones* Rocl *and* Lin *are incomparable. Neither* Rocl *nor* Lin *is closed under product.*

Thus in particular $D_1'^*$ is not a generator of Alg.

Proof. In view of Examples 4.6 and 4.7, neither $D_1'^*$ nor $S_2$ has an independent system of iterative pairs of type $(e\bar{e})^2$. Thus by Corollary 6.5 the language

$(S_=)^2$ is neither in Rocl nor in Lin. This proves the second part of the proposition. We already know that $D_1'^*$ is not a linear language (Example V.6.5). To show that Lin is not contained in Rocl, consider the language $L=\{x^n y^p z^p t^n \mid n, p \geqslant 0\}$. Clearly $L$ is linear and there is an independent system of type $e^2\bar{e}^2$ in $L$. Thus in view of Corollary 6.7 and Example 4.7, $L$ is not a restricted one counter language. ∎

The next proposition extends Example 6.1 and shows the existence of an ascending chain of rational cones which are all contained in the least full AFL Lin $\hat{\Gamma}$ containing the linear languages and which are all, excepted the first, neither contained in Rocl nor in Lin.

**Proposition 7.2** *The rational cones $(S_=)^n\Gamma$ $(n \geqslant 1)$ form a strictly increasing chain of cones.*

Proof. Obviously $(S_=)^{n+1} \geqslant (S_=)^n$ for $n \geqslant 1$. Next $(S_=)^n$ has an independent system of iterative pairs of type $(e\bar{e})^n$, but has no such system of type $(e\bar{e})^{n+1}$. Thus by Corollary 6.7, $(S_=)^{n+1}$ does not dominate $(S_=)^n$. ∎

There is another interesting chain built of cones of linear languages. Define $A_n \subset (X_n \cup \bar{X}_n)^*$, where $X_n = \{x_1, \ldots, x_n\}$ and $\bar{X}_n = \{\bar{x}_1, \ldots, \bar{x}_n\}$ by

$$A_n = \{x_1^{k_1} \cdots x_n^{k_n} \bar{x}_n^{k_n} \cdots \bar{x}_1^{k_1} \mid k_1, \ldots, k_n \geqslant 0\}.$$

Clearly

$$A_n = S_n \cap x_1^* \cdots x_n^* \bar{x}_n^* \cdots \bar{x}_1^*,$$

where $S_n$ is the symmetric language over $X_n \cup \bar{X}_n$.

**Proposition 7.3** *The cones $A_n\Gamma$, $(n \geqslant 1)$ form a strictly increasing chain of rational cones.*

Proof. Clearly $A_n$ has independent very strict systems of iterative pairs of type $e^n\bar{e}^n$, but none of type $e^{n+1}\bar{e}^{n+1}$. Thus the conclusion follows from Corollary 6.7. ∎

The rational cone $\mathscr{L} = \bigcup_{n \geqslant 1} A_n\Gamma$ is not principal, thus $\mathscr{L}$ is strictly contained in Lin. It can even be shown that $\mathscr{L} \subsetneq \mathrm{Ng}(\mathrm{Lin}) = \{L \in \mathrm{Lin} \mid L\Gamma \neq \mathrm{Lin}\}$ (see Exercise 7.1). The cones $A_n\Gamma$ and $(S_=)^m\Gamma$, $(n, m \geqslant 2)$ are pairwise incomparable. For a generalization, see Exercise 7.2. In order to pursue the classification of families of languages, we now consider systems of iterative pairs in principal full AFLs, and in the syntactic substitution of two languages. We need a definition.

**Definition** A system of iterative pairs $\theta$ in a language $L$ is called prime if either $\theta$ is an iterative pair or $\theta = \pi * \theta'$.

Clearly $\theta$ is prime iff its type $\chi(\theta, L)$ is a Dyck prime in the language $D_4'^*$ over $\Xi$. Thus any system $\theta$ can be written in the form

$$\theta = \theta^{(1)} \times \cdots \times \theta^{(r)}$$

where $\theta^{(1)}, \ldots, \theta^{(r)}$ are prime systems.

**Proposition 7.4** *Let L be a language, $d \notin X_L$. If $\theta$ is a prime system of iterative pairs in $(Ld)^+$, then there is a prime system $\bar{\theta}$ of iterative pairs in L such that* $\mathrm{Exp}(\bar{\theta}, L) = \mathrm{Exp}(\theta, (Ld)^+)$.

Proof. We consider first iterative pairs. Let $\pi = (a, u, b, v, c)$ be a nondegenerated iterative pair of a word

$$w = h_1 d h_2 d \cdots h_m d \qquad (h_i \in L)$$

in $L$. Then $w = aubvc$. We claim that neither $u$ nor $v$ contains an occurrence of the letter $d$. Assume indeed for instance that $|u|_d > 0$. Then there are integers $i < j$, and factorizations

$$h_i = h_i' h_i'', \qquad h_j = h_j' h_j''$$

such that

$$a = h_1 d \cdots h_{i-1} d h_i', \qquad u = h_i'' d h_{i+1} d \cdots h_{j-1} d h_j'.$$

Since $au^n bv^n c \in (Ld)^+$, necessarily $h_j' h_i'' \in L$, $h_i' h_j'' \in L$. Consequently, $au^n bv^n c \in (Ld)^+$ implies $au^p bv^n c \in (Ld)^+$ for all $p \geqslant 0$, whence $au^* bv^* c \in (Ld)^+$ and $\pi$ is degenerated contrary to the assumption. Thus there is an index $i$, $(1 \leqslant i \leqslant m)$ and a factorization

$$h_i = \bar{a} u h_i', \quad \text{with} \quad a = h_1 d \cdots h_{i-1} d \bar{a}.$$

Then two cases are possible: a) There is an index $j > i$ such that $h_j = h_j' v h_j''$, and $b = h_i' d \cdots h_{j-1} d h_j'$. But then $au^n bv^n c \in (Ld)^+$ implies $\bar{a} u^n h_i' \in L$, $h_j' v^n h_j'' \in L$, whence again $au^* bv^* c \in (Ld)^+$. Therefore we have: b) $h_i'$ factorizes into $h_i' = bv\bar{c}$ (and $\bar{c} d h_{i+1} d \cdots h_m d = c$). Then $au^n bv^r c \in (Ld)^+$ iff $\bar{a} u^n bv^r \bar{c} \in L$, thus $\pi = (\bar{a}, u, b, v, \bar{c})$ is an iterative pair of $h_i$ in $L$, and $\mathrm{Exp}(\bar{\pi}, L) = \mathrm{Exp}(\pi, (Ld)^+)$. Consider next a prime system

$$\theta = (a_0, u_1, a_1, \ldots, u_{2n}, a_{2n})$$

of iterative pairs of $w$ in $(Ld)^+$, and let $\theta = \pi * \theta'$, with $\pi = \langle u_1, u_{2n} \rangle$. Then by the first part of the proof, there are factorizations $a_0 = w' \bar{a}_0$, $a_{2n} = \bar{a}_{2n} w''$ with $w', w'' \in (Ld)^*$, such that $\bar{a}_0 u_1 a_1 \cdots u_{2n} \bar{a}_{2n} \in L$. Thus

$$\bar{\theta} = (\bar{a}_0, u_1, a_1, \ldots, u_{2n}, \bar{a}_{2n})$$

is a system of iterative pairs in $L$ and clearly $\mathrm{Exp}(\bar{\theta}, L) = \mathrm{Exp}(\theta, (Ld)^+)$. ∎

**Example 7.1** The symmetric language $S_2$ is not in $\mathrm{Ocl} = D_1'^* \hat{\Gamma}$. Assume indeed that $(D_1'^* d)^+$ dominates $S_2$. Since by Example 4.6, $S_2$ has independent very strict systems of type $e^p\bar{e}^p$ for any $p \geq 1$, the language $(D_1'^*d)^+$ must have independent systems of these types by Corollary 6.7. But then Proposition 7.4 implies that $D_1'^*$ itself has independent systems of these types, and this contradicts Example 4.7.

We now extend the previous result to the syntactic substitution $L \uparrow M$, with the aim to prove that $S_2$ is not in the substitution closure of the one counter languages. Consider two disjoint alphabets $X$, $Y$, and let $L \subset X^*$, $M \subset Y^*$. Let $\varphi$ be the projection $(X \cup Y)^* \to X^*$.

We first consider two special cases.

**Lemma 7.5** *Let $\theta = (a_0, u_1, \ldots, u_{2n}, a_{2n})$ be a system of iterative pairs in $L \uparrow M$, and suppose that $u_1 a_1 \cdots a_{2n-1} u_{2n} \in Y^*$. Then there exists a system of iterative pairs $\bar{\theta}$ in $M$ with* $\mathrm{Exp}(\bar{\theta}, M) = \mathrm{Exp}(\theta, L \uparrow M)$.

Proof. Set $w = a_0 u_1 \cdots u_{2n} a_{2n}$. Since $w \in L \uparrow M$, $w = h_1 x_1 \cdots h_p x_p$ for some $h_1, \ldots, h_p \in M$, $x_1, \ldots, x_p \in X$, $x_1 \cdots x_p \in L$. The assumption implies that there is an index $i$, $(1 \leq i \leq p)$ and a factorization

$$h_i = \bar{a}_0 u_1 a_1 \cdots u_{2n} \bar{a}_{2n}$$

$$a_0 = h_1 x_1 \cdots x_{i-1} \bar{a}_0, \qquad a_{2n} = \bar{a}_{2n} x_i h_{i+1} x_{i+1} \cdots h_p x_p.$$

Then $\bar{\theta} = (\bar{a}_0, u_1, a_1, \ldots, u_{2n}, \bar{a}_{2n})$ clearly satisfies $\mathrm{Exp}(\bar{\theta}, M) = \mathrm{Exp}(\theta, L \uparrow M)$. ■

**Lemma 7.6** *Let $\theta = (a_0, u_1, \ldots, u_{2n}, a_{2n})$ be a system of iterative pairs in $L \uparrow M$, and suppose that $|u_i|_X > 0$, $(i = 1, \ldots, 2n)$ and $|a_i|_X > 0$, $(i = 1, \ldots, 2n-1)$. Then $\varphi(\theta)$ is a system of iterative pairs in $L$ and* $\mathrm{Exp}(\varphi(\theta), L) = \mathrm{Exp}(\theta, L \uparrow M)$.

Proof. Clearly $\mathrm{Exp}(\theta, L \uparrow M) \subset \mathrm{Exp}(\varphi(\theta), L)$. In order to prove the converse inclusion, observe first that $L \uparrow M = \varphi^{-1}(L) \cap (MX)^*$. Next $\mathrm{Exp}(\varphi(\theta), L) = \mathrm{Exp}(\theta, \varphi^{-1}(L))$. Set $K = a_0 u_1^* a_1 \cdots u_{2n}^* a_{2n}$. Suppose proved already that $K \subset (MX)^*$. Then

$$\mathrm{Exp}(\theta, \varphi^{-1}(L)) = \mathrm{Exp}(\theta, \varphi^{-1}(L) \cap K) \subset \mathrm{Exp}(\theta, L \uparrow M),$$

whence $\mathrm{Exp}(\varphi(\theta), L) = \mathrm{Exp}(\theta, L \uparrow M)$ as desired. Thus it suffices to verify the inclusion $K \subset (MX)^*$. For this, set $m = 2n$,

$$w = a_0 u_1 \cdots u_m a_m = h_1 x_1 \cdots h_p x_p,$$

with $h_1, \ldots, h_p \in M$, $x_1, \ldots, x_p \in L$. The conditions $\varphi(u_i) \neq 1$, $\varphi(a_i) \neq 0$ imply that there exist indices $1 \leq r(1) < s(1) < r(2) < \cdots < r(m) < s(m) \leq p$ such that

$$u_i = h''_{r(i)} x_{r(i)} \cdots x_{s(i)-1} h'_{s(i)} \qquad i = 1, \ldots, m$$

$$a_i = h''_{s(i)} x_{s(i)} \cdots x_{r(i+1)-1} h'_{r(i+1)} \qquad i = 1, \ldots, m-1$$

$$a_0 = h_1 x_1 \cdots x_{r(1)} h'_{r(1)}, \qquad a_m = h''_{s(m)} x_m \cdots h_p x_p$$

for factorizations $h_{r(i)}=h'_{r(i)}h''_{r(i)}$, $h_{s(i)}=h'_{s(i)}h''_{s(i)}$, $(i=1,\ldots,m)$. Since $a_0a_1a_2\cdots a_m\in L\uparrow M$, we have $h'_{r(i)}h''_{s(i)}\in M$ for $i=1,\ldots,m$. Similarly, since $a_0u_1^2a_1u_2^2\cdots a_{m-1}u_m^2a_m\in L\uparrow M$, we have $h'_{s(i)}h''_{r(i)}\in M$ for $i=1,\ldots,m$. Consequently $a_0u_1^{k_1}a_1\in(MX)^*h'_{r(2)}$ for all $k_1\geq 0$, and by induction

$$a_0u_1^{k_1}a_1u_2^{k_2}a_2\cdots u_i^{k_i}a_i\in(MX)^+h'_{r(i+1)}$$
$$\text{for}\quad i\leq m-1\quad\text{and}\quad k_1,\ldots,k_i\geq 0,$$

and the desired result follows. ∎

**Proposition 7.7** *Let $\pi=(a,u,b,v,c)$ be a nondegenerated iterative pair in $L\uparrow M$. Then*

(i) *either $|u|_X>0$, $|v|_X>0$ and $\varphi(\pi)$ is an iterative pair in $L$;*

(ii) *or $ubv\in Y^+$, and there is an iterative pair $\bar\pi$ in $M$ such that* $\operatorname{Exp}(\bar\pi,M)=\operatorname{Exp}(\pi,L\uparrow M)$.

Note that we do not claim that $\varphi(\pi)$ is nondegenerated. Indeed this stronger result is false in general.

**Example 7.2** Let $X=\{x,y\}$, $Y=\{a,b,c\}$, $L=x^*y\cup S_<$, and $M=\{1,a,bc,c,ab\}$. Consider the iterative pair $\pi=(a,x,b,cy,y)$ in $L\uparrow M$. Then $\operatorname{Exp}(\pi,L\uparrow M)=\{(n,m)\mid 0\leq n\leq m\}$. Next $\varphi(\pi)=(1,x,1,y,y)$ and $\operatorname{Exp}(\varphi(\pi),L)=\mathbb{N}\times\{0\}\cup\{(n,m)\mid 0\leq n\leq m\}$. Thus $\pi$ is left strict and $\varphi(\pi)$ is degenerated.

However, the previous example is somewhat artificial as is shown by the

**Corollary 7.8** *Let $\pi=(a,u,b,v,c)$ be a nondegenerated iterative pair in $L\uparrow M$ and suppose $|u|_X>0,|v|_X>0$. Then $\pi'=(a,u,ub,v,vc)$ is an iterative pair in $L\uparrow M$ with $\chi(\pi',L\uparrow M)\geq\chi(\pi,L\uparrow M)$, and $\varphi(\pi')$ is an iterative pair in $L$ satisfying* $\operatorname{Exp}(\varphi(\pi'),L)=\operatorname{Exp}(\pi',L\uparrow M)$.

Proof. Clearly $(k,l)\in\operatorname{Exp}(\pi',L\uparrow M)$ iff $(1+k,1+l)\in\operatorname{Exp}(\pi',L\uparrow M)$. Thus $\chi(\pi',L\uparrow M)\geq\chi(\pi,L\uparrow M)$. Next by Proposition 7.7, $\varphi(\pi')$ is an iterative pair in $L$. Finally $\pi'$ satisfies the conditions of Lemma 7.6, and this lemma gives the required equality. ∎

Proof of Proposition 7.7. If $|u|_X>0$ and $|v|_X>0$, then $\varphi(\pi)$ is an iterative pair in $L$ since $|\varphi(uv)|>0$ and $\operatorname{Exp}(\pi,L\uparrow M)\subset\operatorname{Exp}(\varphi(\pi),L)$. Next if $ubv\in Y^+$, then the conclusion follows from Lemma 7.5. Thus it remains only to prove that no other possibilities arise. For this, we show that $u\in Y^+$ implies $ubv\in Y^+$. The symmetrical implication: $v\in Y^+\Rightarrow ubv\in Y^+$ is proved in the same way. Set

$$w=aubvc=h_1x_1\cdots h_px_p,$$

with $h_1,\ldots,h_p\in M$, $x_1,\ldots,x_p\in X$, $x_1\cdots x_p\in L$. If $u\in Y^+$, then there is an index $i$, $(0\leq i\leq p)$ such that

$$a=h_1x_1\cdots h_{i-1}x_{i-1}h'_i,\qquad h_i=h'_iuh''_i,\qquad bvc=h''_ix_i\cdots h_px_p.$$

We claim that $bv$ is a left factor of $h_i''$, showing thus that $ubv$ is in $Y^+$. Assume the contrary. Then $h_i''$ is a proper left factor of $bv$, and for $n \geqslant 1$,

$$au^n bv^n c = h_1 x_1 \cdots h_{i-1} x_{i-1} h_i' u^n h_i'' x_i f_n,$$

where $f_n \in (MX)^*$ is such that $x_1 \cdots x_i \varphi(f_n) \in L$. Since $h_i' u^n h_i'' \in M$ for all $n \geqslant 1$, we also have

$$h_1 x_1 \cdots h_{i-1} x_{i-1} h_i' u^m h_i'' x_i f_n = au^m bv^n c \in L \uparrow M$$

for all $m$, $n > 0$. Thus $\pi$ is degenerated, contrary to the assumption. ■

**Theorem 7.9** *The symmetric language $S_2$ is not a finite counter language. The cones* Fcl *and* Lin *are incomparable.*

Thus the relation Lin $\not\subset$ Rocl of Proposition 7.1 is strengthened to: Lin $\not\subset$ Fcl. The "dual" result, namely Rocl $\not\subset$ Qrt, will be proved below. Recall that $\mathrm{Fcl} = \bigcup_{n \geqslant 1} \mathrm{Fcl}(n)$, where $\mathrm{Fcl}(1) = \mathrm{Rocl}$ and $\mathrm{Fcl}(n+1) = \mathrm{Rocl} \mathbin{\square} \mathrm{Fcl}(n)$ for $n \geqslant 1$. The theorem is a direct consequence of the following proposition.

**Proposition 7.10** *Let $A \in \mathrm{Fcl}(n)$ and suppose that $A$ has an independent system of iterative pairs of type $e^p \bar{e}^p$. Then $p \leqslant n$.*

Proof of Theorem 7.9. The symmetric language $S_2$ has independent systems of iterative pairs of type $e^p \bar{e}^p$ for any $p \geqslant 1$ (Example 4.6). Thus by the previous proposition, $S_2$ is not in Fcl($n$) for any $n \geqslant 1$. Thus Lin $\not\subset$ Fcl. Since Rocl $\not\subset$ Lin, the converse relation is obvious. ■

Proof of Proposition 7.10. (i) We first consider the following situation: Let $A = D_1'^* \uparrow M$, where $D_1'^* \subset X^*$ with $X = \{x, \bar{x}\}$, $M \subset Y^*$, $X \cap Y = \emptyset$, and $\varphi$ is the projection of $(X \cup Y)^*$ onto $X^*$. Suppose that

$$\theta = (a_0, u_1, a_1, u_2, a_2, u_3, a_3, u_4, a_4) = \pi_1 * \pi_2$$

is an independent system of type $e^2 \bar{e}^2$ in $A$. We claim that then $|u_1|_X = 0$ or $|u_2|_X = 0$. Indeed, assume the contrary. Then by Proposition 7.7, $|u_3|_X > 0$ and $|u_4|_X > 0$. Consider the system

$$\theta' = (a_0, u_1, u_1 a_1, u_2, u_2 a_2, u_3, u_3 a_3, u_4, u_4 a_4).$$

Since $\mathrm{Exp}(\theta, A) = \Delta * \Delta$, we have $\mathrm{Exp}((\theta', A) = \Delta * \Delta$. In view of Lemma 7.6, $\varphi(\theta')$ is an independent system of iterative pairs of type $e^2 \bar{e}^2$ in $D_1'^*$ which is impossible by Example 4.7.

(ii) Next we claim that if $A = D_1'^* \uparrow M$ has an independent system of type $e^p \bar{e}^p$ with $p \geqslant 2$, then $M$ has an independent system of type $e^{p-1} \bar{e}^{p-1}$. Let indeed

$$\theta = (a_0, u_1, \ldots, u_{2p}, a_{2p}) = \pi_1 * \pi_2 * \cdots * \pi_p$$

be an independent system of type $e^p \bar{e}^p$ in $A$. Then $\pi_1 * \pi_2$ is an independent system of type $e^2 \bar{e}^2$ in $A$, and by part (i), we have $u_1 a_1 \cdots a_{2p-1} u_{2p} \in Y^+$. In

view of Lemma 7.5, there is an independent system of type $e^{p-1}\bar{e}^{p-1}$ in $M$.
(iii) We now define languages $A_n$, $(n \geq 1)$ as follows: $A_1 = D_1'^*$, $A_{n+1} = D_1'^* \uparrow (A_n' d_n)^+$, where $A_n'$ is a copy of $A_n$ over an alphabet disjoint from $X$, and where $d_n$ is a new letter. By Theorem VI.5.7, $A_n$ is a generator of the rational cone Fcl$(n)$. (Note that the use of Exercise VI.5.7 would supply generators of simpler form.) If $A \in \text{Fcl}(n)$, then $A_n \geq A$, and by Corollary 6.7, $A_n$ has an independent system of iterative pairs of type $e^p \bar{e}^p$. Thus we have to prove that if $A_n$ has an independent system of type $e^n \bar{e}^n$, then $p \leq n$. This will be done by induction, the case $n = 1$ being true in view of Example 4.7. Thus suppose $n > 1$ and $p > 1$. Then by part (ii), the language $(A_{n-1}' d_{n-1})^+$ has an independent system of type $e^{p-1}\bar{e}^{p-1}$. Since this system is prime, we apply Proposition 7.4. Consequently, $A_{n-1}'$, whence also $A_{n-1}$, has an independent system of type $e^{p-1}\bar{e}^{p-1}$. By induction, $n-1 \geq p-1$. ■

We now want to prove a result "dual" to Theorem 7.9, namely that $D_1'^*$ is not a quasi-rational language. The proof cannot use independent systems of iterative pairs, since such systems in $D_1'^*$ are all of type $e\bar{e}$. Instead of independent systems, we shall use grammatical systems and apply Proposition 6.2. Recall that

$$\text{Qrt} = \bigcup_{n \geq 1} \text{Qrt}(n),$$

where

$$\text{Qrt}(1) = \text{Lin}, \qquad \text{Qrt}(n+1) = \text{Lin} \,\square\, \text{Qrt}(n).$$

The idea of the proof is the following: We observe first that for a fixed grammar generating a fixed generator of the cone Qrt$(n)$, any grammatical system of iterative pairs of some special type satisfies an additional constraint (Proposition 7.12). Then we construct a restricted one counter language having systems violating this constraint, whatever is the grammar generating it (Lemma 7.13. This is the main part of the proof). By an application of Proposition 6.2, we conclude that this language is not in Qrt$(n)$. Consequently, the cone Rocl is contained in none of the cones Qrt$(n)$, and therefore $D_1'^*$ is in none of the Qrt$(n)$.

We begin with construction of an algebraic grammar generating the language $S_2 \uparrow (Md)^+$, where $S_2 \subset Z_2^*$ is the symmetric language, $M \subset Y^*$ is some algebraic language. $Z_2 \cap Y = \emptyset$ and $d \notin Z_2 \cup Y$ is a new letter. This will be used below with $M$ a generator of Qrt$(n)$, and thus yields a grammar for a generator of Qrt$(n+1)$. As for the Dyck language $D_1'^*$, the application of Exercise VI.5.6 could simplify the construction.

Let $G_1$ be the grammar generating $S_2$ with productions

$$\xi \to x_1 \xi \bar{x}_1 + x_2 \xi \bar{x}_2 + 1,$$

and let $G=\langle V, Y, P\rangle$ be an algebraic grammar such that $L_G(\sigma)=M$ for some $\sigma\in V$. We assume $\xi\notin V$, and consider a new symbol $\sigma_+$. Define a grammar

$$G_1\uparrow G=\langle \xi\cup\sigma_+\cup V, Z_2\cup Y\cup d, P\cup P'\rangle$$

by $$P'=\{\xi\to\sigma_+x_1\xi\sigma_+\bar{x}_1, \xi\to\sigma_+x_2\xi\sigma_+\bar{x}_2, \xi\to 1, \sigma_+\to\sigma d\sigma_+, \sigma_+\to\sigma d\}.$$

Then a straightforward verification shows that

$$L_{G_1\uparrow G}(\xi)=S_2\uparrow (Md)^+.$$

**Lemma 7.11** *If $S_2\uparrow(Md)^+$ has a system of iterative pairs grammatical with respect to $G_1\uparrow G$ of type $\beta=e\alpha\alpha\bar{e}$, where $\alpha\in\{e,\bar{e}\}^+$ is the type of a prime system of iterative pairs, then $M$ has a system of iterative pairs of type $\alpha$ grammatical with respect to $G$.*

Proof. Let $\theta=(a_0, u_1, \ldots, u_{2n}, a_{2n})$ be a grammatical system of type $\beta$. Then by definition, there are variables $\eta, \eta_1, \eta_2\in V\cup\xi\cup\sigma_+$ such that, in $G_1\uparrow G$,

$$\xi\xrightarrow{*} a_0\eta a_{2n},\qquad \eta\xrightarrow{*} u_1\eta u_{2n},\qquad \eta\xrightarrow{*} a_1\eta_1 a_n\eta_2 a_{2n-1} \tag{7.1}$$

and since $\alpha$ is a Dyck-prime

$$\eta_1\xrightarrow{*} u_2\eta_1 u_n,\qquad \eta_2\xrightarrow{*} u_{n+1}\eta_2 u_{2n-1} \tag{7.2}$$

Observe that $\eta_1\neq\sigma_+$ since the iterative pair $\langle u_2, u_n\rangle$ is nondegenerated and consequently $u_n\neq 1$. Similarly $\eta_2\neq\sigma_+$. Thus $\eta_1, \eta_2\in V\cup\xi$. If $\eta_1=\eta_2=\xi$, then by (7.1), $\xi\xrightarrow{*} a_0a_1\xi a_n\xi a_{2n-1}a_{2n}$. This is impossible since $\xi$ appears at most once in the right side of each production of $G_1\uparrow G$. Thus either $\eta_1\in V$ or $\eta_2\in V$. In the first case, the first of the derivations (7.2) holds also in $G$, and $\langle u_2, \ldots, u_n\rangle$ is a system of iterative pairs of type $\alpha$ grammatical with respect to $G$. In the second case, $\langle u_{n+1}, \ldots, u_{2n-1}\rangle$ is grammatical with respect to $G$. (Note that we do not claim that both $\eta_1, \eta_2$ are in $V$.) ■

We now define languages $B_n$, $(n\geqslant 1)$ as follows:

$$B_1=S_2,\qquad B_{n+1}=S_2\uparrow(B'_nd_n)^+\qquad (n\geqslant 1),$$

where $B'_n$ is a copy of $B_n$ over an alphabet disjoint from $Z_2$, and where $d_n$ is a new letter. Next let $G_{n+1}=G_1\uparrow G'_n$, where $G'_n$ is the grammar generating $B'_n$ derived from $G_n$ in the obvious way. Then $L_{G_n}(\xi)=B_n$, $(n\geqslant 1)$. Define a sequence $(\alpha_n)$ of types by

$$\alpha_1=e\bar{e},\qquad \alpha_{n+1}=e\alpha_n\alpha_n\bar{e}\qquad (n\geqslant 1).$$

**Proposition 7.12** *If $B_n$ has a system of iterative pairs of type $\alpha_p$ grammatical with respect to $G_n$, then $p\leqslant n$.*

Actually, it is easy to show that $B_n$ has systems of type $\alpha_n$, grammatical with respect to $G_n$ (Exercise 7.3).

Proof. The result is true for $n=1$ by Example 4.6. Assume that $B_n$ has a system of iterative pairs of type $\alpha_p = e\alpha_{p-1}\alpha_{p-1}\bar{e}$, $(p \geq 2)$ grammatical with respect to $G_n$. Then by the previous lemma, $B_{n-1}$ has a grammatical system with respect to $G_{n-1}$ of type $p-1$. By induction $p-1 \leq n-1$. ■

The main lemma for the proof of the announced result is the following:

**Lemma 7.13** *Define* $L(k) = D_1'^* \cap (x^+\bar{x}^+)^{2^{k-1}}$, $(k \geq 1)$. *For any grammar H generating L(k), there exists a system of iterative pairs in L(k) of type $\alpha_k$ which is grammatical with respect to H.*

Take the lemma for granted for a moment. Then we can prove

**Theorem 7.14** (Yntema [1967]) *The language $D_1'^*$ is not quasi-rational. The cones* Rocl *and* Qrt *are incomparable.*

Proof. If $D_1'^* \notin$ Qrt, then Rocl $\not\subset$ Qrt. Since Lin $\not\subset$ Rocl by Proposition 7.1, the cones Rocl and Qrt are incomparable. Thus it suffices to show that $D_1'^*$ is not quasi-rational. Assume the contrary. Then $D_1'^* \in \text{Qrt}(n)$ for some $n \geq 1$. Let $k > n$. Since $D_1'^* \geq L(k)$ and Qrt$(n)$ is a cone, we have $L(k) \in \text{Qrt}(n)$, whence $L(k) \leq B_n$. Next in view of Lemma 7.13 and of Corollary 6.4, there is a system of iterative pairs $\theta$ in $B_n$ grammatical with respect to $G_n$ and such that $\chi(\theta, B_n) \geq \alpha_k$, whence $\chi(\theta, B_n) = \alpha_k$. By Proposition 7.12, we have $k \leq n$. This yields the desired contradiction. ■

Thus it remains to prove Lemma 7.13. The proof which follows is similar to a proof given by Gruska [1971b]. Fix an integer $k \geq 1$, set $X = \{x, \bar{x}\}$, $L = L(k)$, $K = (x^+\bar{x}^+)^{2^{k-1}}$, and let $H = \langle V, X, P\rangle$ be a grammar such that $L = L_H(\sigma)$ for some $\sigma \in V$. We first give two preliminary lemmas.

**Lemma 7.15** *One can assume that H is strictly reduced in $\sigma$, i.e. satisfies the two following conditions:*

(i) *For any $\xi \in V$, there are $a, b \in X^*$ such that $\sigma \xrightarrow{*} a\xi b$.* (7.3)

(ii) *For any $\xi \in V$, $L_H(\xi)$ is infinite.* (7.4)

Proof. In view of Lemma II.2.6, a strictly reduced grammar $H'$ generating $L$ exists. By the construction given in Section II.2, it is straightforward that any system of iterative pairs which is grammatical with respect to $H'$ is also grammatical with respect to the original grammars, since all pivots satisfy condition (ii). ■

For the rest of the proof, we assume that $H$ satisfies (7.3) and (7.4). Then we have

**Lemma 7.16** (i) *For all* $\xi \in V$, $u, v \in L_H(\xi)$ *implies* $\|u\| = \|v\|$.
(ii) *For all* $\xi \in V$, $L_H(\xi) \not\subset x^* \cup \bar{x}^*$.
(iii) *If* $\eta \xrightarrow{*} u\eta v$ *for* $u, v \in (V \cup X)^+$, *then* $u = x^p$, $v = \bar{x}^p$ *for some* $p > 0$.

By (i), there is an integer $\delta(\xi)$ for each $\xi \in V$ such that $\|u\| = \delta(\xi)$ for all $u \in L_H(\xi)$.

Proof. (i) By (7.3), there are $a, b \in X^*$ such that $\sigma \to a\xi b$. Thus $aub, avb \in D_1'^*$, whence $0 = \|aub\| = \|avb\|$. Consequently $\|u\| = \|v\|$.
(ii) If $L_H(\xi) \subset x^* \cup \bar{x}^*$, then $L_H(\xi)$ consists of a single word $x^{\delta(\xi)}$ or $\bar{x}^{-\delta(\xi)}$ according to $\delta(\xi)$ being positive or negative. This contradicts (7.4).
(iii) In view of (7.3) and (7.4), there are $a, b, c \in X^*$ such that $\sigma \xrightarrow{*} a\eta c$, $\eta \xrightarrow{*} b$. Further $\sigma \xrightarrow{*} au^n bv^n c$ for $n \geq 0$. If $u \notin X^+$, then $|u|_\xi > 0$ for some $\xi \in V$. By (ii) $L_H(\xi)$ contains a word in $X^*(x\bar{x} \cup \bar{x}x)X^*$. Consequently $u \xrightarrow{*} u'$ for some $u' \in X^*(x\bar{x} \cup \bar{x}x)X^*$. Then $u'^n \in X^*(x\bar{x})^n X^* \cup X^*(\bar{x}x)^n X^* \subset X^*(x\bar{x})^{n-1}X^*$, and $au^n bv^n c$ derives into a word which is not in $K$ for sufficiently large $n$. Thus $u \in X^*$ and by the same argument as above $u \in x^+ \cup \bar{x}^+$. Since $au^n$ is left factor of a word in $D_1'^*$ for $n \geqslant 0$, we have $\|au^n\| = \|a\| + n\|u\| \geqslant 0$ for all $n$ whence $\|u\| \geqslant 0$. Thus $u \in x^+$. Symmetrically $v \in \bar{x}^+$. Finally $\|u\| + \|v\| = 0$. ■

Now let $T$ be the integer associated to the grammar $H$ by Ogden's Lemma II.2.5, set $D = \max\{|d(\xi)| : \xi \in V\}$ and set $N = D + T$. The two main lemmas of the proof are the following.

**Lemma 7.17** *Suppose that* $\sigma \xrightarrow{*} w = gx^N\bar{x}^N h$ *for some* $g, h \in (V \cup X)^*$. *Then there exist* $\xi \in V$, *and integers* $r, s \geqslant 0$, $p > 0$ *such that*

(i) $\quad r + p, s + p \leqslant N$;

(ii) $\quad \sigma \xrightarrow{*} gx^r\xi\bar{x}^s h, \qquad \xi \to x^p\xi\bar{x}^p, \qquad x^{p+r}\xi\bar{x}^{p+s} \xrightarrow{*} x^N\bar{x}^N.$

Thus the lemma asserts that $w$ admits an iterative pair (in the language $\hat{L}_H(\sigma)$ of sentential forms) such that both iterating elements are in the central factor $x^N\bar{x}^N$ of $w$.

Proof. Mark the $T$ rightmost positions in the left factor $gx^N\bar{x}^N$ of $w$, and apply Ogden's Lemma II.2.5 to $w$. Then there exist a factorization $w = aubvc$, and a variable $\xi$ such that $\sigma \xrightarrow{*} a\xi c$, $\xi \xrightarrow{*} u\xi v$, $\xi \xrightarrow{*} b$. By Lemma 7.16 (iii), $u = x^p$, $v = \bar{x}^p$ for some $p > 0$. Further $c = \bar{x}^s h$ for some $s \geqslant 0$, and $p + s \leqslant T \leqslant N$. Two localizations are possible for the iterating element $u$:
(i) $|a| \geqslant |g|$. Then $a = gx^r$ for some $r \geqslant 0$, and clearly $r + p \leqslant N$. Further $b = x^{N-p-r}\bar{x}^{N-p-s}$.
(ii) $|a| < |g|$. Then $w' = ubv = g'x^N\bar{x}^{N-S}$, where $g'$ is some right factor of $g$. Mark the $T$ left most positions in the right factor $x^N\bar{x}^{N-s}$ of $w'$, and apply

Ogden's Lemma, noting that $\xi \xrightarrow{*} w'$. Then there are a factorization $w' = a'u'b'v'c'$ and a variable $\eta$ such that $\xi \xrightarrow{*} a'\eta c'$, $\eta \xrightarrow{*} u'\eta v'$, $\eta \xrightarrow{*} b'$. As above, $u' = x^q$, $v' = \bar{x}^q$ for some $q > 0$, and $a' = g'x^{r'}$ for some $r' \geqslant 0$. Moreover $r' + q \leqslant T$. Further $c' = \bar{x}^{s'}$ for some $s'$ such that $q + s' \leqslant N - s$, whence $(s'+s) + q \leqslant N$. Since $\sigma \xrightarrow{*} aa'\eta c'c = gx^{r'}\eta\bar{x}^{s'+s}h$ and $b' = x^{N-r'}\bar{x}^{N-(s'+s)}$, the lemma is verified also in this case. ∎

**Lemma 7.18** *Suppose that*

$$\sigma \xrightarrow{*} w = gx^{N+r_1}\eta_1\bar{x}^{s_1}x^{r_2}\eta_2\bar{x}^{s_2+N}h$$

*for some* $g, h \in (V \cup X)^*$, *and that further*

(i) $r_1, r_2, s_1, s_2 \leqslant N$;

(ii) $x_1^{r_1}\eta_1\bar{x}^{s_1} \xrightarrow{*} x^N u_1 \bar{x}^N, \qquad x^{r_2}\eta_2\bar{x}^{s_2} \xrightarrow{*} x^N u_2 \bar{x}^N$ *for some* $u_1, u_2 \in D_1'^*$.

*Then there exist* $\xi \in V$, $t_1, t_2 \geqslant 0$, $p > 0$ *such that*

(iii) $t_1 + p, \quad t_2 + p \leqslant N$;

(iv) $\sigma \xrightarrow{*} gx^{t_1}\xi\bar{x}^{t_2}h; \qquad \xi \xrightarrow{*} x^p\xi\bar{x}^p, \quad x^{t_1+p}\xi\bar{x}^{t_2+p} \xrightarrow{*} x^{N+r_1}\eta_1\bar{x}^{s_1}x^{r_2}\eta_2\bar{x}^{s_2+N}$;

(v) $x^{t_1+p}\xi\bar{x}^{t_2+p} \xrightarrow{*} x^N u_3 \bar{x}^N$ *for some* $u_3 \in D_1'^*$.

The lemma claims the existence of an iterative pair of $w$ whose iterating elements are contained in the $N$ first $x$'s of the block $x^{N+r_1}$ and in the $N$ last $\bar{x}$'s of the block $\bar{x}^{s_2+N}$ of $w$ respectively.

Proof. The conclusion (v) is a direct consequence of (iv) and (ii), since $x^{t_1+p}\xi\bar{x}^{t_2+p} \xrightarrow{*} x^N x^N u_1 \bar{x}^N x^N u_2 \bar{x}^N \bar{x}^N$, whence $u_3 = x^N u_1 \bar{x}^N x^N u_2 \bar{x}^N \in D_1'^*$. We proceed as in the proof of the previous lemma: Mark the $T$ rightmost positions of the left factor $gx^{N+r_1}\eta_1\bar{x}^{s_1}x^{r_2}\eta_2\bar{x}^{s_2+N}$ of $w$, and apply Ogden's Lemma to $w$. Then there are a factorization $w = aubvc$, and a variable $\xi$ such that $\sigma \xrightarrow{*} a\xi c$, $\xi \xrightarrow{*} u\xi v$, $\xi \xrightarrow{*} b$. By Lemma 7.16, $u = x^p$, $v = \bar{x}^p$ for some $p > 0$, and $c = \bar{x}^{t_2}h$ for some $t_2$ such that $t_2 + p \leqslant T \leqslant N$. Three localizations are possible for the iterating element $u$:

a) $|a| \geqslant |gx^{N+r_1}\eta_1\bar{x}^{s_1}|$ and $|bvc| \geqslant |\eta_2\bar{x}^{s_2+N}h|$.

b) $|a| \geqslant |g|$ and $|au| \leqslant |gx^{N+r_1}|$.

c) $|a| < |g|$.

We first verify that case a) is impossible. Assume the contrary. Then $ubv = x^r\eta_2\bar{x}^{s_2+N-t_2}$ for some $r \leqslant r_2$. Next, by (ii), $ubv \xrightarrow{*} x^{r-r_2+N}u_2\bar{x}^{2N-t_2}$. Since $\xi \xrightarrow{*} ubv$, we have $-D \leqslant d(\xi) = r - r_2 - N + t_2 \leqslant -N + t_2 < -N + T$, whence $N < D + T = N$.

Consider now case b). Then $a = gx^{t_1}$ for some $t_1 \geqslant 0$, whence $ubv = x^{N+r_1-t_1}\eta_1\bar{x}^{s_1}x^{r_2}\eta_2\bar{x}^{N+s_2-t_2}$. Since

$$\xi \xrightarrow{*} ubv \xrightarrow{*} x^{2N-t_1}u_1\bar{x}^N x^N u_2 \bar{x}^{2N-t_2},$$

$-D \leqslant d(\xi) = t_2 - t_1 \leqslant T - (t_1 + p)$, whence $t_1 + p \leqslant T + D = N$.
Finally assume that c) holds. Then $ubv = w' = g'x^{N+r_1}\eta_1\bar{x}^{s_1}x^{r_2}\eta_2\bar{x}^{s_2-t_2+N}$, where $g'$ is a right factor of $g$. Mark the $T$ leftmost positions of the right factor $x^{N+r_1}\eta_1\bar{x}^{s_1}x^{r_2}\eta_2\bar{x}^{s_2-t_2+N}$ of $w'$, and apply Ogden's Lemma to $w'$. Remembering that $\xi \xrightarrow{*} w'$, we obtain a factorization $w' = a'u'b'v'c'$ and a variable $\eta$ such that $\xi \xrightarrow{*} a'\eta c'$, $\eta \xrightarrow{*} u'\eta v'$, $\eta \xrightarrow{*} b'$. Then $a' = g'x^{t'}$, and $u' = x^q$, $v' = \bar{x}^q$ for $t' \geqslant 0$, $q > 0$ and $t' + q \leqslant T$. Next either $|c'| \geqslant |x^{r_2}\eta_2\bar{x}^{s_2-t_2+N}|$ or $c' = \bar{x}^l$ for some $l$. In the first case, $u'b'v' = x^{N+r_1-t'}\eta_1\bar{x}^{s'}$ for some $s' \leqslant s_1$, and since $\eta \xrightarrow{*} u'b'v' \xrightarrow{*} x^{2N-t'}u_1\bar{x}^{N+s'-s_1}$, we obtain $d(\eta) = N - t' - (s' - s_1)$. Consequently $N = d(\eta) + t' + s' - s_1 \leqslant d(\eta) + t' < D + T = N$, which is impossible. Thus the second case holds. Then

$$\sigma \xrightarrow{*} aa'\eta c'c = gx^{t'}\eta\bar{x}^{l+t_2}h,$$

$$x^{t'+q}\eta\bar{x}^{q+l+t_2} = x^{t'}u'\eta v'\bar{x}^{l+t_2} \xrightarrow{*} x^{N+r_1}\eta_1\bar{x}^{s_1}x^{r_2}\eta_2\bar{x}^{s_2+N}.$$

Since $t' + q \leqslant T \leqslant N$, it remains only to show that $q + l + t_2 \leqslant N$.
By (ii), $\eta \xrightarrow{*} u'b'v' \xrightarrow{*} x^{2N-t'}u_1\bar{x}^N x^N u_2\bar{x}^{2N-(l+t_2)}$. Thus $d(\eta) = l + t_2 - t'$, whence $q + l + t_2 = d(\eta) + q + t' \leqslant D + T = N$. ■

Proof of Lemma 7.13. Define a sequence $f_n$, $(n = 1, \ldots, k)$ of words by

$$f_1 = x^N\bar{x}^N, \qquad f_{n+1} = x^N f_n f_n \bar{x}^N, \qquad n = 1, \ldots, k-1.$$

Then $f_1 \in x^+\bar{x}^+ = (x^+\bar{x}^+)^{2^0}$, and assuming by induction that $f_n \in (x^+\bar{x}^+)^{2^{n-1}}$, we have $f_{n+1} \in x^+(x^+\bar{x}^+)^{2^n}\bar{x}^+ = (x^+\bar{x}^+)^{2^n}$. Thus $f_k \in L(=L(k))$. We shall prove by induction on $n$ that $f_k$ admits a grammatical system of iterative pairs of type $(\alpha_n)^{2^{k-n}}$. The desired conclusion results for $n = k$.
For sake of simplicity, we set $z(n) = 2^{k-n}$. Then for $n = 1, \ldots, k$, $f_k$ can be written as

$$f_k = g_{n,0}f_n g_{n,1}f_n \cdots g_{n,z(n)-1}f_n g_{n,z(n)} \tag{7.5}$$

where the words $g_{n,i}$ are defined by:

$$g_{k,0} = g_{k,1} = 1,$$

and for $1 \leqslant n \leqslant k-1$:

$$g_{n,0} = g_{n+1,0}x^N, \qquad g_{n,z(n)} = \bar{x}^N g_{n+1,z(n+1)}; \tag{7.6}$$

$$g_{n,2i} = \bar{x}^N g_{n+1,i}x^N \quad (0 < 2i < z(n));$$
$$g_{n,2i+1} = 1 \quad (0 < 2i+1 < z(n)). \tag{7.7}$$

The verification of these formulas is straightforward. We claim that for $n = 1, \ldots, k$,

$$\sigma \xrightarrow{*} h_n = g_{n,0} w_{n,1} g_{n,1} w_{n,2} \cdots g_{n,z(n)-1} w_{n,z(n)} g_{n,z(n)}, \tag{7.8}$$

where for $i = 1, \ldots, z(n)$, $w_{n,i}$ has the form

$$w_{n,i} = x^{r(n,i)} \xi_{n,i} \bar{x}^{s(n,i)}, \tag{7.9}$$

with $r(n, i)$, $s(n, i) \leq N$, and where $\xi_{n,i}$ is the pivot of a grammatical system of iterative pairs of type $\alpha_n$ of the word $f_k$. If this assertion holds, then by (7.8), $f_k$ admits a system of type $(\alpha_n)^{z(n)}$ as desired. We verify first the assertion for $n = 1$, by induction on $i = 1, \ldots, z(1)$. Assume that

$$\sigma \xrightarrow{*} w = g_{1,0} w_{1,1} \cdots w_{1,i-1} g_{1,i-1} f_1 g_{1,i} f_1 \cdots f_1 g_{1,z(1)},$$

where $w_{1,1}, \ldots, w_{1,i-1}$ have the form (7.9). Since $f_1 = x^N \bar{x}^N$, Lemma 7.17 applies to $w$, and consequently

$$\sigma \xrightarrow{*} g_{1,0} \cdots g_{1,i-1} w_{1,i} g_{1,i} f_1 \cdots f_1 g_{1,z(1)},$$

where $w_{1,i}$ has the form (7.9), and further $\xi_{1,i}$ is the pivot of an iterative pair (of type $e\bar{e} = \alpha_1$) of $f_k$. This proves the case $n = 1$.
Suppose now that the assertion is true for some $n < k$, and consider the case $n + 1$. By the induction hypothesis, (7.8) holds.
Write $h_n = h'h''$, with $h' = g_{n,0} w_{n,1} g_{n,1} w_{n,2} g_{n,2}$. Then in view of (7.6), (7.7) and (7.9),

$$h_n = g_{n+1,0} x^{N+r(n,1)} \xi_{n,1} \bar{x}^{s(n,1)} x^{r(n,1)} \xi_{n,2} \bar{x}^{s(n,2)+N} g_{n+1,1} h''' \tag{7.10}$$

with $h''' = x^N h''$ if $n + 1 < k$, and $h''' = h''(= 1)$ if $n + 1 = k$. By (7.10), Lemma 7.17 applies to $h_n$, and consequently

$$\sigma \xrightarrow{*} g_{n+1,0} w_{n+1,1} g_{n+1,1} h'''$$

with $w_{n+1,1}$ described by (7.9). Assume now that

$$\sigma \xrightarrow{*} h = g_{n+1,0} w_{n+1,1} \cdots w_{n+1,i-1} g_{n+1,i-1} x^N w_{n,2i-1} g_{n,2i-1} \cdots w_{n,z(n)} g_{z(n)},$$

for some $i \leq z(n+1)$. Then $h = h'w'h''$, with

$$\begin{aligned} h' &= g_{n+1,0} \cdots g_{n+1,i-1}, \qquad w' = x^N w_{n,2i-1} w_{n,2i} \bar{x}^N \\ h'' &= g_{n+1,i} x^N w_{n,2i+1} \cdots w_{n,z(n)} g_{z(n)} \quad \text{or} \quad h'' = 1, \end{aligned} \tag{7.11}$$

according to $i < z(n+1)$ or $i = z(n+1)$. In view of (7.11) and (7.9), Lemma 7.17 can be applied to $h$, and consequently

$$\sigma \xrightarrow{*} g_{n+1,0} \cdots g_{n+1,i} w_{n+1,i} h'',$$

where $w_{n+1,i}$ satisfies (7.9). Thus the induction step is verified and the proof is achieved. ■

## Exercises

**7.1** (Boasson) Let $S_2$ be the symmetric language over $Z=X\cup\bar{X}$, where $X=\{x, y\}$, $\bar{X}=\{\bar{x}, \bar{y}\}$. Let $\sigma: Z^*\to Z^*$ be the rational substitution defined by $\sigma(x)=x$, $\sigma(y)=y$, $\sigma(\bar{x})=\bar{x}\bar{X}^*$, $\sigma(\bar{y})=\bar{y}\bar{X}^*$ and let $\hat{S}_2=\sigma(S_2)$.

a) Show that any nondegenerated iterative pair in $\hat{S}_2$ is left strict but not right strict.
b) Show that there are independent systems of iterative pairs of type $l^p\bar{l}^p$ in $\hat{S}_2$ for any $p\geq 1$.
c) Show that $\hat{S}_2$ is not a generator of Lin, and that $\hat{S}_2\notin \bigcup_{n\geq 1} A_n\Gamma$, where the $A_n$ are the languages of the text. (A similar argument is used in Boasson [1973b] to show that the family Gre is properly contained in Ng(Alg).)

**7.2** Let $S=\{x^ny^n \mid n\geq 1\}$. For two languages $A, B$, define $A\circ B=A'B$, where $A'$ is a copy of $A$ such that $X_{A'}\cap X_B=\emptyset$. Define $S*A=\{x_1^nhx_2^n \mid n\geq 1, h\in A\}$, where $x_1, x_2$ are letters not in $X_A$. Let $\mathcal{R}$ be the least family of languages containing $S$ and closed under the operations $\circ$ and $*$. Next define a mapping $\alpha:\mathcal{R}\to D_1'^*\subset\{e, \bar{e}\}^*$ by: $\alpha(S)=e\bar{e}$, $\alpha(A\circ B)=\alpha(A)\alpha(B)$, $\alpha(S*A)=e\alpha(A)\bar{e}$.

a) Show that $A\in\mathcal{R}$ has an independent system of iterative pairs of type $\alpha(A)$.
Let $w', w\in D_1'^+$. Say that $w'$ is a Dyck-subword of $w$ if $w'=x_1\cdots x_r$, $w=u_0x_1u_1\cdots u_{r-1}x_ru_r$, $x_1,\ldots,x_r\in\{e, \bar{e}\}$, $u_0,\ldots,u_r\in\{e, \bar{e}\}^*$ and if for $1\leq i<j\leq r$, $x_ix_{i+1}\cdots x_j\in D_1'\Rightarrow x_iu_ix_{i+1}\cdots u_{j-1}x_j\in D_1'$.
b) Show that for $A, B\in\mathcal{R}$, $A\leq B$ iff $\alpha(A)$ is a Dyck-subword of $\alpha(B)$.

**7.3** Show that the language $B_n$ has independent systems of very strict iterative pairs of type $\alpha_n$.

**7.4** Define languages $A_k$, $(k\geq 1)$ by $A_1=S_2$, and $A_{k+1}=[(a_k, b_k)^*, A_kA_k']$, where $A_k'$ is a disjoint copy of $A_k$, and $a_k$, $b_k$ are new letters. Show that $A_k\in \mathrm{Qrt}(k)\setminus\mathrm{Qrt}(k-1)$ for $k\geq 2$.

**7.5** For $k\geq 1$, let $\mathrm{Lin}^{(k)}$ be the family of all languages which are finite unions of products of at most $k$ linear languages.

a) Show that $\mathrm{Lin}^{(k)}$ is a principal cone for $k\geq 1$.
b) Show that $\mathrm{Lin}^{(k)}\subsetneq\mathrm{Lin}^{(k+1)}$ for $k\geq 1$, and that the cone $\mathrm{Mlin}=\bigcup_{k\geq 1}\mathrm{Lin}^{(k)}$ is nonprincipal (Mlin is the family of metalinear languages).
c) Show that $\mathrm{Mlin}\subsetneq\mathrm{Lin}\,\hat{\Gamma}$, and that $\mathrm{Mlin}\subsetneq\mathrm{Ng}(\mathrm{Lin}\,\hat{\Gamma})$. (Hint. Show that $(S_{\leq}\#)^+$ is not in Mlin.)

# IX Open Problems, Further Developments

Fig. IX.1 summarizes the results concerning cones of context-free languages proved in the preceding chapters. Inclusion is represented by containment. A principal cone is represented by a sharp peak, a nonprincipal one by a rounded peak. There are several question marks in the figure. They refer to open problems which we are going to discuss now:

**1.** It is unknown whether the full AFL Ng(Alg) of nongenerators of Alg is principal or not. A positive answer to this question would imply, by Theorem VII.3.5, that Ng(Ng(Alg)) is a full AFL closed under substitution. Thus the same question would hold for that full AFL. Thus there may exist a finite or infinite decreasing chain of substitution closed full principal AFLs Alg $\supsetneq$ Ng(Alg) $\supsetneq$ Ng(Ng(Alg)) $\supsetneq \ldots$, having some similarity with the sequence of powers of the maximal ideal in a local ring. However, it is conjectured (Greibach [1970]) that Ng(Alg) is nonprincipal.

**2.** Is there a cone $\mathscr{L}$ of context-free languages containing no nonregular linear (or one counter) language, i.e. such that $\mathscr{L} \cap \mathrm{Lin} = \mathrm{Rat}$ (resp. $\mathscr{L} \cap \mathrm{Rocl} = \mathrm{Rat}$)? More generally, say that two cones $\mathscr{L}$ and $\mathscr{M}$ are disjoint if $\mathscr{L} \cap \mathscr{M} = \mathrm{Rat}$. Then the question is: Are there disjoint cones of context-free languages?

**3.** Call a cone $\mathscr{L}$ minimal if Rat is the only cone properly contained in $\mathscr{L}$. A minimal cone is principal, and $\mathrm{Ng}(\mathscr{L}) = \mathrm{Rat}$. Are there minimal cones of context-free languages? Note that two distinct minimal cones are disjoint. Only a few partial results concerning this question are known. Thus Latteux [1978] proves that the cone generated by $\bar{D}_1^* = \{x, \bar{x}\}^* \backslash D_1^*$ is minimal among all cones generated by commutative languages. See also Berstel and Boasson [1974].

**4.** The previous question is closely related to the following. Say that a cone $\mathscr{L}$ covers a cone $\mathscr{M}$ if $\mathscr{M}$ is the maximal cone properly contained in $\mathscr{L}$. Thus a cone is minimal iff it covers Rat. For a principal cone $\mathscr{L}$, $\mathscr{L}$ covers $\mathrm{Ng}(\mathscr{L})$. Are there principal cones $\mathscr{L}$ and $\mathscr{M}$ such that $\mathscr{L}$ covers $\mathscr{M}$? In other terms, is there a principal cone $\mathscr{L}$ such that $\mathrm{Ng}(\mathscr{L})$ is principal? A negative answer would imply a negative answer to both problems 1 and 3. A related question is: Is there a cone $\mathscr{L}$ which contains only finitely many incomparable principal cones?

There are other open problems which are even more closely related to context-free languages.

**5.** Is the cone $S_{=}\Gamma$ properly contained in the intersection Lin$\cap$Rocl? A related question is: Is the intersection of two principal cones of context-free languages always a principal cone? (For a negative answer to a related problem, see Ullian [1971].) Another question is: Is there a least principal cone containing a given family of context-free languages?

**6.** Consider a language $L$ which dominates all quasi-rational languages. Is $L$ a generator of Alg? In other terms, let $\mathscr{L}$ be a principal cone containing Qrt. Is it true that then $\mathscr{L}$ contains Alg? Same question for the family Fcl of finite counter languages.

There are some results concerning the modification of the hierarchies represented in Fig. IX.1 when the attention is restricted to some cutset of the

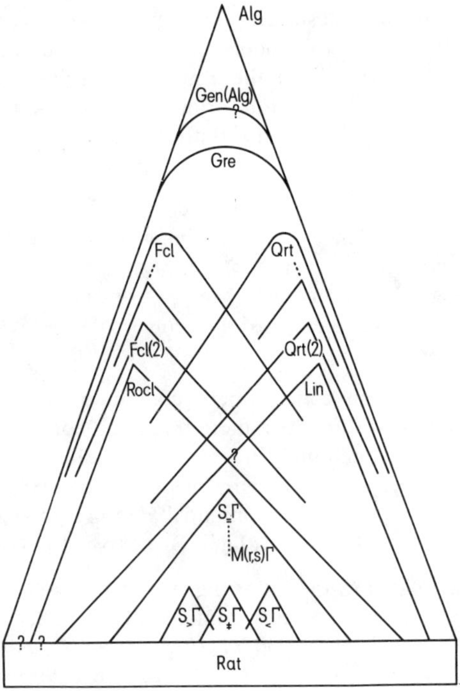


Fig. IX.1

context-free languages characterized by an additional restriction. Thus consider the family Irs of context-free languages that contain no infinite regular set (i.e. satisfying the IRS-condition described in Section V.3). S. Greibach [1975] proves that Irs$\cap$Gre$=$Irs$\cap$Qrt. This implies in particular that any generator of Rocl contains an infinite regular set (since such a generator is not quasi-rational by Theorem VIII.7.14). Along these lines, let Idk be the family of these languages which are rationally equivalent to a language of the form $D_n'^* \cap K$ for some $n \geq 1$, and some regular language $K$. Then Beauquier [1973] shows that for $L \in \text{Idk}$, either $\text{Rocl} \subset L\Gamma$ or $L\Gamma \subset \text{Qrt}$. Note that this result is related to problem 6.

We now discuss some developments which were not included in the preceding chapters, and which are closely related to the topics we are concerned with. Observe that the classification of context-free languages via cones and full AFLs is a classification of languages, in opposition to a classification of machines (acceptors) or of generating devices (grammars). In fact, there exists a theory of AFAs (Abstract Families of Acceptors) which is entirely parallel to the theory of full AFLs. For these results, the reader is referred to the book of Ginsburg [1975]. Up to now, no characterization of the family of context-free grammars generating the languages of a cone is known. The notions of grammar form and of family of grammars introduced in Cremers and Ginsburg [1974] mainly apply to subfamilies of quasi-rational languages.

For some purposes, the concept of cone is too large. Thus the deterministic and the unambiguous context-free languages are closed under inverse morphism and intersection with regular sets, but are not closed under morphism. Therefore, these important families are not cones. This leads to the following weakened version of a cone: A cylinder is a family of languages closed under inverse morphism and intersection with regular languages. There is another reason why cylinders are interesting: cylinders preserve complexity of recognition. More precisely, consider a language $L$. If one has an algorithm $A$ for deciding membership in $L$ (say on a nondeterministic Turing machine) then for a language $L'$ of the form $L' = \varphi^{-1}(L)$ or $L' = L \cap K$, where $\varphi$ is a morphism and $K$ is a regular language, one can easily obtain an algorithm $A'$ for deciding membership in $L'$ such that the order of running time of $A'$ is the same as the order of running time of $A$. (Thus for $L' = \varphi^{-1}(L)$, $A'$ is simply the following algorithm: For a word $f'$, compute first $f = \varphi(f')$ and then apply $A$ to $f$.) Similar considerations hold for space complexity. In this sense, $L$ is "harder" to analyze than $L'$ (since there may exist other algorithms for $L'$ which are faster than $A'$). In other terms, the language $L$ is the "hardest" language in the cylinder generated by $L$. The question whether a given cylinder $\mathcal{L}$ is principal thus is equivalent to whether there exists a "hardest" language in $\mathcal{L}$. Greibach [1973] has given a hardest context-free language. Autebert [1977b] and Boasson and Nivat [1977] have proved that neither the cone of restricted one counter languages, nor the cone of linear languages is a principal cylinder. For a general method, see Autebert [1977a]. The techniques are similar to those exposed in the previous chapters. Thus the theory of transductions and of abstract families of languages serves as starting point for further developments.

## Bibliography

Expository texts on formal languages include Aho and Ullman [1972], Autebert and Cousineau [1976], Becker and Walter [1977], Eilenberg [1974], Ginsburg [1966], Hopcroft and Ullman [1969], Hotz and Walter [1968/1969], Hotz and Claus [1972], Maurer [1977], Salomaa [1969a], [1973]. Salomaa [1973] contains a list of books on formal languages and automata theory up to March 1972. Ginsburg [1975] is a treatise on AFL theory. For rational transductions and rational functions, see Eilenberg [1974].


Aho, A. V.; Ullman, J. D. [1972]: The Theory of Parsing, Translation and Compiling, Vol. I. Englewood Cliffs, N.J. 1972

Alt, H. [1977]: Eine untere Schranke für den Platzbedarf bei der Analyse beschränkter kontextfreier Sprachen. In: Theoretical Computer Science, 3rd GI Conference. Berlin–Heidelberg–New York 1977. = Lecture Notes in Computer Science, vol. 48, 123–131

Alt, H.; Mehlhorn, K. [1976]: Lower bounds for the space complexity of context-free recognition. In: Michaelson, S., Milner, R. (eds.): Automata, Languages and Programming, Third International Colloquium, Edinburgh 1976, 338–354

Amar, V.; Putzolu, G. [1965]: Generalizations of regular events. Inform. and Control **8** (1965) 56–63

Anissimov, A. W.; Seifert, F. D. [1975]: Zur algebraischen Charakteristik der durch kontext-freie Sprachen definierten Gruppen. Elektron. Inform. Verarb. u. Kybern. **11** (1975) 695–702

Autebert, J. M. [1973]: Quelques générateurs des langages algébriques. In: Böhling, K. H.; Indermark, K. (eds.): 1. Fachtagung über Automatentheorie und Formale Sprachen, Bonn. Berlin–Heidelberg–New York 1977. = Lecture Notes in Computer Science, vol. 2, 124–131

— [1977a]: Push-down automata and families of languages generating cylinders. In: Gruska, J. (ed.): Mathematical Foundations of Computer Science 1977. Berlin–Heidelberg–New York 1977. = Lecture Notes in Computer Science, vol. 53, 231–239

— [1977b]: Non-principalité du cylindre des langages à compteur. Math. Syst. Theory **11** (1977), 157–167

Autebert, J. M.; Beauquier, J. [1974]: Une caractérisation des générateurs standard. Rev. franç. d'Automat. Informat. et Rech. Opérationn. 8è Ann. (1974) R-1, 63–83

Autebert, J. M.; Cousineau, G. [1976]: Théorie des automates et des langages formels: I. Les langages algébriques. Institut de Programmation, Université de Paris 1976

Beauquier, J. [1973]: Une caractérisation des langages quasi-rationnels standards. In: Böhling, K. H.; Indermark, K. (eds.): 1. Fachtagung über Automatentheorie und Formale Sprachen, Bonn. Berlin–Heidelberg–New York 1973. = Lecture Notes in Computer Science, vol. 2, 132–138

— [1976a]: Une propriété de non-ambiguité des générateurs algébriques. C. R. Acad. Sci. Paris, Sér. A, **282** (1976) 1203–1205

— [1976b]: Générateurs algébriques non-ambigus. In: Michaelson, S.; Milner, R. (eds.): Automata, Languages and Programming. Third International Colloquium, Edinburgh 1976, 66–73

— [1977]: Contribution à l'étude de la complexité structurelle des langages algébriques. Thèse d'Etat, Université Paris VII, 1977

— [1978a]: Générateurs algébriques et systèmes de paires itérantes. Theor. Comput. Sci. (to appear)

— [1978b]: A remark about the syntactic lemma (submitted for publication)

Becker, H., Walter, H. [1977]: Formale Sprachen. Braunschweig-Wiesbaden 1977

Benois, M. [1969]: Parties rationnelles du groupe libre. C. R. Acad. Sci. Paris, Sér. A, **269** (1969) 1188–1190

Benois, M.; Nivat, M. [1972]: Congruences parfaites et quasi-parfaites. Sémin. Dubreil, 25è Ann., 1971/1972, Université de Paris

Berstel, J. [1973]: Une hiérarchie des parties rationnelles de $N^2$. Math. Syst. Theory **7** (1973) 114–137

— [1974]: Sur une conjecture de S. Greibach. Acta Informatica **3** (1974) 293–295

Berstel, J.; Boasson, L. [1974]: Une suite décroissante de cônes rationnels. In: Loeckx, J. (ed.): Automata, Languages and Programming, 2nd Colloquium. Berlin–Heidelberg–New York 1974. = Lecture Notes in Computer Science, vol. 14, 383–397

Blattner, M.; Head, T. [1977]: Single valued $a$-transducers. J. of Comput. and Syst. Sci. **15** (1977) 310–327

Boasson, L. [1973a]: Two iteration theorems for some families of languages. J. Comput. and Syst. Sci. **7** (1973) 583–596

— [1973b]: The inclusion of the substitution closure of linear and one-counter languages in the largest sub AFL of the family of CFL's is proper. Inform. Process. Let **2** (1973) 135–140

— [1976]: Langages algébriques, paires itérantes et transductions rationnelles. Theor. Comput. Sci. **2** (1976) 209–223

Boasson, L.; Crestin, J. P.; Nivat, M. [1973]: Familles de langages translatables et fermées par crochet. Acta Informatica **2** (1973) 383–393

Boasson, L.; Nivat, M. [1977]: Le cylindre des langages linéaires. Math. Syst. Theory **11** (1977) 147–155

Boë, J. M. [1976]: Représentations des monoïdes: Applications à la théorie des codes. Thèse 3e cycle, Montpellier 1976

Boë, J. M.; Boyat, J.; Cesari, Y.; Lacheny, A.; Vincent, M. [1978]: Automates et monoïdes syntaxiques des sous-monoïdes libres. Inform. and Control (to appear)

Cesari, Y. [1974]: Sur l'application du théorème de Suschkewitsch à l'étude des codes rationnels complets. In: Loeckx, J. (ed.): Automata, Languages and Programming, 2nd Colloquium. Berlin–Heidelberg–New York 1974. = Lecture Notes in Computer Science, vol. 14, 342–350

Choffrut, Ch. [1977]: Une caractérisation des fonctions séquentielles et des fonctions sous-séquentielles en tant que relations rationnelles. Theor. Comput. Sci. **5** (1977), 325–338

— [1978] Contribution à l'étude de quelques familles remarquables de fonctions rationnelles, Thèse d'Etat, Université Paris VII, Paris 1978

Cochet, Y.; Nivat, M. [1971]: Une généralisation des ensembles de Dyck. Israel J. of Math. **9** (1971) 389–395

Conway, J. H. [1971]: Regular Algebra and Finite Machines. London 1971

Cremers, A.; Ginsburg, S. [1975]: Context-free grammar forms. J. Comput. and Syst. Sci. **11** (1975) 86–117

Crestin, J. P. [1978]: Langages quasi-rationnels. In: Crestin, J. P.; Nivat, M. (eds.): Langages algébriques, Ecole Nationale Supérieure des Techniques Avancées, Paris, 1978, 123–166

Davis, M. [1958]: Computability and Unsolvability. New York 1958

Eilenberg, S. [1967]: Algèbre catégorique et théorie des automates. Institut H. Poincaré, Université de Paris, 1967

— [1974]: Automata, Languages and Machines, vol. A. New York–San Francisco–London 1974

— [1978]: Automata, Languages and Machines, vol. C (in preparation).

Elgot, C. C.; Mezei, G. [1965]: On relations defined by generalized finite automata. IBM J. of Res. and Dev. **9** (1965) 47–65

Fischer, P. C.; Rosenberg, A. L. [1968]: Multitape one-way nonwriting automata. J. of Comput. and Syst. Sci. **2** (1968) 88–101

Fliess, M. [1971]: Deux applications de la représentation matricielle d'une série rationnelle non commutative. J. of Algebra **19** (1971) 344–353

Ginsburg, S. [1966]: The Mathematical Theory of Context-Free Languages. New York 1966

— [1975]: Algebraic and Automata-Theoretic Properties of Formal Languages. Amsterdam–New York–Oxford 1975

Ginsburg, S.; Greibach, S. A. [1969]: Abstract families of languages. In: Studies in Abstract Families of Languages. Mem. of the Amer. Math. Soc. **87** (1969) 1–32

— [1970]: Principal AFL. J. of Comput. and Syst. Sci. **4** (1970) 308–338

Ginsburg, S.; Greibach, S. A.; Hopcroft, J. E. [1969]: Pre-AFL. In: Studies in Abstract Families of Languages. Mem. of the Amer. Math. Soc. **87** (1969) 41–51

Ginsburg, S.; Rose, G. F. [1966]: A characterization of machine mappings. Can. J. of Math. **18** (1966) 381–388

Ginsburg, S.; Spanier, E. H. [1968]: Derivation-bounded languages. J. of Comput. and Syst. Sci. **2** (1968) 228–250

— [1970]: Substitutions in families of languages. Inform. Sci. **2** (1970) 83–110

Goldstine, J. [1972]: Substitution and bounded AFLs. J. of Comput. and Syst. Sci. **6** (1972) 9–29

Greibach, S. A. [1969]: An infinite hierarchy of context-free languages. J. of the Assoc. of Comput. Mach. **16** (1969) 91–106

— [1970]: Chains of full AFLs. Math. Syst. Theory **4** (1970) 231–242

— [1972]: Syntactic operators on full semi-AFLs. J. of Comput. and Syst. Sci. **6** (1972) 30–76

— [1973]: The hardest context-free language. SIAM J. of Computing **2** (1973) 304–310

— [1975]: One-counter languages and the IRS condition. J. of Comput. and Syst. Sci. **10** (1975) 237–247

Greibach, S. A.; Hopcroft, J. E. [1969]: Independence of AFL operations. In: Studies in Abstract Families of Languages. Mem. of the Amer. Math. Soc. **87** (1969) 33–40

Gruska, J. [1971a]: A characterization of context-free languages. J. of Comp. and Syst. Sci. **5** (1971) 353–364

— [1971b]: A few remarks on the index of context-free grammars and languages. Inform. and Control **19** (1971) 216–223

Hopcroft, J.; Ullman, J. [1969]: Formal Languages and Their Relation to Automata. Reading, MA–Don Mills, Ont.–London–Amsterdam–Sydney 1969

Hotz, G.; Claus, V. [1972]: Automatentheorie und formale Sprachen III. Mannheim–Wien–Zürich 1972. = B.I. Hochschultaschenbücher, vol. 823

Hotz, G.; Messerschmidt, J. [1974]: Dyck-Sprachen sind in Bandkomplexität log $n$ analysierbar. Tech. Ber. Universität des Saarlandes, 1974.

Hotz, G.; Walter, H. [1968/69] Automatentheorie und formale Sprachen I, II. Mannheim 1968, 1969. = B. I. Hochschultaschenbücher, vol. 821, 822

Kleene, S. C. [1956]: Representation of events in nerve nets and finite automata. In: Shannon, C. E.; McCarthy, J. (eds.): Automata Studies. Princeton, N.J. 1956, 3–40

McKnight Jr., J. D. [1964]: Kleene quotient theorems. Pac. J. of Math. **14** (1964) 1343–1352

McKnight Jr., J. D.,; Storey, A. J. [1969]: Equidivisible semigroups. J. of Algebra **12** (1969) 24–48

Latteux, M. [1977]: Produit dans le cône rationnel engendré par $D_1^*$. Theor. Comput. Sci. **5** (1977) 129–134

— [1978]: Cônes rationnels commutatifs. J. of Comput. and Syst. Sci. (to appear)

Lewis, D. J. [1970]: Closure of families of languages under substitution operators. In: Proc. of the 2nd Annual ACM Symp. on Theory of Computing 1970, 100–108

Maeda, F.; Maeda, S. [1974]: Theory of Symmetric Lattices. Berlin–Heidelberg–New York 1974. = Grundlehren der math. Wiss., vol. 173

Magnus, W.; Karrass, A.; Solitar, D. [1966]: Combinatorial Group Theory. New York–London–Sydney 1966

Maurer, H. [1977]: Theoretische Grundlagen der Programmiersprachen. Repr. Mannheim–Wien–Zürich 1977

Nivat, M. [1967]: Transductions des langages de Chomsky. Thèse d'Etat, Université de Paris, 1967

— [1968]: Transductions des langages de Chomsky. Ann. de l'Inst. Fourier **18** (1968) 339–456

— [1970]: On some families of languages related to the Dyck language. In: Proc. of the 2nd Annual ACM Symp. on Theory of Computing. 1970, 221–225

— [1975]: Opérateurs sur les familles de langages. Rapp. 106, IRIA, France, 1975

Ogden, W. [1968]: A helpful result for proving inherent ambiguity. Math. Syst. Theory **2** (1968) 191–194

Perrin, D.; Schützenberger, M. P. [1977]: Codes et sous-monoïdes possédant des mots neutres. In: Theoretical Computer Science, 3rd GI Conference. Berlin–Heidelberg–New York 1977. = Lecture Notes in Computer Science, vol. 48, 270–281

Perrot, J. F.; Sakarovitch, J. [1977]: A theory of syntactic monoids for context-free languages. In: Gilchrist, B. (ed.): Information Processing 77, IFIP. Amsterdam–New York–Oxford 1977, 69-72

Rosenberg, A. L. [1967]: A machine realization of the linear context-free languages. Inform. and Control **10** (1967) 175–188

Sakarovitch, J. [1977]: Sur les groupes infinis, considérés comme monoïdes syntaxiques des langages formels. In: Séminaire P. Dubreil 1975/76. Berlin–Heidelberg–New York 1977. = Lecture Notes in Mathematic, vol. 586, 168–179

Salomaa, A. [1969a]: Theory of Automata. Oxford–New York–Toronto–Sydney 1969

— [1969b]: On the index of a context-free grammar and language. Inform. and Control **14** (1969)474–477

— [1973]: Formal Languages. New York–San-Francisco–London 1973

Salomaa, A.; Soittola, M. [1978]: Automata-Theoretic Aspects of Formal Power Series. Berlin–Heidelberg–New York 1978

Schnorr, C. P. [1974]: Rekursive Funktionen und ihre Komplexität. Stuttgart 1974. = Teubner Studienbücher Informatik. Leitfäden der angewandten Mathematik und Mechanik, vol. 24

Schützenberger, M. P. [1961a]: Some remarks on Chomsky's context-free languages. M.I.T. Res. Lab. Electron. Quart. Prog. Rept. 63, 1961

— [1961b]: A remark on finite transducers. Inform. and Control **4** (1961) 185–196

— [1973] Sur un langage équivalent au langage de Dyck. In: Suppes, P. et al. (ed.): Logic, Methodology and Philosophy of Science IV. Amsterdam–New York–Oxford 1973, 197–203

— [1975]: Sur les relations rationnelles. In: Brakhage, H. (ed.): Automata Theory and Formal Languages, 2nd GI Conference. Berlin–Heidelberg–New York 1975. = Lecture Notes in Computer Science, vol. 33, 209–213

— [1976]: Sur les relations rationnelles entre monoïdes libres. Theor. Comput. Sci. **3** (1976) 243–259

— [1977a]: Sur une variante des fonctions séquentielles. Theor. Comput. Sci. **4** (1977) 47–57

— [1977b]: Une propriété de Hankel des relations fonctionnelles entre monoïdes libres. Advances in Math. **24** (1977) 274–280

Ullian, J. S. [1971]: Three theorems concerning principal AFL. J. of Comput. and Syst. Sci. **5** (1971) 304–314

Vogel, H. [1972]: Zur Theorie der rationalen und deterministischen Mengen. Mitt. der Gesell. f. Math. Datenverarb. **18** (1972)

Walljasper, S. J. [1970]: Nondeterministic automata and effective languages. Ph. D. Thesis, University of Iowa, 1970

Yntema, M. K. [1967]: Inclusion relations among families of context-free languages. Inform. and Control **10** (1967) 572–597

# List of Symbols


$\mathbb{N}$, $\mathbb{Z}$ 9
$\mathfrak{P}(E)$ 9
$A \backslash B$ 9
$P * Q$ 232
$|u|$, $|u|_x$, $|u|_Y$ 10
$\tilde{u}$ 10
$u \wedge v$, $\bigwedge A$ 104
$\|u, v\|$ 104
$\oplus$ 75
$\mathbb{k}$ 75, 169
$\sqcup\!\sqcup$ 74
$A^+$, $A^*$ 10
$B^{-1}A$ 11
$X^{\oplus}$ 15
$X_n^{(*)}$ 43
$[A, M]$ 154
$L \uparrow M$ 148
$A \geqslant B$, $A > B$, $A \approx B$ 135
$\delta_n$, $\delta'_n$, $\delta_I$ 36
$D_n$, $D'_n$, $D^*_n$, $D'^*_n$ 36
$D_I$, $D^*_I$ 36
Ł 47
$E$ 49
$E_n$ 188
$G_\theta$ 138
$S_n$ 151
$S_=$, $S_<$, $S_\leqslant$, $S_>$, $S_\geqslant$, $S_\neq$ 151
$\rightarrow$, $\xrightarrow{p}$, $\xrightarrow{*}$, $\xrightarrow{+}$ 23
$\vdash_I$, $\overset{*}{\vdash}_I$ 37
$\models$ 201
$\overset{k}{\models}$, $\overset{+}{\models}$ 202
$L_G$, $\hat{L}_G$ 23
$\psi(G)$, $\varphi^{-1}(G)$ 31
$G_K$ 32
$\Delta$ 219
$\mathrm{Exp}(\pi, L)$ 219
$\chi(\pi, L)$ 220
$\mathrm{Exp}(\theta, L)$ 233
$\chi(\theta, L)$ 233
$\mathrm{Diag}(\theta, L)$ 233
$\pi * \theta$, $\theta_1 \times \theta_2$ 233
$\Gamma$ 136
$\hat{\Gamma}$ 145
$\Phi$, $\Phi^{-1}$ 164
$\Phi_a$, $\Phi_a^{-1}$, $\Phi_\pi$, $\Phi_\pi^{-1}$ 165
$I_R$ 165
$\overline{\mathcal{M}}$, $\underline{\mathcal{L}}$ 145
$\underline{\mathrm{Rat}}$ 139
$\mathcal{L} \square \mathcal{M}$ 145
$\mathcal{L} \vee \mathcal{M}$ 138
$\mathcal{L} \wedge \mathcal{M}$ 177
$\mathring{\mathcal{L}}$ 184
Gen($\mathcal{L}$), Ng($\mathcal{L}$) 198
Rec($M$) 52
Rat($M$) 55
$\Sigma$ 134
$\hat{\Sigma}$ 135
$\mathbb{F}$ 162
Alg 134
Elm 134
Fcl, Fcl($k$) 199
Fin 134
Gre 200
Ind($k$) 211
Irs 266
Lin 150
Mlin 264
Nexp 209
Ocl 202
Qrt, Qrt($k$) 209
Rat 134
Rocl 202

# Index


**a**bstract family of languages (AFL) 141
accessible state 18
addition 74, 99
algebraic grammar 22, 71
— language 24
— subset 71
— transduction 71
alphabet 10, 134
alphabetic morphism 13
automaton, accessible part 18
—, finite 16, 55
—, minimal 18
—, nondeterministic 17
—, semi- 18, 108

**b**ifix code 11
bimachine 123
—, state complete 124
bimorphism 62
bounded variation 105
bracket 154, 217

**c**losure operator 164
code, prefix, suffix, bifix 11
computation 78, 201
cone, disjoint 265
—, generator 137, 198
—, minimal 265
—, nongenerators 198
—, principal 137
—, rational 136
configuration 201
congruence 14
—, saturated 52
—, syntactic 15
context-free grammar 22
context-free language 24
— — substitution 31
continuous morphism 13
— transduction 87
copy 13, 62
cross-section 111
cylinder 267

**d**educed iterative pair 225
— system of iterative pairs 241
derivation 23
— bounded language 199, 208
—, index of a 211
—, length of a 23
—, proper 23
diagonal 219
distance of languages 15
— of words 104
domination, rational 135
—, strict 135
Dyck congruence 36
— language 36, 187, 203, 228
— —, restricted 36, 184, 199, 221, 233, 236, 251, 259
— primes 36
— reduction 37
— -subword 264

**e**quations, algebraic 26
—, equivalent 29
—, right linear 64
—, solution of 27
—, system of 26
extensive operator 164

**f**actor, left, right, proper 11
faithful transduction 87

family of languages 134
— — —, abstract (AFL) 141
— — —, finite 135
— — —, incomparable 135
— — —, rationally closed 138
finite counter language 199
free commutative monoid 15
— group 43
— monoid 10
full AFL 141
— — generator 142
— —, principal 142
— semi-AFL 141
— trio 136
function, (left) sequential 97
—, length preserving 123, 127
—, rational 92
—, right sequential 98
—, subsequential 100
— with bounded variation 105

generator of a cone 137, 198
— — — full AFL 142
grammar, algebraic, context-free 22, 71
—, index of a 211
—, linear 150
—, proper 30
—, reduced 34
—, strict 28
—, strictly reduced 34
grammatical iterative pair 223
— system of iterative pairs 238
— type 248

idempotent operator 164
incomparable families of languages 135
independent system of iterative pairs 234
index of a derivation 211
— — — grammar 211
— — — language 211
input label 78
inverse transduction 65
IRS-condition 140, 266
Iteration lemma for algebraic grammars 34
— — — — languages 33
— — — linear languages 155
— — — rational relations 63
— — — regular languages 20
iterative pair, bounded 222
— —, deduced 225
— —, degenerated 220
— —, grammatical 223
— —, pivot of an 223
— —, strict, left strict, right strict 220
— —, underlying 233
— —, very strict 220

label 78
—, input 78
—, output 78
language 10, 134
—, algebraic, context-free 24
—, derivation bounded 199, 208
—, finite counter 199
—, index of a 211
—, linear 150
—, local regular 18
—, Lukasiewicz 47, 202
—, marked 178
—, metalinear 264
—, nonexpansive 199, 209
— of arithmetic expressions 49, 188
— — palindromes 156
— — sentential forms 23
—, one counter 202
—, quasi-rational 199, 209
—, rational 56
—, recognizable 16
—, regular 16
—, restricted one counter 199, 202
—, standard matching choice 199, 208

language, symmetric 151, 184, 221, 235, 256
length of a system of iterative pairs 233
— — — word 10
— preserving function 123, 127
linear grammar 150
— language 150
local regular language 18
locally finite 163
Lukasiewicz language 47, 202

**m**arked language 178
— $\mathcal{M}$-substitution 178
marking operator 178
matrix representation 81
—, normalised 93
—, trim 86
— with twinning property 128
metalinear language 264
minimal automaton 18
— cone 265
morphism 12
—, alphabetic 13
—, continuous, $\varepsilon$-free 13
—, semigroup 12, 81
—, semimonomial 117
—, strictly alphabetic 13
—, summed up 121
—, unambiguous 117
—, (0, 1)- 117
$\mathcal{M}$-substitution 145

**n**ondeterministic automaton 17
nonexpansive language 199, 209
nongenerators 198
norm of a language 15
normalized matrix representation 93

**o**ne counter language 202
operator 163
—, closure 164
—, extensive 164
—, idempotent 164
operator, marking 178
output label 78

**p**ath, successful 78
pivot of an iterative pair 223
— — a system of iterative pairs 238
prefix code 11
prime system of iterative pairs 252
principal cone 137
— full AFL 142
production 22
projection 13
pushdown automaton, one counter 201

**q**uasi-rational language 199, 209
— — relation 217
quotient, left, right 11

**r**ank of a quasi-rational language 199, 209
rational domination 135
— expression 55
— function 92
— image 135
— relation 61, 63
— subset 55
— transduction 65, 67
recognizable language 16
— relation 61
— subset 52
— transduction 72
reduced grammar 34
regular language 16
— substitution 17
relation 14
—, quasi-rational 217
—, rational 61, 63
—, recognizable 61
restricted Dyck language 36, 184, 199, 221, 233, 236, 251, 259
— one counter language 199, 202
reversal of a transduction 66

right linear system of equations 64
— sequential function 98
— — transducer 98
— — transduction 98
rule 201

segment 11
semiautomaton 18, 108
semigroup 9
semilinear subset 56
semimonomial morphism 117
semiring 11
sequential function 97
— transducer 96
— transduction 97
set of exponents 219
shuffle 74
standard matching choice language 199, 208
starheight 58
strict grammar 28
strictly alphabetic morphism 13
— reduced grammar 34
subsequential function 100
— transducer 100
subset, algebraic 71
—, rational 55
—, recognizable 52
—, semilinear 56
—, unambiguous rational 123
substitution 13
—, context-free 31
—, $\mathcal{M}$- 145
—, marked $\mathcal{M}$- 178
—, regular 17
—, syntactic 148
subword 11
successful path 78
suffix code 11
symmetric language 151, 184, 221, 235, 256
syntactic congruence 15
— monoid 15
syntactic substitution 148
system, deduced 241
—, grammatical 238
—, independent 234
—, length of a 233
—, of iterative pairs 233
—, pivot of a 238
—, prime 252
—, type of a 233
—, very strict 249

transducer 77
—, right sequential 98
—, sequential 96
—, subsequential 100
—, unambiguous 114
transduction 65
—, algebraic 71
—, continuous 87
—, faithful 87
—, inverse 65
—, recognizable 72
—, reversal 66
—, right sequential 97
—, sequential 98
transition, authorized, forbidden 19
trim matrix representation 86
twinned states 128
twinning property 128
type of an iterative pair 220
— — a system of iterative pairs 233

unambiguous morphism 117
— rational subset 123
— transducer 114
underlying pair 233
unitary operator 163

variable 22
very strict system of iterative pairs 249
word 10
—, conjugate 12
—, primitive- 12